ARISTOTLE'S
PHYSICS

ARISTOTLE'S PHYSICS

newly translated by
RICHARD HOPE
with an
Analytical Index
of Technical Terms

University of Nebraska Press, Lincoln

1961

FOREWORD

Aristotle's thought has furnished successive ages and cultures with stimulus to their own attempts to reflect upon the world in which man finds himself. Today, one of the most suggestive parts of his writing is his careful analysis of the concepts involved in the notion of natural process, to be found in his *Physics*. Central in his vision is the great panorama of living things, ceaselessly unrolling their life cycle of birth, growth, maturity, and death. Descendant of an Asklepiad family and rooted in the Greek medical tradition, he sees man taking his place at the summit of this spectacle of life, united with the rest of animate nature in so much of his own manner of living, yet capable of a distinctive way of acting shared by none of the rest. Man alone displays the power of knowing, reason and intelligence, which the Greeks called *nous*. This power enables man to communicate with his fellows through speech, to dwell together with them in a community or *polis,* to share a sense of right and justice, to foresee the consequences of his acts, and to live intelligently. Above all, it makes it possible for him to share in the deathless and divine through knowing. All these distinctive human functions, whose importance for Aristotle was reinforced by what he had learned from Plato's vision, form a pervasive theme in his thought which he strove to interweave with his own vision of the processes of life in general.

The nineteenth century found most suggestive Aristotle's embracing of human life within the domain of life in general, his showing of how the life of reason can be a natural human life. It cherished his *Ethics* and *Politics,* whose careful analysis displays the way in which all the functions of a rational and political animal can be performed in accordance with reason. These treatments of human life are still among the most fruitful of Aristotle's writings for present-day thinking. The *Ethics* in particular plays a central role in much contemporary moral philosophy.

But Aristotle himself applied the concepts and the scheme of analysis he found so fruitful for understanding living processes in general not only to human actions in particular. He also generalized them to hold for all natural processes. He was convinced that there is a fundamental continuity between the simplest and the most complex motions of things. The scheme of understanding he worked out for living processes he pushed in the one direction to include human actions, and in the other to include all natural motions. This attempt to bring nature, life, and human practice and art within a common scheme of analysis takes Aristotle close to one of the major enterprises of twentieth-century philosophizing. It enables our own philosophies of process to find much that is suggestive and fruitful in his concepts. It has brought the *Physics*, among the major Aristotelian writings, into the center of attention once more. And it also makes the *Physics* the best source for Aristotle's analysis of art. For since he held that in art, human production, the factors present in natural processes are most clearly and distinctly differentiated, he normally uses art as both a clue to and a contrast to natural operations. Hence the *Physics* is as much about the processes of art as about other natural processes. Together with Book Theta of the *Metaphysics*, it supplies the general philosophy of art against which the analysis of poetry in the *Poetics* has to be understood.

Historically, in his philosophy of nature Aristotle is trying to revive and reinstate in precise analytical form the older Ionian notion of "nature" or *physis*, the first philosophical idea of the Greeks of which we have any record. The earliest Greek thinkers, doubtless trying to make sense out of Mesopotamian cosmogonies, saw the world as a career or *physis*, a process of growth involving generation, maturity, decay, and death. This vision accorded well with Aristotle's own focus upon life as the central process in the world. It provided also for that serial order and that achievement of goals which impressed him so much in living processes. But Aristotle lived after Empedocles and Democritus, and their atomistic and mechanistic analysis of the manner in which processes are effected. He accepted their view of this manner, but he was convinced they left out something essential. An egg is a chemical process, but it is one that grows into a chicken. It reaches a goal after a fixed order of development. So he set out to revise the mechanistic view of Democritus, in the light of the earlier notion of process of "nature." This makes his many criticisms and supplementations of Democritus' mechanism quite pertinent to our own revisions and enlargements of Newtonian mechanism, and brings the concepts he formulated in criticism of the Greek atomists into the context of the enterprise of transcending our inherited Newtonian concepts. Facing a problem very similar to that of our own natural philosophy, he offers much that challenges our careful consideration.

For the *Physics* is not, of course, what we should call physical science itself. It is rather an attempt at the philosophy of science: it is an analysis of the concepts involved in such a scheme of understanding. Aristotle called such concepts *archai*, principles or beginnings of understanding. In the *Physics* he analyzes such fundamental concepts as the natural, chance, necessity, and natural end, and those involved in motion, continuity, infinite, place, the void, and time. His conception of time as not absolute, but rather a dimension, making possible a system of measurement, and of place as an element in a coordinate system, and hence relative, have come down in the Western tradition side by side with the Platonic conceptions of time and place as absolute. Here again Aristotle is suggestive in the attempt to escape the difficulties of Newton's Platonistic views.

Much recent scholarly concern with the *Physics*, as with the other Aristotelian writings, has been stimulated by the genetic studies of Werner Jaeger to try to find different stages of Aristotle's intellectual development in the document. This involves the dissection of the text into several parts, and the attempt to fit these dissevered members into a scheme of the growth of Aristotle's thought, buttressed by reference to a similar treatment of other texts in the Corpus. This is a fascinating enterprise, and with an obviously composite work like the *Metaphysics* doubtless suggestive enough. The kind of results that may be hoped for in this case may be found judiciously set forth in Sir David Ross's edition of the *Physics*. With this text, it may be doubted that the order of composition of its parts, could it be recovered by such a genetic analysis, would throw much light on the philosophic content and issues involved. It might clarify the relation between the last two books. But, historically important as they have been, they hardly embrace those parts of Aristotle's analysis that are likely to strike the modern thinker as most significant.

This new English version of the *Physics* is the last contribution to the understanding of Greek thought of Richard Hope, long a teacher of philosophy at the University of Pittsburgh. An able classical scholar, he was early caught by Frederick J. E. Woodbridge's vision of the inexhaustible fertility and the ever-renewed modernity of Greek philosophizing. His *Book of Diogenes Laertius* is a penetrating study of that sole ancient textbook in the history of philosophy to come down to us entire. It emphasizes what we can learn from that heir to centuries of textbook writing of the philosophic tastes and preferences of the late Hellenistic generation for whom it was compiled. Hope's translation of the *Metaphysics* signalized his long devotion to the study of Aristotelian thought, inspired by the philosophic analysis of his teacher Woodbridge, a man for whom the ideas of the master of them that know were no mere exhibits in an historical museum, but genuine keys to the understanding of the world man finds spread out before him. Shortly before his untimely death, Richard Hope completed the revision of his version of the *Physics*. This writing he had always seen as embodying many of Aristotle's most enduring insights.

In his translations Hope attempted to have them make sense to the English reader, and above all to make philosophic sense to anyone trying to understand not only Aristotle but the world as well. With Aristotle this often involves a quite literal rendering into plain English of what the text says. But it often involves also a transformation of the Greek syntax, adherence to which is most helpful if the English is to be used side by side with the Greek, but hardly otherwise. Above all, to grasp Aristotle's scheme of analysis, and to follow his terms through the various shades of meaning they assume in different contexts—to see how the many subject matters he explores, distinctive as each is, are all caught up in a single scheme of intelligibility and way of understanding which, for all their variety, they make possible—it is necessary to make clear the single Greek term which must often be rendered differently in different places. This Hope has tried to do through his apparatus.

Any translation of Aristotle involves a philosophic interpretation. If Hope's can scarcely claim to set forth at last what Aristotle "really" meant, it at least endeavors to arrive at something like what Aristotle might have said, were he speaking in their own tongue to English readers today. As such, presented in the form in which he left it, it can stand as a monument to the thinking of a learned and penetrating philosophical mind.

John Herman Randall, Jr.

PREFACE

Aristotle's *Physics* constitutes a basic work in natural science. Others had written treatises *On Nature;* but these have not come down to us as such, and Aristotle found them deficient in a sense of natural structure, genesis, and development. Emphasis on nature's structural, productive, and historical aspects accordingly affords an important clue to Aristotle's contributions to the subject. This emphasis also accounts in large part for his uses of such terms as nature, change, transformation, movement, process, beginning, end, place, time, potentiality, actuality, activity, passivity, principle, explanation, life, art, search, knowledge, speech, and ways of being.

The present translation has been planned especially for the benefit of college and graduate students of philosophy. Instead of aiming at a literal reproduction of Aristotle's words and sentences, this version undertakes to present his ideas as we would express them, yet with as much fidelity to his text as the exigencies of a clear English rendering will allow. Bracketed numbers inserted into the translation refer to the "Analytical Index of Technical Terms." This Index also contains terms which have been derived from the two Latin versions reprinted by the Musurgia Publishers in Volume XVIII of the Works of St. Thomas Aquinas (New York, 1949). Footnotes include quotations from the Commentary of St. Thomas Aquinas which have been selected for their historical significance in the interpretation and utilization of Aristotle.

It is a privilege to acknowledge my indebtedness to helpful friends. A typewritten draft of this work was used and gone over in a Seminar on Aristotle's *Physics* at Columbia University in the City of New York.

Richard Hope

CONTENTS

I. *Book Alpha:* NATURAL SCIENCE AND ITS PRINCIPLES   3
1. Order of Procedure in Natural Science
2. Nature Not One Unchanging Being
3. Arguments against Being as One
4. Contraries and Other Principles in Natural Philosophies
5. Arguments for Contraries as Principles
6. The Number of First Principles
7. Subject and Contrary Terms of Change
8. Coming from Being and from Nonbeing
9. Matter, Form, and Privation

II. *Book Beta:* NATURAL SCIENCE AND EXPLANATION   23
1. Nature and Art
2. Natural Science and Related Sciences
3. Material, Formal, Efficient, and Final Explanation
4. Opinions Concerning Luck and Chance
5. Facts and Traits of Luck and Chance
6. Luck, Chance, and Explanation
7. Relations of Explanatory Factors
8. Natural Processes and Their Ends
9. Necessity as Related to Ends

III. *Book Gamma:* MOVEMENT AND THE INFINITE   41
1. Definition of Movement
2. Movement and Moved Movers
3. Actualization of Mover and Moved
4. The Being of the Infinite
5. The Infinite as Not Actual
6. The Infinite as Potential
7. The Infinite and Its Subjects
8. The Infinite and the Finite

IV. *Book Delta:* PLACE, THE VOID, AND TIME   58
1. Place and Its Classification
2. Place Not Form or Matter
3. Place Not in Another Place

4. Definition of Place
5. Ways of Being in a Place
6. Arguments against and for a Void
7. Definition and Nonbeing of a Void
8. No Independent or Occupied Void
9. No Void within Bodies
10. Time Not a Whole or Change
11. Time as Dependent on Events
12. Quantitative Time and Things in Time
13. Definitions of Temporal Terms
14. Temporal Distinctions as Relative and Objective

90   V. *Book Epsilon:* CLASSIFICATION OF MOVEMENTS
1. Types of Changes and Processes
2. Kinds of Movement and the Immovable
3. Succession, Contact, Contiguity, and Related Distinctions
4. Movements as Unified and Diversified
5. Movements Contrary to Each Other
6. Movement and Rest, Natural and Violent

105   VI. *Book Zeta:* CONTINUITY OF MOVEMENT
1. The Continuous as Infinitely Divisible
2. Time and Spatial Magnitudes as Continuous
3. Moments, Movement, and Rest in Time
4. Subjects and Kinds of Divisibility
5. Ends and Beginnings of Changes
6. Continuity as Not Divisible into First Parts
7. Finite and Infinite Time and Distance
8. Coming to Rest, Rest, and Stages of Change
9. Fallacies of Taking Divisibility as Prior to Movement
10. The Indivisible as Motionless and Change as Definite

129   VII. *Book Eta:* SERIES OF MOVEMENTS
1. Moved Movers and the First Mover
2. Togetherness of Agents and Things Acted upon
3. Objects and the Senses in Qualitative Alterations
4. Comparability and Incomparability of Movements
5. Proportions of Forces, Objects, Distances, and Durations

143   VIII. *Book Theta:* ETERNITY OF MOVEMENT
1. Movement as Ungenerated and Indestructible
2. Refutation of Objections to Eternal Movement
3. Rest, Movement, and Kinds of Beings
4. Movements as Due to Agents

5. The First Mover Taken as Self-Moved or Unmoved
6. The First Mover as Eternal, One, and Immovable
7. Local Motion as Primary and Continuous
8. Circular Motion as Continuous and Infinite
9. Circular Motion as Primary; Recapitulation
10. The First Mover without Parts or Magnitude

Analytical Index of Technical Terms

   I. Being
  II. Categories
 III. Principles of Explanation
 IV. Nature and Change
  V. Human Nature and Soul
 VI. Mind
VII. The Divine

Index of Names

ARISTOTLE'S
PHYSICS

I. BOOK ALPHA

Natural Science and Its Principles

1. Order of Procedure in Natural Science

Getting scientific knowledge has conditions. Any science deals with 184a10
matters which can be systematically explored. Such affairs, moreover,
call for interpretation in terms of their principles, basic factors, or elements. For only then may we claim to comprehend our subject matter
when we have mastered its fundamental facts and principles and resolved it into its elements.

This is clearly true also of the science of nature.* In exploring nature,
therefore, we must begin by trying to determine its first principles.†

There is a natural path [or order] for us to follow. It leads from
what is familiar [181b] or evident [59c] *to us* to what is *by nature*
clear or conclusive. The reason for this is that what is intelligible
relatively to ourselves and what is inherently [105] intelligible are not
the same. Hence it is also necessary for us to conduct our investigation
in this manner. We must start with what is naturally obscure, though 20
apparent to us; and we must advance to what is naturally manifest and
determinate.‡

---

* Thomas Aquinas: "Cujus subjectum est ens mobile simpliciter." Duns Scotus:
"natural substance." Zabarella: "natural body" (cf. ii.1.192b14); other parts of
"natural science," on "species of natural bodies."
† Simplicius ascribes to Aristotle a treatise *On Principles* (*Physics* i-iv, but later
identified with i-v) and a treatise *On Movement* (v-viii, or vi-viii). Cf. *Meteor-
ology* i.1.338a20.
‡ Thomas Aquinas: "quae plus habent de entitate," "quae sunt magis in actu."
Simplicius distinguishes the formal and material "contributing factors" (i) from
the efficient and telic "factors" (ii) and attributes to Plato the recognition, in
addition, of an instrumental "contributing factor" and of an exemplary "factor."

Now, what is at first plain and obvious to us is a confused situation to be analyzed [59e]. This is also the source from which the elements and principles in question later become known to us on analysis [75]. From generalities [43], then, we must proceed to their particular aspects [40].§ Thus, we are acquainted with the situation which we "sense" as a whole;‖ and a generality is [analagously] a kind of "whole" contain-184b10 ing [65g] many [varieties] as parts.¶ Again, names are related in some such manner to definitions. The name "circle," for example, designates a certain "whole" without further determination [72g]; definition of the circle analyzes it into its various features or meanings [40]. So, too, children at first call men indiscriminately [150] as they do their fathers and women as they do their mothers; when they become older, they distinguish [72e] each individual explicitly.

2. Nature Not One Unchanging Being

There must be [in our scientific exploration of nature] either one or more than one principle or beginning [82]. If there is a single principle, it is either independent of movement as Parmenides and Melissus allege,° or subject to movement as the natural philosophers say; some of the latter identify the first principle with air, whereas others identify it with water.† If there is a plurality of principles, they are either limited or unlimited in number: if they are numerically limited but more 20 than one, there are two, three, four, or some other definite [4] number of them; if they are numerically infinite, they are either, as Democritus describes them, homogeneous, though different in shape or in kind [20], or even contraries.‡ Just so, even those who want to find out how many *beings* there are undertake to inquire whether there is one or more than one principle or element of "beings" and whether, in the latter case, such principles or elements are limited or unlimited in number; what they, too, are therefore seeking to discover is whether there is one or more than one *principle or element*.

---

§ i.7.189b31,32; iii.1.200b24,25; *On Generation and Corruption* ii.9.335a24-28; *On the Parts of Animals* i.1.639a23,24; 4,644a23-28.

‖ Repeated perceptions and memories lead to unified experience and knowledge (*Posterior Analytics* ii.19).

¶ Universals become clarified by application to fresh instances.

° Eudemus (cited by Simplicius): Plato contrasted "supra-natural" considerations (like those presented by the Pythagoreans and the Eleatics) with the "elements" and other "principles" in things "natural and generated."

† Thales held "water" to be the generating element. Anaximander taught the "eternal movement" of the "infinite"; Anaximenes described it as "condensation and rarefaction" of "air."

‡ i.4.187a25, 5.188a22.

However, to consider whether being is a unity and is independent of movement, is to turn one's eyes away from nature! How, then, could we reason about principles—any more than we can in geometry with one 185a who subverts *its* principles unless, to be sure, we appeal to a super-ordinate [16] or universal [92] science? Indeed, if there is nothing but unity (like that suggested), then there is no longer any principle, since a principle is a principle "of" some fact [4] or facts!§ To investigate whether there is a unity of the sort proposed would therefore be like launching out on an elaborate dialectic against any other arbitrary view [64] for the sake of having a "discussion": for example, against the Heraclitean position; or against a debater who would declare "being" to be "one *man.*" Or it would be like trying to resolve a contentious argument: for example, the arguments of both Melissus and Parmenides, which conclude wrongly from false premises; or rather especially the 10 argument of Melissus, which from a single absurdity simply deduces the rest with no grace or effort. We, on the other hand, must regard it as basic [64h] that all or at least some natural beings are changeful, as is evident from induction. We need not, then, refute all arguments but only those which in a "demonstration" are erroneously derived from basic principles—not those not so derived. Thus, a geometer would properly refute the attempt [made by Hippocrates of Chios] to square the circle with the use of "segments," but not so Antiphon's attempt to square the circle [by "exhaustion," with a merely "approximate" result]. Never-theless, since those of whom we are speaking do‖ not deal with nature, yet do happen to raise physical problems,¶ we may do well by briefly discussing them; for our inquiry has philosophic import.   20

To begin with the question which seems most appropriate: since "to be" [1]°° has various meanings, let us try to make out what those who declare all things to "be" one are saying. Does "all things" mean "what is" [26] or quantities or qualities? And are all things "one" primary being [26] (for example, one man or one horse or one living being) or "one" quality (for example, whiteness or warmth), and so forth? All these alternatives differ considerably among themselves and are impos-sible to sustain. If all things are not only "what is" but also in any amount and of some sort, then, whether these are separable from one another or not, beings will be many. But if all things are either "such" or "so much," then, whether there is primary being or not, we are con- 30 fronted with an absurdity if not an impossibility: for [qualities and quantities and] other considerations predicted of a primary being or

§ i.1.184a21-23.
‖ Or: "deal with nature, although they happen not to raise physical problems."
¶ i.2.185a33, 3.186a17, 187a2.
°° "Being" is to be understood as a verb, in the sense of *esse* rather than of *ens.*

subject matter [85] cannot be detached from such primary being as if they were independent [74] of it. In particular, Melissus says that "being is infinite." Is "being," then, "so much of—"? Since "infinite" comes under the category of quantity, and a primary being or a quality or attribute can be infinite only indirectly [3], namely, in a quantitative respect, we take recourse, in defining the "infinite," to quantitative operations, not to a primary being ["the infinite"] or to a quality or trait ["infinity"]. Accordingly, if "being" is both primary being and quantitative, then it is a duality, not a unity; but if it is primary being only, then "being" is not infinite and does not have magnitude (or else it would somehow have to be quantitative).

185b

Next, since "unity" (like "being") has various meanings, we must also examine in what sense everything is declared to be "one." To be "one" means to be (1) continuous or (2) indivisible or (3) one and the same in a definition stating "what-it-meant-to-be-something" [88] (for example, vine-culture and wine-growing). But (1) if things are "one" in the sense of being continuous, their "one" is "many" inasmuch as anything continuous is infinitely divisible. Also, part and whole would involve a difficulty (even if not for the argument, but on its own account): Do part and whole constitute a unity or a [duality or] plurality, and in what sense? If a [duality or] plurality, what kind? What if the parts are not continuous [as in a quantitative whole]? And if each of two unremovable parts is one with the whole [organism or situation], are they also one with each other? Then (2) if things are "one" in the sense of being indivisible, there will be no quantity or quality; and "being" will not be infinite, as Melissus describes it, or limited, as Parmenides describes it (since it is a *limit* rather than the *limited*†† that is indivisible). And (3) if all things are one in definition (like "clothes" and "garment"), the advocates of unity will have to become adherents of the Heraclitean view that to be good and to be bad or to be good and not to be good [88a] are the same; and then the same thing may be both good and not good or, for that matter, both a man and a horse. Their view will have to be, not that "beings are one," but that they are "no anything." Also, being a "this-such" [88a] will turn out to be the same as being "so much."

10

20

The "problem of the one and the many" continued to worry even the more recent of the older thinkers, who protested against letting the "same" thing become "one" and "many." Some, including Lycophron, wanted to eliminate "is" [from "*A* is *B*"]. Some even proposed recasting ordinary expressions like "the man is white" or "the man is walking" into such forms as "the man has received [the attribute] whiteness" or

†† A terminating point rather than the line.

"the man [in motion] walks"; they feared that, by putting in the word 30
"is," they would present one [primary being] as *being* many [circum-
stances]—as if "one" or "being" had but a single meaning! However,
anything [1a] may be many things differently defined, as when the
same thing is both white and musical despite the difference between
"being white" and "being musical"; so that what is "one" is in this sense
"many."‡‡ Or a thing may be many by division; namely, in the sense in
which a whole is its parts. Confronted with such facts, the thinkers 186a
referred to became thoroughly perplexed and began to acknowledge
that the one is many—as if [it had ever been credible that] the same
thing cannot be "one" and "many," except (to be sure) when the "one"
and the "many" are themselves mutually opposed;§§ at any rate, a
thing‖‖ may be potentially [many] and actually [one].

3. Arguments against Being as One

Not only is it apparent on this approach [namely, from the meanings
of the terms in the alleged principle of unitary being] that "beings"
cannot be "one" [being]; but there is also no difficulty in refuting the
arguments for the view under discussion. Both Melissus and Parmenides
reason contentiously: their arguments conclude wrongly from false premises; or rather especially the argument of Melissus, which from a single
absurdity simply deduces the rest with no grace or effort.   10

The fallacy of Melissus is obvious: he thinks that if "whatever has
originated has a starting-point," then "anything which has *not* originated
does not have a starting-point" [but is spatially infinite]! It is strange,
too, that *everything* which *has* originated should have a starting-point
(that is, a starting-point not of the time but of the *thing*, and not only
of absolute generation but also of qualitative change), as though there
were no change which comes about all at once!* Furthermore, why
should the All, if it is one, be therefore exempt from movement? Why
should it not move in the way a unitary part of it (such as a body of
water) moves, namely, within itself? And why should there be no qualitative change? Besides, the All cannot by any means be one in kind 20
[20]; but the All can be one, if at all, only with respect to the material
of which [86] it consists. Even some of the natural philosophers maintain the latter sort of unity, but not the former, since man differs from
horse in kind, and so does one contrary from another.

---

‡‡ Against Menedemus of Eretria. Cf. also Plato *Philebus* 14, 15; *Sophist* 251.
§§ *Metaphysics* x.6.
‖‖ MSS. EIJ and Philoponus: "the one"; MS. F: "being"; Alexander: "being and
one."
 * vi.5.236a27; viii.3.253b23.

The same objections may be brought forward, together with others even more pertinent, against the view of Parmenides: his premises are false, and his conclusions do not follow. He not only falsely assumes that "to be" [1] has but one meaning, whereas it has several; but he also reasons wrongly. Even on the assumptions that "white" has but one meaning and that there is nothing but "white," there will none the less be many white things, not one only: the "white" will not be "one," either by being continuous† or in definition. Thus, in definition, "to be *what white is*" has one meaning; "to be *what is white*" [12a], quite

30 another. Yet there will not [on this account] be something separate from and independent of the "white": whiteness differs from its possessor [82f] not by being *separate* from the latter but in its [distinctive way of] *being* [23]. However, Parmenides was unable to command a view of both [of these ways of being in their distinctness and in their togetherness]. His argument therefore requires more than that "to be" have but one meaning whenever it is *predicated;* but "to be" would also have to refer to an *identifiable* "being" [26a] with an identifiable "unity" [24e]. Otherwise, since an attribute is predicated of *something,* and since the subject is *different* from the "being" it happens to have, that

186b subject would not "be"; there would thus [contrary to the thesis of Parmenides] "be" something that "is-not." For this reason, too, "what primarily is" [26a] cannot be an attribute of something else: if it were, its possessor could not itself be an existing *something* unless, indeed, "being" has many meanings such that each instance of it *is something.*‡

However, we are at present assuming [with Parmenides] that "to be" has but one meaning: if, then, "what primarily is" [26a] does not belong to something else, but other modes of being belong to it, why after all would "what primarily is" signify "what *is*" [1] rather than "what is-not" [1b]? Consider: if "what primarily is" not only "is" but also is white, and if "being what white is" is not "what primarily is" (since "being" cannot belong to it, on the assumption that there "is" nothing but "what primarily is"), then there "is" nothing white; indeed, not only is white not [an opposite] *something,* but white would not be at

10 all. Then, too, since we declared our subject to be white, and white to mean what is-not, even *what primarily is* is-not! But if, in consequence, we take even "white" as denoting "what primarily is," then [contrary to the position of Parmenides] "being" has more than one meaning. Besides, if "being" is [immediately identical with] "what primarily is," then "being" will not even have magnitude [contrary to the views of Melissus

† i.2.185b10.
‡ Simply "to be," ontologically, characterizes anything whatever most generally; more significantly, "to be" is "to be something" and is therefore an affair of distinctions.

and Parmenides]. A reason for this is that if "being" has magnitude, then each of its parts will have a *different* "being" from that of the other [which, again, contradicts the original assumption].§

It is evident that "what primarily is" is even in definition [90] divided into "what primarily is" something [26b] else. Thus, if "man" exemplifies "whatever is something," so must also "animal" and "biped." Otherwise the latter would be accidents [3a] belonging either to "man" or else to some other referent [85]. But either of the latter alternatives is impossible. On the one hand, an "accident" may or may not belong [82f] to a subject, or else an accident is defined by means of the defini- 20 tion of its possessor: for example, "sitting" is a separable accident; but "snubness" cannot be defined without reference to the "nose" which we describe as "snub." Moreover, the parts or elements in a definition do not contain, as an inherent part [82h] of their own definitions in turn, the definition of the whole: thus, a definition of "biped" does not contain a definition of "man"; and in defining "white," we do not define "a white man." If all this is so and if a "man" only happened to be "twofooted" [as he may happen to be "sitting"], then "biped" would have to be a separable accident, so that "man" might not be a "biped";‖ or else the definition of "biped" would contain the definition of "man" 30 [as "snub" is identical with "snub nose"], but this is of course impossible since it is rather the definition of "man" that contains "biped."¶ On the other hand, if "biped" and "animal" are accidents of something other than "man," and if neither of them exemplifies "whatever is something," then even "man" will also be an accident of something else.**

However, let it be so that "whatever is something" cannot be an accident of anything, and let the subject of "animal" and of "biped" be also the subject of both combined:†† is the All, then, composed of things that do not lend themselves to analysis [75a]? Some men did, indeed, 187a at this point yield too much to two arguments:‡‡ to the argument that

---

§ i.2.185b11-19, 3.187a2.

‖ If "man" is different from "biped" or "animal," then "man" cannot be "biped" or "animal."

¶ If "man" is identical with "biped" or "animal," then "man is man," "biped" is "biped," "animal is animal."

** If "man" just "happens" to be "biped" or "animal" and either of these is "something else," then "man" and "biped" or "animal" happen to be in some respects alike and in other respects different. But "man" is said by definition to *be* a "twofooted animal," since "to be anything" is an affair of logic (186b14).

†† Yet it is not something else which is said to be a "man" as well as an "animal" and a "biped."

‡‡ Themistius: Plato, against Parmenides; Xenocrates, against Zeno of Elea. *On Generation and Corruption* i.8.324b25-325a32. Cf. also Plato *Sophist* 251-259; Aristotle *Metaphysics* xiv.2.1088b35-1089a19.

on a single meaning of "being" all things are "one,"§§ by replying that there "is" what-is-not [or the "void"]; and to the argument from bisection, by positing indivisible [41] magnitudes.|||| But even if "being" has only a single meaning and cannot at the same time mean its opposite [49], it is evidently not true that there can therefore be *nothing* which is-not: even if there cannot be anything which absolutely [105] is-not, what is there to prevent what-is-not from being what-is-not-[an-opposite-]*something?* Then, too, as for the statement that, if there is nothing besides "being itself," then all things will be "one," this is absurd: who would understand [184] "being itself" to be anything but *whatever-is-something;* and what is there in this understanding of the situation,
10 as we have just shown, to prevent "being" from being "many"? Clearly, then, it is impossible for "being" to be "one" in the [undifferentiating] way in which it has been characterized.

4. Contraries and Other Principles in Natural Philosophies

The accounts which the natural philosophers give fall into two types [55b]. (1) Some assume a single body persisting in change [85], namely, either one of the well-known three* or something else which is both denser than fire and rarer than air; and they interpret all other things as many and as generated by the thickening and thinning of the one. The latter, moreover, are "contraries" consisting (to use general terms) of an "excess" and a "deficiency." Thus, Plato speaks of the "great-and-small" [or "quantity"]; but he presents these as material and the "one" as form, whereas the former men present the one persisting
20 thing as material and the contraries as differentiae, that is, as forms. (2) Others, however, say that the contraries are present in [23b] the "one" and that they come to be separated [164d] out of it. This is what Anaximander says and is also the view of all those who declare the "one" to be "many," as in the case of the "mixture" from which, according to Empedocles and Anaxagoras, other things become separated. But there are also differences in the views of these men: Empedocles believes that the process is cyclical, whereas Anaxagoras thinks of it as nonrecurrent; and Anaxagoras supposes that his "things with similar parts" [22a] as well as the contraries are infinite in number, whereas Empedocles suggests his socalled "elements" only.

---

§§ Thomas Aquinas: "vel substantiam tantum vel accidens tantum." "Dicebat enim Plato quod accidens est non ens."

|||| Is "to be anything," then, exclusively an affair of atoms moving in the void? i.2.185b26, 186a23, 5.188a23.

* Water (Thales, Hippo), air (Anaximenes, Diogenes of Apollonia), fire (Heraclitus, Hippasus).

Anaxagoras seems to have conceived "infinities" as he did because he accepted as true the doctrine which the natural philosophers hold in common: "nothing can come from what is-not." Here is the reason why these men say that "all things were together" and that, when any 30 kind of thing [5] comes into being, what happens is but a qualitative change or, according to some, an affair of combination and separation. Furthermore, each contrary comes "from" the other; hence, they reasoned, each was inherent [82h] in the other. Since everything that is originated must come from what is or from what is-not, and since what is cannot come from what is-not (as all natural philosophers agree), they hold the remaining alternative to be necessarily the true one, namely, that things come into being from inherent [82h] beings; but, they explain, we cannot perceive the latter because of their minuteness. 187b Accordingly, they declare everything to be mixed up in everything, and for an understandable reason: they have been seeing everything as coming from everything.† The different appearances and names of things, corresponding to their preponderating parts, they explain by the infinitely numerous components in the mixture: no concrete whole, they think, is purely [and wholly] white or black or sweet, flesh or bone; but the most numerous of such components in anything determine the "nature" of that "thing."

However, what is infinite is unknowable in so far as it is infinite: thus, what is infinite in number and in magnitude is some unknowable quantity, and what is infinite in kind is some unknowable quality. But if the principles themselves are infinite both in number and in kind, 10 then, since we get to know a composite thing by distinguishing its elements and their number, we cannot get knowledge of any concrete objects! Again, a whole may be indifferently of any size if its constituent parts may be so. But since a living being cannot be indifferently large or small, neither can any of its constituent parts: flesh and bones of animals or fruits of plants cannot grow or shrink indifferently to 20 any extent we please; if they could, so could the organism as a whole! Again, let all such things be in one another, and let them not come into being but be constituents which become separated [out of the mixture]; let things get their names from the constituents predominant in them, and let anything come from anything (for example, water separated from flesh, and flesh from water); then, since every limited body is at last broken up [118] by a limited body, each thing obviously cannot be in each! As the flesh removed from the water increases in amount by continued separation out of what is left, even if the individual portions of flesh separated out become ever smaller, there is none the less some 30

---

† i.1.184a23-26.

minimal bit of flesh such that no smaller bit can be taken. Consequently, if the separation of the flesh comes to an end, there will be some water remaining in which there is no flesh, in which case everything is evidently not in everything; or if the separation of the flesh does not come to an end but the subtraction of flesh continues indefinitely, there will be in the limited piece of flesh an infinite number of equal limited pieces, which is impossible. Besides, since every body from which something is taken away necessarily becomes smaller, and since the maximum and minimum quantity of flesh is determinate, it is evi-
188a dent that no body can be separated out of the minimum quantity of flesh, for there would then be [a piece of flesh left which is] less than the least possible.

Again, the original infinite bodies would already have infinite flesh and blood and brain present in them, distinct [73] from one another, yet no less existing, and each infinite, which is unreasonable. But when it is said that there will never be a complete separation, this pronouncement is made inadvertently; still, the statement is true, since attributes do not have independent being. Suppose, then, colors and states of being to have been mixed together and subsequently separated: in consequence, "white" and "healthy" would be something by themselves, would be nothing but "white" and "healthy," and would not even be attributes of anything! The "mind" in this theory therefore foolishly undertakes
10 an impossibility if it wants to separate [completely] what cannot be separated either in quantity, where there is no least magnitude, or in quality, since attributes have no separate being. But the theory does not even grasp rightly the formation [116a] of things similar in form [57a]: mud can be divided into chunks of mud, but also in other ways; water and air have their existence and origination from each other, but not as do bricks from a house or as does a house from bricks; and it is better to assume fewer principles, as Empedocles assumes a limited number of them.

5. Arguments for Contraries as Principles

All men [with whom we are here concerned] set up contraries as
20 principles. Even in the company of those who say that the All is one and independent of movement, we find Parmenides treating hot and cold as principles, though he calls them "fire" and "earth." In another group, there are those who speak of the rare and the dense. Then, too, Democritus talks about the full and the empty, calling the former "being" and the latter "nonbeing," and about differences in position, shape, and arrangement, each of the latter being a genus which includes con-

traries: in position, we distinguish up and down, before and behind; in shape, we distinguish being with and without angles, straight and curved. Clearly, then, all these men treat contraries somehow as principles, and justifiably so: principles are not to be derived from one another or from anything else, but they are themselves the beginnings of everything; and these conditions are satisfied by primary contraries which, being primary, are not derived from anything else and, being 30 contraries, not from each other.

However, we must also consider the reason [90] underlying this fact [3b]. Let us assume at the outset a natural functioning [101c] of all beings such that not any chance thing acts upon, is acted upon by, or comes from any chance thing, except incidentally. The process whereby something becomes white does not arise from the circumstance that that thing is musical, unless the musical thing happens to be black or at least not white: only what is *not white* becomes white; that is to say, not *anything* which is "not white," but what is black or intermediate 188b between black and white. In another example, only something nonmusical becomes musical; yet not *anything* which is other than musical, but the "unmusical" or something intermediate between the unmusical and the musical if there is any such intermediate state. By the same token, neither does anything pass [117] into any first chance thing: "white" does not pass into "musical," except perhaps incidentally, but into "nonwhite"; yet not into any chance thing that is nonwhite, but into "black" or into something intermediate between black and white. So, too, "musical" passes into "nonmusical"; yet not into any chance thing that is "nonmusical," but into "unmusical" or into something intermediate between the unmusical and the musical if there is any such intermediate state.

All processes, the simple and the complicated ones alike, follow the 10 same pattern [90]; but we readily overlook this fact when we fail to name the opposite [13] stages [64c] explicitly as such. Thus, it must be from a state of being out of tune that a state of being in tune arises, and vice versa; and what a state of being in tune must pass into is a state of being out of tune—not at all into any chance disharmony but into the corresponding opposite state. It makes no difference whether we speak of harmony or arrangement or composition: all these processes exhibit the same pattern. A house, a statue, and any other products arise in similar ways: a house, from materials not yet put together but relatively unordered, and a statue or anything shaped, from the respec- 20 tive unshaped state; the things of which a house consists constitute a certain arrangement, and the things of which a statue consists constitute a certain composition. This being so, everything that comes into being or passes away comes from or passes into one of a pair of contrary states or a state intermediate between them; and since the intermediate states are composed of contraries (colors, for example, of light and dark shades), therefore all the things that are naturally produced are contraries or are composed of them.

As we have already indicated, most other philosophers would perhaps go along with us up to the point we have just reached: truth compels all of them, even if they fail to give any reason [90] for this position,*
30 to describe their "elements" or "beginnings" (as they call them) in terms of contraries. They differ in the precise contraries to which they give priority in their accounts [83] of the origination [116a] of things: some select such as are more familiar in the order of sense perception, for example, hot and cold, wet and dry; others select such as are more intelligible in the order of reason [90], for example, odd and even, love and strife.† The contraries selected are thus somehow both the same and different. Although to most people they appear to be different,
189a yet they are the same by analogy. Each of them is taken from the same list‡ of positive or corresponding negative terms, some of the pairs named being more inclusive than others. Not only are the contraries singled out [36] by the philosophers therefore at once somehow the same and different, but they are also by the respective criteria already stated superior or inferior. Some take the contraries which are more intelligible in the order of reason, namely, the universal (since reason grasps the universal), for example, the "great-and-small"; whereas others take the contraries more accessible in the order of sense perception, namely, the particular (since sense perception grasps the particular),
10 for example, the dense and the rare. At any rate, it is evident that our principles must be contraries.

6. The Number of First Principles

We take up next the question: Are there two or three or more first principles? There cannot be one principle only, for contraries are not one [of their terms without the other]. But neither can there be an infinite number of first principles. In that case, "being" would be unknowable. Another reason is that there is in any one genus [19] but one

---

* Thomas Aquinas: "interdum intellectus hominis quadam naturali inclinatione tendit in veritatem, licet rationem veritatis non percipiat."

† Simplicius: Parmenides (fire and earth), Anaximenes or Xenophanes (earth and water), Plato (great and small), Empedocles (love and strife).

‡ *Metaphysics* i.5.986a23.

[fundamental] contrariety, and primary being is but one genus.* Besides, derivation from a limited number of principles not only suffices but, as is apparent from the principles used by Empedocles, excels derivation from an infinite number of principles; thus, Empedocles claims to derive from his principles all that Anaxagoras derives from his infinite principles. Finally, some contraries have priority over others; derived contraries (for example, sweet and bitter, white and black) differ [from the primary ones]; but first principles must always remain [first principles].† Clearly, then, there is not a single principle only, and the 20 number of first principles is not infinite.

The number of first principles being limited, it would seem reasonable [90] not to present them as two only. For one may very well ask: How can density naturally function so as to convert rarity into anything, or vice versa? So with any other pair of contraries: friendship, for example, does not attract strife and produce something out of it; or vice versa. We must take a third something of quite a different sort along with a pair of contraries. Indeed, some‡ believe that a larger number of sources [86] is required to constitute the nature of beings. Again, if one does not suppose [64h] a different nature basic to contraries, there is the added difficulty that we do not observe contraries functioning as the primary being of any beings or take the first princi- 30 ple to be an attribute of any subject; if we did, there would be a principle of that principle, since the claim of what persists [85] in a change to being a first principle apparently takes precedence over the claim of what is predicated of that subject. Again, a primary being is not contrary to a primary being. How, then, can nonprimary beings [like contraries] constitute a primary being? Or how can nonprimary being have a status of priority to primary being?

Accordingly, if we accept as true both the previous argument [that contraries are principles] and the present argument [that the contraries do not constitute primary being], doing justice to both of these truths 189b requires that we assume a third something. This is what those do who declare the All to be a single nature such as water or fire or something intermediate between them. Now, since fire, earth, air, and water involve contraries, an intermediate nature seems preferable to them. It is with good reason, therefore, that some characterize the persistent being as different from those [four] natures; second to them in order of preference would be those who identify the persistent being with air, since air is the element which has the fewest sensible differences; third would come those who identify the persistent being with water.

---

* Simplicius: one subject-in-process.
† "Form" and "privation."
‡ Democritus, Empedocles.

At any rate, all of these men elaborate the accounts they give of their
10 "one" by drawing upon contraries such as "dense and rare" or "more
and less" which, as has been said,§ are in each case an excess and a
deficiency. Indeed, no less ancient than the treatment of contraries as
principles seems to be the opinion that the "one," together with "excess" and "deficiency," are the first principles of beings. But this opinion
has taken various forms: the earlier men deal with the latter two principles as active and the "one" as passive; some of the more recent
thinkers prefer to employ the "one" as the active principle and the
other two as the passive principles.

In the light of these and other considerations, the view seems reasonable that our principles are three in number, in contrast to the view
that there are more than three. Not only does one passive element
20 suffice; but if there are two pairs of contraries (as in the theory of four
elements), there will be two distinct intermediate natures to go along
with the two pairs of contraries, or the two pairs of contraries will
be able to generate from each other,‖ and then one of the two pairs
of contraries will be superfluous. Moreover, there cannot be more than
one pair of primary contraries, since primary being is but one [and the
same] genus of being: its principles will therefore differ from one another in priority and subsequence only; these will not differ in genus,
since there is in any single genus a single pair of contraries to which
all the pairs of contraries may be reduced.

It is evident, then, that there is not one element only and that there
are not more than two or three; but whether there are two or whether
there are three is, as we have suggested, very difficult to determine.

7. Subject and Contrary Terms of Change

30   Let us now formulate our own view, beginning with "becoming" in
general and thus proceeding in a natural way from the features which
all changes have in common to the features which are peculiar to the
several kinds of change. We designate the "terms" of a change *from*
"something" *to* "something else," or "something different," by means of
expressions which are either "simple" or "complex." I mean that a "man"
may become "educated" and that the "noneducated" may become "edu-
190a cated" or that a "noneducated man" may become an "educated man."
I call "simple" the initial term "man" or "noneducated" and the final
term "educated"; I call "complex" the initial and the final terms when

---

§ i.4.187a16.
‖ That is to say, the two pairs of contraries will have the same intermediate nature
as their subject matter, so that a contrary in one pair can serve as a contrary
in the other.

we say that "a noneducated man becomes an educated man." Also, we say either that "the noneducated becomes educated," or that "it is *from* the noneducated that the educated comes to be"; but we do not say that "it is from a man that the educated comes to be," but rather that "a man becomes educated." Of the subjects simply designated as such in these assertions of "becoming," the latter endures, whereas the former 10 does not. For a "man" endures as such, that is, he is a "man" even when he has become "educated"; but the "noneducated" or "uneducated" does not endure as such, and neither does the "noneducated man" or the "uneducated man."

In the light of the distinctions uncovered in the various ways of describing any change, we find that change always presupposes [85a] something which changes [116a] and that this is numerically but not formally one. That is to say, it is not one in definition, since "to be a man" is not the same as "to be uneducated." The nonopposite, "man," endures; but the "noneducated" or "uneducated" does not, and neither 20 does the "uneducated man." The term "from" which the change is said to start is chiefly what does not endure, as when we say that "it is from the uneducated" (not: "from the man") "that the educated comes to be"; although we sometimes use this description also for what does endure, as when we refer to the bronze "out of" which the statue comes (rather than to the bronze which "becomes" a statue). In the case of the nonenduring opposite, we say either that "it is from the uneducated that the educated comes to be," or that "the uneducated becomes educated." Hence, we also say either that "it is from an uneducated man 30 that an educated man comes to be," or that "an uneducated man becomes an educated man."

However, "becoming" has more than one meaning. Some things do not "come to be" but "become this or that" [4b]. Only primary beings simply "come to be." In other categories there must evidently be some subject which becomes so much or such or related to something else or to some time or to some place: these ways of being are predicated of primary being; only primary being is not in turn predicated of some other subject. Still, reflective consideration shows that even primary or 190b independent [105] beings come into being from some existing source [85]: such a source is always present whenever anything arises; for example, plants and animals come from seeds. In general, things come about in different ways: a statue, by shaping; growth, by addition; the Hermes, by extraction from a stone; a house, by being put together; and things whose material is converted into some other state, by qualitative alteration. Since all these changes evidently arise out of an exist- 10 ing source, it is clear from what has been said that anything involved in "becoming" is always complex: there is what comes into being [for example, one "educated"]; and there is, in what undergoes such a change, a double aspect, namely, the persistent being (for example, a "man") and an opposite (for example, the "uneducated"). Other examples of an "opposite" are the unshaped, the unformed, the unordered; other examples of a "being persisting through a change" are bronze, stone, gold.

It is evident, therefore, in what sense every product is composed "of" a persistent being [85] and a form [91]. These are the fundamental principles or factors which explain what natural beings are or have come
20 to be; not incidentally, however, but in what we call their "essential being" [26]. So an "educated man" may be analyzed into the components "man" and "educated." Thus it is clear in what sense the things which have come into being are composed "of" such elements. Moreover, the initial subject in process [85] is numerically one, although it has two formal aspects. What is to be counted is the "man" or the "gold" or, in general, the "material"; this is a positive "something" and is not an accidental feature of the source out of which the product arises. The "privation," on the other hand, which is "contrary" [to the form in which the change culminates], is incidental [to the subject]. Also, the [final] form is one: for example, the "order," "musical knowledge," or any other similar predicate. Accordingly, fundamental princi-
30 ples are in one sense two and in another sense three: in the former sense, they are the contrary terms of a change (for example, educated and uneducated, hot and cold, in tune and out of tune); in the latter sense, they are not the contraries, since one contrary cannot be acted upon by another. This difficulty, too, is readily resolved: what persists in change is other than the contraries; it is not itself contrary to anything. In one sense, therefore, the principles are not more numerically than are the contraries, that is to say, not more than two; but in another
191a sense they are not two at all but three, because to be a subject in process (for example, to be a man, to be bronze) and to be a contrary (for example, to be uneducated, to be unshaped) are different [aspects of the starting-point of the change].

We have now stated how many principles there are in the case of natural beings which are implicated in "becoming": clearly, there must be a subject of contraries, and the contraries are two; yet in another sense this is not necessary, since one of the contraries suffices to account for [34] a change [115] by its absence or presence. The persistent nature* is known by analogy: as bronze is to the statue, wood to the

---

* Thomas Aquinas: ". . . materia prima . . . consideratur subjecta omni formae . . ." Zabarella: Out of "nude matter," nothing can be generated.

bed, or relatively unformed material to something having a certain form, 10
so is the persistent nature to the primary being which is an existent
"this-something." The persistent nature, then, is one beginning or principle (though as a principle it is not "one" or a "being"† in the same
sense in which a "this-something" is); another principle is [the object
of] the definition; the third is the "privation" which is contrary to it.
We have already explained in what sense these principles are two and
in what sense more than two.

To recapitulate: we have shown, first, that contraries are principles;
next, that there must also be something else [persisting through the
change *from* one contrary *to* the other], so that there are three principles; finally, how the contraries differ, how the princples are interre-
lated, and what the subject in process is. It is not clear as yet whether 20
the primary being [26] is the form [20] or the subject-in-process [85];‡
but it is clear that and how there are three principles and in what way
each of them is a principle. So much, then, for the question: how many
and what are the principles?

8. Coming from Being and from Nonbeing

The analysis of change affords the only way in which we can follow
out the fundamental difficulty of the ancients to a solution. In their
inexperience, those who first sought philosophic truth and the natural
development [101] of beings were diverted into a wrong course of
reasoning. "Nothing comes into being or passes out of being," they said,
"because whatever comes into being would have to come from what
is or from what is-not; and both of these alternatives are impossible."
They went on to explain: "What is does not become anything, since it 30
already is; and nothing comes from what is-not, since something must
underlie." Thereupon they even went beyond this opinion as they progressively amplified its consequences until they came to the conclusion:
"There cannot be many beings; only being itself is."*

We are now in a position to present a solution of the view they held
for the reasons stated. What does it mean "for anything to come from
what is or from what is-not"? Or what does it mean "for what is-not
or for what *is* to act upon anything or to be acted upon by anything or
to become anything"? Nothing essentially different from what it means 191b
"for a physician to act upon anything or to be acted upon by anything
or to become anything"! What is said in the example of the physician

---

† Thomas Aquinas: ". . . dicitur ens et unum inquantum est in potentia ad
formam."
‡ ii.1. *Metaphysics* vii.3.
* i.2.184b16, 4.187a21,22, 5.188a22.

has two meanings; so that we must distinguish two meanings also in such expressions as "coming from what is" and "what is, acts and is acted upon." On the one hand, the physician does not build a house in his capacity as a physician, but as a house-builder; and he does not turn white in so far as he is a physician, but in so far as he has been dark. On the other hand, he effects or fails to effect a cure in so far as he is a physician; and it is chiefly in this latter sense that we speak of a "physician" acting upon anything or being acted upon by anything or becoming anything, namely, *as a physician*. Clearly, then, [to deny] that anything "comes from what is-not" means, properly, [to deny] that

10 anything "comes from what is-not *in so far as it is-not*."† Because they failed to make this distinction, the early philosophers left their problem without a solution; and because of this perplexity [190] they even went so far as to deny the becoming and the being of anything else [that is, other than being itself], and thus to abolish all becoming. We for our own part agree with them that nothing comes from what "is-not" *absolutely*, but insist that a thing does come from what "is-not" *in an incidental sense*: it comes from its "privation," and this is, by itself, what "is-not";‡ hence, from something that does not remain in any product of a change. But the early men found this "mysterious" and therefore held it impossible for anything to come from what is-not.

Just so, we agree that nothing comes from what "is" or that what "is" does not become anything, except (to be sure) *incidentally;* but we insist that this, too, does occur in this latter sense. The point at issue

20 is as if we argued about an animal coming from an animal, a particular animal from a particular animal, a dog from a horse: the dog would come not only from a particular animal but from "an animal," but would not therefore come into being *as an animal* since this [character] is already there; if a particular animal is to come *into* being not incidentally but *absolutely*, what it will come *from* is not an "animal." Similarly, if any being is to come into being [in an absolute sense], it will not come from what is any more than it will come from what is-not (namely, as we have said, in so far as the latter is-not). Moreover, we are not denying that "anything either is or is-not" [which is implicitly denied in the opinion we have been examining].

We have thus presented one solution of the difficulty raised by our predecessors; but there is another solution. The same terms may be analyzed with the aid of the distinction between the potential and the actual: [a product comes from what "is not" that product actually but

---

† Cf. Plato *Sophist* 258E.

‡ i.9.192a5. Zabarella: "privation," the contrary on whose removal a state of affairs comes into being; absence of a form in a material having proximate capacity for the form.

from what "is" that product *potentially*]. However, we have elaborated this distinction more accurately elsewhere.§

We have now resolved the difficulties which prompted our predecessors to deny some of the things we have been trying to explain, so that they were driven far away from the course which leads to an understanding of generation and destruction and of change generally. Had they but given attention to the "nature" [which persists in change],‖ they would have found in it the answer to their whole perplexity.¶ 30

9. Matter, Form, and Privation

Unlike the men whose views we have just examined,* certain others acknowledge the nature which persists in change; yet they have not grasped it adequately.† In the dilemma of Parmenides, they accept the alternative that a product arises out of what absolutely is-not. Besides, 192a they think that, since the nature in which a change starts is numerically single, it has but a single import [11]; but this is a very different consideration.‡ We ourselves distinguish a "material" and a "privative" aspect: the material factor *incidentally* is-not [what it becomes], whereas what we call the "privation" is *essentially* what is-not-[yet]; also, a material is in some sense almost even if not quite a primary being, whereas a "privation" is not a primary being in any way at all. In contrast, the other men declare their "great-and-small" (whether taken together or separately but in any case *both aspects alike*) to be "what is-not," with the consequence that their triad [the "one" and the "great-and-small"] differs considerably from ours [form, material, and privation]. They have progressed far enough to recognize that some nature 10 must be present, but they still portray it as single. Although at least one of them characterizes it as a "dyad" of the "great-and-small," his account nevertheless comes to the same result. The reason is that he overlooks the other nature [the "privation"]. The material which persists is jointly responsible [83] with the form (like Plato's "mother") for the products of changes; yet that member of the pair of contraries which we call the "privative" aspect often seems to those who think only of its baneful character, not to be at all. Hence, whereas we, in

---

§ *Metaphysics* ix.

‖ Simplicius: the nature of privation and matter, or of the absolute and incidental, and the potential and actual, or the composite, nature.

¶ The subject-in-process endures while its initial form becomes replaced by the contrary form which it initially lacks; in other words, a material "is not" actually but "is" potentially what it becomes.

* i.8.191a24-33.

† Plato *Timaeus* 50.

‡ iii.1.201a34-b3.

acknowledging a divine and good goal of movement, distinguish between what is contrary to it and what in its natural activity has a natural impulse to tend in its direction, their view has the consequence that
20 [matter as presumably] one of the contraries has an impulse to its own destruction! Surely, the form cannot tend towards itself, for it does not come short of itself; and a contrary cannot tend to it, for contraries are mutually destructive. But as the female or the ugly inclines to the male or to the beautiful (albeit not essentially but incidentally), so what [naturally] tends to a form is matter.§

We must also distinguish a sense in which "matter" is destroyed or is produced and a sense in which it is not. As that "in which" [there is a privation], a material really ceases to be; what ceases to be is the privative aspect in it; but as a potentiality matter itself [2] does not really pass away, but must be indestructible and unproducible. If matter were produced, some first constituent [82h] would have to be
30 present [85a] out of which matter would arise; but to be such a constituent, is matter's own nature. Were matter produced, matter would therefore have been before it arose! But "matter" is by definition the "first" persistent being out of which anything arises and which inheres in the product in a way that is not incidental. So, too, if matter were destroyed, it would pass into matter in the end; hence, matter would have perished before it perished!‖

There is also the question whether the formal principle is one or more than one and what it is or what they are. However, the accurate determination of this issue falls within the province of first philosophy, to
192b which we therefore leave this subject.¶ In the expositions [63] which follow here, we shall deal with natural and perishable forms.

What we have undertaken to show in our preceding analysis is that there are first principles, what they are, and how many there are. But let us resume our subject and, in so doing, start at another starting-point.

---

§ Thomas Aquinas (against Avicenna): "Nihil igitur est aliud materiam appetere formam, quam eam ordinari ad formam; ut potentia ad actum. Et . . . est ei semper appetitus formae; . . . quia est in potentia ad alias formas dum unam habet in actu. Nec etiam utitur hic figurata locutione, sed exemplari. . . ."

‖ Thomas Aquinas: "Sed ex hoc non excluditur quin per creationem in esse procedat."

¶ *Metaphysics* vii, xii.7-9.

II. BOOK BETA

---•◦ ✦▌◥✦ ◦•---

Natural Science and Explanation

1. Nature and Art

Among beings, some are formed by nature, some by other causes.[*]
Among those formed by nature, we may name animals[†] and their parts,
plants, and the simple bodies (earth, fire, air, and water); all of these,   **10**
together with beings like them, we call "formed by nature." Observation
discloses [173b] how they differ from things not constituted [111i] by
nature: each of them has *within itself* a beginning of movement and
rest, whether the "movement" [or specific type of behavior] is a local
motion, growth or decline, or a qualitative change. Such is not the case
with things like beds and clothes: that is to say, to the extent that these
come within the classification [25] of "products of art" [171], they do
not have implanted within themselves any tendency [155] to change
[115]; nevertheless, in so far as they happen to consist of stone or earth   **20**
or a composite material, they do have such a beginning of movement
and rest, but only in this respect. But even this circumstance gives evidence that the nature of a thing[‡] is in some sense the factor [83] which
initiates [82] movement and rest within that thing in which it is itself
immediately, not incidentally, present [82f]. The reason for saying "not
incidentally" may be illustrated by a physician who "incidentally" heals
himself; but since even he cannot as a patient practice the medical art,
physician and patient are usually two separate individuals, although

---

[*] ii.6.198a9,10. Cf. Plato *Timaeus* 30A,47E,52D,53B; *Philebus* 59; *Laws* x.886B- √
899D.

[†] *On the Parts of Animals* i. presents numerous parallels to *Physics* ii. √

[‡] *Metaphysics* v.4. ⟩

under certain conditions the same individual may happen to be his own physician and his own patient. What applies to the restoration of a patient's health, applies also to the products of any other art: not one of them has the source of its own production within itself; rather is this 30 source in an agent external to the product (as in the case of a house or of any other product of manual labor) or, when the thing happens incidentally to act [83] upon itself, the source is in some distinct aspect of the product itself.

These remarks bring out what "nature" is. Accordingly, a thing may be said to "have a nature" if it has within itself the sort of "beginning" described: every such thing is a "primary being," since it is a "subject" [of change]; and a "nature" always involves a "subject" in which it inheres. Such things are also "according to nature." So, too, are their 193a essential attributes; for example, the upward motion of fire neither *is* nor *has* a "nature" but happens "*by* nature" or, in other words, "*according to nature*." Thus we have explained what "nature" is and what it is to be "by nature" and to happen "according to nature." It would be ridiculous, however, to try to prove *that* nature is: it is obvious that there are many such natural beings; but to want to prove the obvious by what is not obvious shows inability to discriminate between what we can ascertain directly and what indirectly.§ Yet as a man born blind may resort to reasoning to convince himself about colors, so this failing, too, is a possible one; but those who have it must be unthinkingly talking about words.‖

10   Now, some hold that the nature or the primary being of natural beings is their proximate constituent by itself, apart from any arrangement of it: the nature of a bed,¶ they say, is wood and, of a statue, bronze. As Antiphon suggests, by way of giving a clue [38a] to this interpretation: bury a bed and let it rot until it gets enough power to send forth a shoot, this shoot would not be a bed but wood; hence, the bed's arrangement by convention and by art is only incidental to it, whereas its primary being is what remains continuously through its changing conditions! Moreover, suppose the relation between an object and its material to hold also between the material and something else (for example, between bronze or gold and water, between bronze or wood and earth, 20 and so forth), then that element is the nature or primary being of the object! This is the reason why some declare earth, others fire or air or water, and still others some or all of these elements, to be the nature

---

§ "Knowing *that* our subject matter is, we inquire *what* it is." *Posterior Analytics* ii.1.89b34. With the approach of *Physics* ii.1, compare that of i.1.

‖ Thomas Aquinas (against Avicenna): "ignorantia principiorum moventium non impedit quin naturam esse sit per se notum."

¶ 192b19, 20.

of beings. Whichever element or elements any of these men chooses, he puts it forward as the whole of a primary being, viewing all its other aspects as its modifications, states, or dispositions; any element, moreover, he contends, is eternal, since it cannot be transformed into anything else, and all other aspects of things he sets forth as coming and going an endless number of times. This, then, is one interpretation of "nature": the immediate persisting material of anything which has within itself a beginning of movement or change.

According to another interpretation, "nature" means "shape" or "form" as expressed in a definition. Analogously to the term "art," which relates to artistic skill and its products, the term "nature" relates to a natural process and its products. But since we do not ascribe the artistic character of an art object to a thing as long as it is only potentially, for example, a bed and has not yet received the actual form of a bed, a corresponding principle is to be maintained in the realm of things constituted by nature: what is potentially flesh or bone has not yet attained to its proper "nature" and is not properly a "natural" being as long as it has not assumed the definite [90] form by means of which we define what flesh or bone is. On this interpretation, therefore, "nature" would be the form of anything which has within itself a beginning of movement, the form being not separable except in thought [90]. To be sure, the being composed of both matter and form (for example, a man) is not *a* "nature," although it is [produced] "*by* nature";** yet, at any rate, form rather than material is "nature," since the term "nature" marks an actual being more appropriately than a potential one.

Again, man generates man. Because, on the other hand, a bed does not produce a bed, its "nature" is said to be not its pattern but wood: if the bed sprouted, we are told, what would come forth would not be another bed but wood. But even if this [pattern] is art, the shape [of man] is [his] nature, for man generates man.

Again, "nature" as genesis†† is a process *towards* [the product's] "nature." By way of contrast, the attempt to heal is a process directed not to the "art" of healing but to a healthy state; the task of healing must start *from* the art of healing,‡‡ instead of leading to it. But "nature" [as productive] is related to [the product's] "nature" in a different way: what grows [101b] out of something proceeds to something or "grows," not towards that from which it starts, but that towards which it tends. Hence, its final shape is its "nature." On the other hand, the

30

193b

10

** 192b32-193a1.
†† Thomas Aquinas: "puta si natura dicatur nativitas."
‡‡ Thomas Aquinas: "actiones denominantur a principiis, passiones vero a terminis."

"shape" or "nature" of anything has two meanings, for there is a sense
20 in which even a "privation" is a "form." But whether or not there is a
"privation" or a contrary [of the final form] in simple generation, we
must consider later.§§

2. Natural Science and Related Sciences

Having distinguished the different meanings of "nature,"* let us see
wherein the mathematician differs from the natural philosopher, for the
solids, planes, lines, and points with which the mathematician deals are,
after all, aspects [33] of natural bodies; at the same time, let us consider
more specifically whether or not astronomy is a part of natural science,
for it would surely be unreasonable to expect the natural scientist to
know what the sun and the moon are if he is not to know any of their
essential attributes, especially when writers on nature patently do dis-
30 cuss what shape the sun and the moon have and whether or not the
earth or the cosmos is spherical. Now, although the mathematician, too,
occupies himself with these things, he does not concern himself with
them as limits of natural bodies or with the properties they have in
this status [23]; what he therefore does is to separate [73] them, particularly since they can be separated in thought [169c] from processes
without any resulting incongruity [76] or falsity. But the advocates of
the "ideas" do this confusedly [205] when in their theory they separate
194a natural objects, which are much less separable than mathematical
objects. This will become clear in any attempt to define these two kinds
of entities and their respective attributes: odd and even, straight and
curved, number, line, and figure can be independently of movement;
not so things like flesh, bone, and man, which are defined like "snub
nose," not like "curved." This difference is also clarified by reference to
the more physical branches of mathematics (for example, optics, har-
monics, and astronomy), which really proceed, in a sense, by a method
10 the inverse of that of geometry: although geometry investigates physical lines, it ignores their physical aspects; optics, on the other hand,
examines lines which are indeed mathematical, but deals with them not
so much in their mathematical as in their physical aspects.

Accordingly, we must inquire [187] into "nature," since it has two
meanings: form and matter, in the same way in which we would consider what "snubness" is: reflection shows that such things are neither
purely immaterial nor considered in their purely material aspect. So one
may here, too, ask: With which of the two aspects of nature does the
natural philosopher properly deal? Or does he deal with that composed

§§ v.1; *On Generation and Corruption* i.3.
* ii.1.192a21, 193a28,b3,6,12.

of both of them together? If so, then he will also deal with each of them separately; but will he do this, then, in the same science or in separate sciences? If we look for light on this problem to the ancients, natural philosophy would appear to be exclusively interested in matter,† for even Empedocles and Democritus touched but lightly on questions of form and of what it means to be anything. If, on the other hand, art imitates nature, and every scientific technique [179] requires knowledge [182] of its distinctive form as well as of the material suited to it—if, for example, a physician understands health as well as the bile and phlegm on which health depends, and a builder understands the form of a house as well as its materials (such as bricks and timber), and so forth, then natural philosophy must take cognizance of both [the formal and the material] aspects of nature.

Again, any art or science takes cognizance both of the "where-for" or "end" and of all the means to this end. So, too, nature [as form] is an end which is correlative to [material nature as] means. In other words, when anything in a continuous passage [from its initial to its final form] comes to its appropriate terminus, this sort of stopping-point is also what the process is for. How absurd, therefore, the poet's line: "He has the consummation of his life!"‡ As if all endings [18a] were completions [100], instead of only the best ones! Then, too, the arts construct material objects, sometimes with intended useful results, but sometimes not. At any rate, we use everything as if it existed in our own interest, since we are ourselves in some sense an "end." In our work On Philosophy, we have distinguished two meanings of the expression "for the sake of": [namely, "for the sake of what" and "for the sake of whom"].§ But there are two sorts of arts which preside [82d] over a material by virtue of their knowledge: one sort uses the material, whereas the other is "architectonic" in directing its construction—although even the former sort is in a sense "architectonic"; but the two differ in that one of them knows the form (for example, the steersman knows and specifies the kind of form a helm is to have), whereas the other sort knows the material (for example, the maker of the helm knows out of what kinds of materials and by means of what kinds of processes the helm is to be made). Thus it is that in a process of art we ourselves make the material with a view to its uses [9c]; in natural processes, on the other hand, the materials are there to begin with [82f].

Again, "material" is one term of a relative distinction, since different forms require different materials. To what extent, then, must the natural philosopher know the form of a material, that is, *what* it is [87]? No

† *Metaphysics* i.3.
‡ Euripides.
§ Cf. *De Anima* ii.4.415b2,20.

doubt, to the extent that a physician knows sinews, and the smith, bronze; namely, to the extent of knowing the function [96a] of the respective materials. The forms [or functions] with which the natural philosopher is concerned are such as may be distinguished from but are embodied in materials; for it is a man that generates a man, no less than the sun. But whether there are forms separate from materials and what they are, it is the task of first philosophy to determine.‖

3. Material, Formal, Efficient, and Final Explanation

On the basis of these distinctions, let us now examine what and how many sorts of explanatory factors [83] there are. All inquiry aims at knowledge; but we cannot claim to know a subject matter until we have

20   grasped the "why" [203a] of it, that is, its fundamental explanation. It must clearly, therefore, be our aim in the present inquiry to get knowledge of the first principles to which we may refer any problem in our exploration of generation and destruction and of any natural transformation.

"An explanatory factor," then,* means (1) from one point of view, the material constituent from which [86] a thing comes; for example, the bronze of a statue, the silver of a cup, and their kinds. From another point of view, (2) the form [20] or pattern [89a] of a thing, that is, the reason [90] (and the kind of reason) which explains what it was to be [88] that thing; for example, the factors in an octave are based on the ratio of two to one and, in general, on number. This kind of factor is found in the parts of a definition [90]. Again, (3) the agent

30   whereby [95] a change or a state of rest is first produced [82]; for example, an adviser is "responsible" [83a] for a plan, a father "causes" his child, and, in general, any maker "causes" what he makes, and any agent causes what it changes. Again, (4) the end [100] or the where-for [96]; so, when we take a walk for the sake of our health, and someone asks us why we are walking, we answer, "in order to be healthy," and thus we think we have explained our action. So any intermediate means to the end of a series of acts: for example, as means of health

195a   there are reducing, purging, drugs, instruments, and so forth; for all these are for an end, though they differ from one another in that some are instruments, and others are actions [9c].

Since what we call an "explanatory factor" may be any one of these different aspects of a process, it follows not only that anything actually has several such factors which are not merely accidental differences of

---

‖ *Metaphysics* xii.6-10.

* The text of 194b23-195b21 is nearly identical with the text of *Metaphysics* v.2.

meaning (as both the sculptor's art and the bronze are needed to explain a statue as a statue, the bronze being its material, and the sculpturing, its agent), but it follows also that these factors are reciprocal: for example, exercise explains good health, and good health explains exercise; 10 though they explain each other differently (good health as end, and exercise as [82] means [109]). And the same thing may explain contraries: for the same thing which by its presence explains a given fact is "blamed" [83b] by its absence for the contrary fact; for example, a shipwreck is "caused" by the absence of the pilot, whose presence is responsible for the ship's safety.

All the factors here mentioned clearly fall under four varieties. From letters come syllables; from building materials come buildings; from fire, earth, and so forth, come bodies; from parts come wholes; and from assumptions come conclusions. The first factor in each of these pairs is the subject matter or the parts; the second is what it meant to be 20 that particular whole, or synthesis, or form. A "cause" in the sense illustrated by a seed, a physician, an adviser, and any agent generally, is the factor whereby a change or state of being is initiated. Finally, there are the ends or the good of the others; for all the others tend toward what is best as toward their end. It makes no difference now whether we say "their good" or "their apparent good."†

These, then, are the kinds of explanatory factors. But they fall into many lesser varieties, which can also be summarized under a few heads. There are several ways in which explanatory factors explain, even when they are of the same general kind [57a]. Thus one factor is prior to 30 another, which is posterior: for example, health is prior to both the physician and the technician; the octave is prior to the ratio of two to one and to number; and so always, the inclusive factor is prior to individual factors.

Then there are accidental factors of various kinds; for example, a statue is, we say, by Polyclitus, but it is also by a sculptor; the sculptor happens to be Polyclitus. And so the kind (sculptor) and the accidental (Polyclitus) it embraces are both factors in the statue; thus a man is responsible for the statue, and so is the more general species "animal"; 195b for Polyclitus is a man, and man is an animal. These accidental factors are sometimes remote and sometimes proximate; for example, between Polyclitus in particular and man in general there would be such intermediate factors as "a white man" and "an artist."

Besides, any factor, whether essential [55] or accidental [3], may be actually in operation [9a] or merely capable of acting [11a]: a house being built is the work of "builders," but more actually of the builder

---

† Thomas Aquinas: "quia quod apparet bonum non movet, nisi sub ratione boni."

who is building it. The same is true of the things to which explanatory factors refer—they may be singled out or referred to more generally: for example, "this statue," or "a statue," or even more generally, "an image"; and "this bronze," or "of bronze," or, generally, "of matter"; and similarly with reference to the accidental factors.

10   Moreover, both accidental and essential factors may be combined: for example, instead of Polyclitus or the sculptor, we say "Polyclitus the sculptor." However, these varieties reduce to but six, each being taken either individually or collectively: the accidental factors (individual or collective); combined or separate factors; and actual or potential factors. There is another difference between them: the operating and individual causes exist and cease to exist simultaneously with their effects (for example, this man actually healing is correlative with this man who is
20   now being healed, and this actual builder, with this thing-being-now-built); but potentially they do not exist together (for the house and the builder do not perish with the act of building).

We must, however, always seek the "highest" [or "principal"] explanatory factor of each case, as in any other investigations [of "reasons why"]; for example, a man builds only because he is a builder, and a builder, only because he has mastered the builder's art, which is therefore the more primary factor; and so in all such cases. Again, generic effects go with generic explanatory factors (for example, a statue with a sculptor), particular effects go with particular explanatory factors (for example, this statue, with this sculptor); so, too, potential effects correspond precisely to potential factors, and things actualized, to factors actually operating.

Let this suffice, then, concerning types of explanatory factors and the
30   ways in which they operate.

4. Opinions Concerning Luck and Chance

In explaining events, we often speak of "luck" and "chance," since many beings and happenings occur "by luck" or "by chance." Accordingly, we must ask: in what sense do luck and chance explain anything? Are they the same or different? In general, what is luck, and what is chance?*

196a   Some question whether there is any such thing. Nothing happens by luck, they say, but everything called "by luck" or "by chance" has some determinate explanation; for example, if someone goes "by luck" to the market-place where he unexpectedly meets someone he has been wishing to meet, this is, after all, a consequence of his decision to go to

---

* These questions are answered in reverse order in ii.5, 6.

market. So for any other event ascribed to luck, they say, it is always possible to find some explanation other than luck; indeed, it would seem very strange, they argue, if luck were really [7] a definite [4] factor. Moreover, they ask, why did none of the ancient sages in explaining generation and destruction develop a theory [72e] of luck? May it not 10 have been because they, too, believed that nothing happens by luck? However, we may reply, is not this, too, very "strange"? Consider: men generally are not unaware of the possibility of tracing many matters of luck and of chance to some definite factor which accounts for what happens (as the old argument [of Democritus] has it which eliminates luck), yet they continue to ascribe some events to luck and others not. Ought not the ancient sages for this reason alone to have given some attention to this subject? Yet they did not even recognize luck along with friendship, strife, mind, fire, or any other factor which they named. Their failure to develop a theory of luck and of chance is unjustifiable, whether they set them aside altogether or simply passed them by; espe- 20 cially since they did not disdain on occasion to take recourse to them. Thus, Empedocles remarks that air is not uniformly drawn upward but, as it were, by luck; as he puts it in his cosmogony, "it happened at that time [when friendship prevailed] to move in that way, but at other times in other ways." Empedocles also asserts that most parts of animals originated by luck.

Again, some attribute [83b] the heavens and all worlds to chance happenings, on the theory that the "vortex" arose "of itself," that is, the motion which separated and arranged the entire universe in its present order. This, too, is very "strange"! It is not by luck, they say, that animals and plants are and come to be, but these are to be explained by 30 nature or mind or something of the sort, since a seed does not develop into any random thing but one kind into an olive and another kind into a man, and so forth; yet the heavens, the most divine of all things visible, have no such determinate explanation as animals and plants have but arise spontaneously. Even if this were the case, it would be a situation that might well give us pause and prompt us to say something 196b of importance about it. Apart from the strangeness of the theory in other respects, it is especially "strange" to assert chance in the case of the heavens where we do not observe it and to deny luck to a region in which many events occur by luck; what happens is the very opposite of what their theory would lead us to expect.

Some, again, hold luck to be an explanatory factor but to be something divine and rather daemonic and therefore obscure to human intelligence.

Hence, we must examine what chance and luck are, whether they are the same or different, and how they are to be classified in relation to the explanatory factors which we have distinguished.

5. Facts and Traits of Luck and Chance

10   We observe that some events always occur in the same way and some usually so. Evidently, we do not ascribe either of these two classes of events to luck; nor do random events happen in the same way either necessarily and always or even for the most part. But everyone distinguishes, besides uniform and typical events, exceptional or nonnormal events. Evidently, then, there is such a thing as luck or chance; and by nonnormal events we mean random events, as by random events we mean nonnormal events.°

Then, too, we distinguish events which happen to some end [96b] and those which do not; and among ends, we distinguish those which are intended [178] and those which are unintended. Clearly, then,
20   things may not only have necessary uses and probable [125] uses, but they may also have at least some other possible [12a] uses. Now, things happen† to some end either by nature or by design [170]. They happen by luck when they come about by accident. Thus, just as anything may be either essentially or accidentally, so explanatory events or factors may also be of either kind. In the building of a house, for example, the builder is an essential factor, but the fact that he may be white or a musician is incidental; the essential factor is determinate, whereas incidental ones are indeterminate, since any individual may have an infinite number of attributes. As we have
30   said, then, when events which happen to some purpose or end come about by accident, we ascribe them to luck or to chance. (How these two differ, we shall explain later; but for the present it is evident that both of them refer to things that happen to some purpose or end.) Suppose, for example,‡ that a creditor would have gone to a market to recover his loan had he known that his debtor was there, but he happened to go there for another purpose with the result that he got his
197a   money, although it was not his usual or invariable practice [as it might have been for someone else] to go to the place where the two men met: the result (getting the money) is, like any object of deliberate choice, a factor external to the agent; and we say that the event happened by luck [relatively to the normal case], for we would not say this if he had gone there regularly or normally for the purpose of soliciting funds. Clearly, then, luck is an accidental factor which may intervene in events otherwise directed to an end in accordance with some intelligent choice

---

° *On Generation and Corruption* ii.6.334b4-19; *Rhetoric* i.10.1369a32-b5; *Posterior Analytics* i.30.87b20,21. Thomas Aquinas (against Avicenna): "nihil quod est ad utrumlibet, exit in actum, nisi per potentiam appetivam determinetur ad unum."

† 196b21-25, 197a5-14, 25-27 are duplicated in *Metaphysics* xi.8.

‡ ii.4.196a1-5.

[178]. Hence luck and thought [170] pertain to the same event, for choice takes thought.

However, the explanatory factors of events that occur by luck are necessarily indeterminate; therefore, luck, too, is held to be indeterminate and obscure to men, so that in a way nothing might be held to 10 happen by luck. All of these opinions have some reasonable justification. In one way there are things which happen by luck, namely, accidentally, so that luck is in some sense an accidental factor; but in another way nothing happens by luck, namely, absolutely. For example, a builder is responsible for a house; that he may be a flute-player, is incidental. Again, there may be innumerable reasons for a man's coming and getting his money when he did not come for that purpose: he may have wanted to see someone or have been following or avoiding someone or have gone to see some spectacle. So, too, luck is rightly said to be unpredictable. Prediction [90] applies to what is [1a] always or for the most part; but luck characterizes a third class of events. Consequently, since factors 20 of this kind are indeterminate, luck is also indeterminate. Still, there are cases in view of which we may ask: may we, then, explain a random occurrence by *any* random factors? We may explain a given case of health, for example, by the blowing of the wind or by the rays of the sun, but hardly by a haircut; for some incidental factors are less remote than others.

Moreover, it is good or bad luck when the result is good or evil; and prosperity or misfortune, when the results are on a grand scale. Even when we fail by but a little to come to some great good or evil, we are fortunate or unfortunate in that we can all the more readily think of 30 the good or evil as present and the slight miss as absent. Again, fortune is rightly held to be unstable, since luck is unstable inasmuch as none of the things due to luck are either invariable or typical.

Both luck and chance, then, are incidental factors (as has been said) in the class of events which may happen, but which do not happen absolutely or usually, yet which happen to some end.

6. Luck, Chance, and Explanation

Chance and luck differ in that the former has a wider extent. Everything that occurs by luck, occurs by chance; but not everything that occurs by chance, occurs by luck. 197b

Luck and the consequences of luck happen only to beings with capacity [82f] for good fortune and for the conduct of life [188b] generally.*

---

* Thomas Aquinas: ". . . vita practica sive activa est eorum quae habent dominium sui actus. . . ."

Hence luck must pertain to practical affairs; as is evidenced by the belief that good fortune is almost though not quite identical with happiness and that happiness is in turn a kind of activity, namely, "doing well." But beings lacking capacity for conduct of affairs cannot achieve anything even by luck; for example, inanimate beings, brutes, or children. They are incapable of shaping their course by intelligent choice [178] or of experiencing good or bad fortune—except by analogy [57],

10 as Protarchus pronounced altar-stones fortunate because they are valued above others which serve as stepping-stones. Yet beings unable to act by luck may indeed in some sense be acted upon by luck, namely, when an agent brings something about by luck in dealing with them.

Chance, on the other hand, is found both among living beings other than man and in many inanimate beings. We say, for example, that a certain horse came by chance to a safe spot because in consequence of his coming he found safety there, though we do not say that he came "with a view to his safety." Similarly, we say that a certain three-legged chair [tossed into the air] fell onto its feet "of itself" because it did so not "in order that" but "with the result that" someone could sit on it.

Evidently, then, we attribute to "chance" events which happen to some

20 end or result, but which do not happen with the end-result in view, and which have an external explanatory factor; and, of the events which happen by "chance," we attribute to "luck" those only which are possible objects of choice [178] for beings who can choose. The meaning of "chance" or "automatism" is reflected [38a] in the phrase "in vain," as applied to an action which does not result in the outcome intended: for example, a walk taken to stimulate bowel-movement is said to have been taken "in vain" if the desired result does not occur; so that any action which naturally functions [101c] as a means to an end other than itself is "in vain" when it does not issue in the end to which it is naturally instrumental—for it would be ridiculous for anyone to say that he had bathed "in vain" because he did not thereby bring about an eclipse of the sun, since no one bathes to such an end. Etymologically,

30 then, the "auto-matic" [or "self-active"] characterizes an event which by "itself" happens "in vain": a stone, for example, does not fall for the purpose of hitting anyone; hence, it falls "automatically" because [contrariwise] it might fall by the action of an agent and for that purpose. However, luck and chance are most clearly differentiated in events which happen by nature but contrary to nature: we attribute such events to chance rather than to luck; yet even such events differ [from chance in the strict sense] inasmuch as the latter is due to an external factor, whereas these are due to an internal factor.

We have explained what chance and luck are and wherein they differ 198a
from each other. Both belong to the type of "explanatory factors"
whence comes the beginning of movement. They are always a sort of
factor operating either by nature or by design, although the number of
these is indeterminate. But in spite of the fact that the results of chance
and of luck are results for which mind or nature might be responsible,
those chance results have come about because of some accidental factor.
It is clear that,† since nothing accidental is prior to the essential, neither
are accidental factors prior to essential factors. Accordingly, chance and 10
luck are posterior to mind and nature; and hence, even if chance is a
factor responsible for the heavens, mind and nature must be prior factors
not only for many other things but also for this universe itself.

7. Relations of Explanatory Factors

Clearly, there are explanatory factors, and they fall into four types
corresponding, as we have analyzed them, to the four meanings of the
question "why." The "why" of anything may ultimately mean: (1) "what
it is," for example, a straight line or the commensurable or anything else
as defined in mathematics or in the realm of unchanging considerations
generally; or (2) "what started a process," for example, a war, as in
revenge for a raid; or (3) "to what end," for example, in order to win 20
rulership; or (4) in the case of products, their "material."* Since these
are the four types of explanation, the natural philosopher must try to
understand them all if he is to deal adequately [101a] with the "why"
of anything in terms of each type of explanatory factor: the material,
the form, the agent, the "where-for."

Often the [first] three factors coincide. "What" something is and "to
what end" it is, may be the same; and a prime mover may be identical
with these factors in form or species, since man generates man, and
so with moved movers generally. As for unmoved movers, they do not
belong to the domain of natural science, since they evoke movement
without having within themselves either movement or a beginning of
movement. In this regard, therefore, we must distinguish three sciences 30
[197]: one, concerning unmoved beings; another, concerning moved
beings which are imperishable; and a third, concerning moved beings
which are perishable.

In giving an account of the "why" of anything, then, we must take
into consideration at once its material, what it is, and its mover. But

---

† Lines 7-13 are duplicated in part at the end of *Metaphysics* xi.8.
* Thomas Aquinas: "Sed quia forma est causa essendi absolute, aliae vero tres sunt
causae essendi secundum quod aliquid accipit esse, inde est quod in immobili-
bus non considerantur aliae tres causae, sed solum causa formalis."

investigators seeking explanations of events [116a] employ the method of asking, chiefly: "What comes after what?" "What started the process? And on what did it act?" Thus they give their chief attention to the sequence of events [136c]. However, the beginnings in control of natural movements are of two sorts. One of them is nonphysical, since it has no tendency to change within itself. This is an unmoved mover: it is (1) the completely immovable or first of all beings; and it is (2) what anything is or its form, which is also at the same time its completion or the end to which it functions. Hence, since nature [in its processes] has uses [96b], we must in our attempts to get knowledge admit the "final" factor along with others.

198b

In short, the question "why?" calls for a comprehensive answer. Thus, [we must explore (1) the efficient factor:] "this must result from *that*," and "from that" either without qualification or in most cases; [(2) the material factor:] "if this is to be, then *that* must be," just as syllogistic conclusions are conditioned by their premises; [(3) the formal factor:] "*that* was what it meant for this to be"; and [(4) the final factor:] "*that* is why it is best for this to be thus and so," not of course absolutely but relatively to its [40] distinctive being [26].

8. Natural Processes and Their Ends

10   Let us now state how end-results [96b] are grounded [83] in nature and then how there is necessity in natural processes [101a]. Writers on nature generally reduce their explanations to "necessity." Since hot and cold (and so forth) naturally function in certain ways [5], they say, it is by necessity that states of affairs [4a] are as they are and arise as they do. If these men speak of any other factor such as friendship and strife or mind,* they touch on it briefly and abruptly bid farewell to it. So, they ask, why should not nature act, not to some preferred end—but as it rains, not in order that crops may grow, but by necessity? Rising vapor must cool and, having become cool, must turn into water and descend, whereupon crops happen to grow; so, too, if crops on the threshing-floor are spoiled, the rain did not fall in order to spoil them, but this is simply the way things come about. Hence, why should not even bodily parts like teeth have developed in the necessary course of nature—sharp front teeth suited for the tearing of food and flat back teeth suited for the crushing of food? May they not have been produced, not to some end, but by coincidence? And may it not be so with all bodily parts supposedly having some inherent end or purpose? Those organic structures, then, which came into the world as if they had been

20

---

* Cp. Plato *Phaedo* 97-99.

produced to some end, survived because they had been automatically 30
organized in a fitting way; all others, like the man-faced offspring of
oxen in the theory of Empedocles, have perished and continue to perish.

However, this or any similar line of reasoning in objection [to natural
ends] cannot be sustained in the sense in which it is usually pursued.
All natural products like those mentioned are either always or for the
most part generated in definite ways, which is not the case with any
products of luck or of chance! Thus, we do not ascribe frequent rain 199a
in winter or heat in summer to chance or to coincidence, as we do
frequent rain in summer or heat in winter. If events, then, presumably
either result from coincidence or else happen to some end, and the
events mentioned cannot be ascribed to coincidence or to chance, then
such events must happen to some end. Even the men who argue as we
have reported agree that events like those in our illustration happen
naturally. Hence, there must be among natural beings and products such
as exist or come into existence to some end. Again, in any procedure
[188a] which has an end, what comes first and what comes next are
performed for that end. But as in human operations, so in natural proc- 10
esses; and as in processes, so in human operations (unless something interferes). Human operations are for an end, hence natural processes are so too. If a house, for example, came into being by nature, it
would come into being [in stages] just as it now does by art; and if
natural objects not only came into being by nature but also by art, they
would come into being [in successive stages] just as they do in the course
of nature; [in either case] each stage is [continuous with and] for the
sake of the next.

In general, moreover, art completes what nature is unable to carry
to a finish; or art imitates nature. If, then, processes by art are to some
end, it is clear that natural processes are too. The earlier [in a
series] are related to the later ones in processes by nature as they are
in processes by art.

This is strikingly evident in those other animals who do not act by 20
[conscious] art [171] or experimentation [194] or deliberation [177a],
which is the reason why some people debate whether or not spiders, ants,
and the like work by intellect [169] or something else. Moreover, as we
pass gradually down the scale, it becomes apparent that there is adaptation to ends also in the growth of plants which, for example, put forth
leaves to shelter their fruit. Hence, if it is *both* by nature *and* to an end
that the swallow builds its nest and the spider spins its web and that
plants put forth leaves for the benefit of the fruit and send their roots
down rather than up for nourishment, it is evident that there is such a
factor [as an "end"] in natural processes and beings. Further, since "na- 30

ture" is double, meaning either "material" or "form," and the latter is the end, everything else being for the sake of the end, the "form" will be the For What aimed at.

Now, mistakes occur even in processes by art, as when a scribe writes incorrectly or a physician gives a wrong dose; so that, clearly, 199b mistakes are possible in processes by nature also. If there are certain processes by art in which right procedures serve their respective ends, and failures occur when the end sought is not achieved, so it is with natural products; among the latter, monstrosities or freaks exemplify a falling short of natural ends.

If, then, among original formations [in the theory of Empedocles], the offspring of [man-faced] oxen were unable to attain to a certain [72b] completion or end, they must have been generated by a principle in which was some corruption, just as now [monsters are generated by some corruption] of the seed. Again, seeds must have come into being first, and not animals all at once; the "undifferentiated" which came "first" [before extant animal forms] must have been seed [not animal].
10 Again, since plants function to some end, only with less organization than animals, were there or were there not among them olive-headed offspring of vines? Absurd! But no more absurd than man-faced offspring of oxen among animals!

Again, any chance products ought [on his theory] to arise from seeds. Such an assertion would completely abolish everything "natural" and "nature"; for that is "natural" which in a continuous "movement" from some "beginning" within itself arrives at a definite end—not indeed the same end for all from any beginning whatsoever, nor indeed any chance end, but always the same end for each [kind] unless something inter-
20 feres. To be sure, ends and means may come about by luck, as in the story of the stranger who came by luck and before departing ransomed a prisoner† as if he had come for that purpose (although he really did not); here the result is incidental [to the purpose of his coming], for luck is according to our previous analysis an incidental factor. But when a certain result occurs always or for the most part, it is not an incidental or a lucky occurrence; and "natural" results are such as always or usually occur in definite ways unless something interferes.

As for those who think that nothing comes into being for any end if they do not see the moving factor deliberating, their argument is absurd; even art does not deliberate. Indeed, if the art of shipbuilding were *in the wood* [instead of being in a visible artist], it would act like *nature;*
30 hence, if art proceeds to some end, so does nature. This becomes espe-cially clear when someone [without deliberation] heals *himself;* nature

----
† Preferable to the reading "bathed."

is like this. In sum, nature is evidently a [genuine] factor [83] which, moreover, operates to some ends [96b].

9. Necessity as Related to Ends

In what sense, then, does anything happen [82f] "necessarily"? "Conditionally" [64h] [in subjection always to ends]? Or also [without any reference to ends, and thus] unconditionally [105]?

Currently, the belief prevails that "necessity" is to be found in generation [by nature], as though it were supposed that a wall had come into being because of the necessity by which what is heavy is naturally borne downward, and what is light, upward; in particular, because the foundation-stones had fallen to the base, the bricks (being lighter) had risen to a higher place, and the wood (being lighter still) had shot up to the top. Yet a wall, although unproducible *without* these things, does not come into being *because* of them, except as its material conditions; but a wall is brought into being to serve as a shelter and as a protection. So with all other cases in which ends are relevant: they do not come into being without means having a necessary nature, yet they do not therefore arise because of them, except as material conditions; but are produced to some end. Why, for example, is a saw such as it is? In order to perform a certain function [4a] to a certain end. This end cannot be brought about unless the saw is made of iron; then, *if* the saw is to function [9c] as a saw, it *must* be made of iron. Necessity, then, is hypothetical [64h], but not as an end. In other words, necessity is in the *material;* the end is in the "logos." 200a

10

Necessity is almost [55b] the same kind [57d] in natural processes as necessity in mathematics. Thus, *since* a straight line is such [that it forms, together with another on which it stands, adjacent angles equal to two right angles], therefore it is *necessary* for a triangle to have its internal angles equal to two right angles. But the converse does not follow; although, indeed, if the "conclusion" concerning the triangle were *not* so, then the straight line would not be as stated in the "premises." On the other hand, in the case of what happens to some end, the relation is inverse: if an "end" will be or is, its "antecedents" must be future or present also; and if they are not, the end will not be achieved —as in geometry, if the "conclusion" does not hold, neither does the "first principle." The end is itself a "beginning," not, to be sure, of the action, but of the reasoning [176]—whereas in mathematics, where there is no action, first principles are the beginning of reasoning only. Thus, we reason: *if* there is to be a house, *then* certain materials *must* be available [82f] or be produced; or, in general, there must be materials for 20

some end, and among such materials are bricks and stones if the end
is a house. But the end is not present or future because of them, except
that they are its material conditions. Yet if there are no materials at
all, if there are no stones and if there is no iron, then there will be no
30 house and no saw either; just as in geometry, if the internal angles of
a triangle were not equal to two right angles, then the first principles
of geometry would not be true either.

It is evident [from this analogy] that necessity in natural productions
refers to what we speak of as their material and its changes. But the
natural philosopher must speak of both factors, more especially of
the final factor: for it is this which accounts for a material [and its
processes], not the material which accounts for the end; and the end
is that for which [the material is necessary]. [That is to say,] the beginning is from a definition or a meaning! In the processes by art,
200b for example, if the house [we want] is such and such [5], then certain
materials must be produced or be available; and if the health [we de-
sire] is such and such [4a], then certain conditions must be brought
about or be present. Just so [in natural processes], if a man is such and
such [4a], then certain things [are necessary]; and if so, then certain
others. Necessity is also in the definition of a thing: if we define the
function of sawing as a certain way of dividing, the dividing cannot
take place unless the saw has teeth of such and such a kind, and these
cannot be such unless they are made of iron. There are also in a definition certain parts which function, so to speak, as the [necessary] material
for the definition.

III. BOOK GAMMA

Movement and the Infinite

1. Definition of Movement

Since we are exploring nature as a beginning of movement and change, 12
we must not neglect to define what movement is, lest unclearness [190]
about movement lead us into inevitable unclearness about nature; and
then, having defined movement, we must in the same manner take
up the terms which follow [136b] from it. Thus, movement being continuous, what is infinite becomes evident primarily in what is continuous,
with the consequence that the continuous is often defined in terms of
the infinite: "that is continuous which is infinitely divisible." In addition, 20
movement is held to be impossible without place, the void, and time.
Clearly, then, we must investigate and find out what we can about each
of these topics, not only for the reasons stated but also because the
features mentioned are general and common to all the things with which
our science deals; and a theory of special [42] traits must come after
a theory of common [92] traits.

First, then, as we have proposed, let us consider movement. Now,
being,* whether primary or quantitative or qualitative, and so forth,
may be only actual, or [only] potential, or both actual and potential.
Moreover, the relative may be "more or less"; or the relative may be
active or passive, that is, a possible mover or something that can be 30
moved, since what can impart movement is relative to what can receive movement, and vice versa. Then, too, there is no movement apart
from things: for when they change, they always change in primary being

---

* Much of iii.1-3 is duplicated in *Metaphysics* xi.9.

41

201a or quantity or quality or place; and there is nothing which, because it is common to all, falls into no one category, so that there is no movement or change any more than there is any being apart from the categories. But the categories apply [82f] to anything in one of two ways: by form or by privation. This is true of any specific primary being [4a] and also of any quality (white or black), any quantity (complete or incomplete), any spatial change (up or down, light or heavy). Consequently, there are as many kinds [20] of movement and change as there are of being.

10   Since any kind [19] of being may be distinguished as either potential or completely realized, the functioning [10] of what is potential [11c] as potential, that is "being in movement": thus, the functioning of the alterable as alterable is "qualitative alteration"; the functioning of that which *can* increase or decrease the contrary (there is no name common to both) is "increase" or "decrease"; the functioning of that which *can* be generated or destroyed is "generation" or "destruction"; the functioning of what *can* change its place is "local motion."

That this is what motion is, can be shown as follows. When building materials, insofar as we say them to be such, are actually functioning as building materials, there is something being built; and this is [the process of] building. Similarly, learning, healing, rolling, leaping, aging, ripening [are movements whenever something is being completed or 20 fulfilled, neither earlier nor later].† Since some things, then, are both potential and actual, though not at the same time and in the same respect—but (for example) what is potentially hot is actually cold, therefore such things will act upon or be acted upon by one another in many ways (for everything is capable both of acting and of being acted upon). Hence, too, every mover which is a physical agent is moved; indeed, every mover of this sort moves by being itself in movement. It seems to some that every mover is a moved mover; no—what is so will become clear from other considerations‡; there is also an unmoved mover.

The functioning,§ therefore, of what is potential, when there is actually something being realized, not as itself, but as movable, is motion. By 30 "as," I mean this: bronze is potentially a statue; nevertheless, the complete functioning [10] of bronze "as" bronze is not [exhausted in] its being changed [109]. For what it is to be [88a] bronze and what it is to be a certain potentiality for being changed are not the same; since if they were strictly the same in definition, the complete functioning of

---

† *Metaphysics* xi.9.1065b19-22.
‡ viii.5.
§ "Fulfillment" and related terms are to be understood in the sense of verbs.

bronze as bronze would be a movement. But they are not the same, as has been said. This is clear in the case of contraries; for to be capable of being healthy is different from being capable of being dis- 201b
eased (if they were not, being healthy and being diseased would be the same). But it is the same subject [85] that may be either healthy or diseased, such as blood or some other bodily fluid. Since, therefore, [to be bronze and to be potentiality] are not the same, as color and the visible are not the same, the complete functioning of the potential as potential, it is evident, is movement.

Clearly, this is movement; and it takes place whenever complete functioning itself does, and not earlier or later. For any power may sometimes function and sometimes not, for example, building materials; their functioning as building materials is the process of building. For the 10
functioning [9] of building materials is either this process of building, or it is the house. But when it has become the house, it is no longer buildable; hence the buildable exists in the process of building. Its functioning [9], accordingly, must be the process of building; and the process of building is a movement. And the same account applies to other movements.

2. Movement and Moved Movers

The truth of this argument is clear also from what others say about movement and from the fact that it is not easy to define it in any other way. For one cannot include movement or change in any other kind [19] of being. And this is clear from what those say who call it 20
"otherness" or "inequality"* or "nonbeing." None of these, however, is necessarily moved; and things do not change toward these or from these more readily than from their opposites. The reason movement is referred to such kinds of being is that it is held to be something indefinite; and the principles governing the privative type of contraries are indefinite, since no privation is a "this" or a "such" or fits into any of the other categories. And another reason why movement is held to be indefinite is that it cannot be classified either as a potentiality or an actuality of things [1a]; for movement does not apply necessarily to, 30
let us say, a quantity, either in its potential or in its actual state. Movement is supposed to be a sort of imperfect actuality, for the reason that the potentiality, whose actuality it is, is incomplete. And therefore it is hard to grasp what movement is; for it must be classified either under "privation" or under "pure actuality," and none of these appears to be possible. Consequently, what has been said remains plausible: 202a

---

* Plato *Timaeus* 52E, 57E, 58A.

movement is the kind of actuality which has been described, which is hard to discern, but which is capable of being.

Then, too, every mover which is capable of being moved is moved, as we have pointed out, and its unmoved state is "rest," which is an absence of movement in anything subject to movement: for to act upon anything movable as such, is precisely to "move" it; and since such a mover does this by contact, it is at the same time acted upon. Hence, movement is the functioning of the movable as movable when the mover touches the movable so that the mover is at the same time acted upon.

10 Moreover, a mover always conveys a definite [4] form, such as a primary being or a quality or a quantity, and it is in terms of this fundamental factor, the form, that the movement which the mover imparts is to be construed; for example, it is an actual man who begets a man from what is potentially a man.

3. Actualization of Mover and Moved

The difficulty is clear also: movement is in the movable; for movement is the perfecting [10] of the movable by some mover [109c], and the functioning of this agent is not different [from the perfecting of the movable].* For movement must be the perfecting of both; since a thing is an agent or mover because it has the power of moving, and is actually moving when that power is functioning. But it is [also] the power to make the movable function. Hence there is a single functioning of both [powers] alike, just as the intervals from one to two and from 20 two to one are the same, and the ascent and descent; these are one, although their definition is not one. Similarly, with the mover and the moved.

There is a dialectical objection: perhaps it is necessary that the complete functioning of the agent and of what it acts upon be different: of the one, acting; of the other, being acted upon. The function and end of the former is something done; but of the latter, something undergone. Since both are "movements," then, in what do they take place if they are different? Are both of them in what is being acted upon and moved; or is the activity in the agent, and the being acted upon in what is acted upon? (If the latter must be called an "activity," it is in another sense.) On the latter alternative [that the activity is in the "mover," and the undergoing, in "what is moved"], the movement will be in the 30 mover; the same reasoning will hold of mover and moved; so that every mover would either be a moved mover, or would have movement in it without being moved. On the former alternative if both are in what

---

* *Metaphysics* ix.1.1046a19.

is being moved and acted upon, both the acting and the being acted
upon, if, for example, both teaching and learning take place in the
learner), one consequence would be that the functioning of each
thing would not be inherent in that thing. Another consequence
would be the absurdity that what is moved would undergo two movements at once; and which two qualitative alterations would one thing
undergo which is moving toward a single form? This would be impossible. But suppose the functioning [of the mover and the moved] to be
one [functioning]; still, it would seem unreasonable to suppose that two 202b
things differing in form should have one and the same function. Hence,
the consequence would be that if teaching and learning, or acting and
being acted upon, were the same, then to teach would be the same as
to learn, and to act would be to be acted upon, so that a teacher would
have to be learning everything he is teaching, and an agent would have
to be undergoing everything he is producing.

However, it is not at all absurd that the functioning of one thing
should be in something else: teaching is the functioning of someone
able to teach; it is certainly in someone, and is not disconnected; but
[the teaching] of the one [teaching] is in the one [taught]. Again, there
is nothing to prevent the functioning of two things from being the
same: not the same as to being; but as what is potential is related to 10
what is actually functioning. Then, too, it is not necessary for one
to be learning while teaching; and even if to act and to be acted upon
are the same, it is not in the sense in which they have the same definition expressing what-it-meant-for-them-to-be [88], like "clothes" and
"garments," but (as we have said before) in the sense in which the road
from Thebes to Athens and the road from Athens to Thebes are the
"same." The illustrations given show what things have all their attributes
the same: namely, not things which are in any way whatever the same;
but only things which have the same definition [23]. To be sure, even
if teaching is the same as learning, to learn need not therefore be the
same as to teach; just so, even if two places are separated by the same
distance, the spanning of the distance from here to there would not
therefore be the same as the spanning of the distance from there to
here. In general, teaching and learning or acting and being acted upon 20
are not the "same" in the strict sense of the term, but what is the
same is the movement to which each of such correlatives belongs; for
the operation of "this" upon "that" and that's being acted upon by this
differ in definition.

We have thus stated in both general and particular terms what movement is, for it is not hard to see how each of its kinds is to be defined.
Thus, qualitative alteration is the functioning of the alterable in so far

as it is alterable; or in plainer language, the functioning of what can act and of what can be acted upon, as such. So, too, both in general and in particular terms, with building, healing, and so forth; and a similar definition may be given of any of the other kinds of movement.

4. The Being of the Infinite

30   The science of nature deals with magnitudes, movement, and time, each of which is necessarily either infinite or finite, even though not everything is "either infinite or finite"; for example, considerations like qualities and points probably need not be considered in either of these ways; hence, it behooves one who explores [197] nature to develop a theory of the infinite. Is there such a thing or not; and if so, what is it?

203a   That the theory of the infinite belongs [55] to natural science, is indicated by the fact that all the recognized philosophers aiming at a theory of nature have worked out reasoned views [90] of the infinite; indeed, all of them treat [64] the infinite as in some sense a *principle* of beings.

Some, like the Pythagoreans and Plato, regard "the infinite" as something independent [2], that is, not as an attribute [3a] of something else, but as being itself a *primary being* [26]. The Pythagoreans, who do not treat numbers as anything separate [or abstract], locate the infinite in sensible things; they also declare what is beyond the boundary of the universe to be infinite.* Plato, however, denies that there is anything beyond the boundary of the universe, either anything bodily or any "ideas" (since "ideas" are "no-*where*"); and he locates the in-

10   finite both in sensible things and in the "ideas."* Moreover, the Pythagoreans identify the infinite with even number: inasmuch as even numbers can be divided [65d] and are also limited by the odd,† they give [33d] to beings [1a] an infinity [of forms]. By way of evidence, the Pythagoreans appeal to what happens to numbers [in geometrical patterns] when gnomons are placed successively round the unit and [analogous constructions are formed] apart from the unit: in the former series,‡ the resulting [square] figures [20] preserve a unity [of shape]; in the latter series,§ the resulting [rectilinear] figures are always diverse [in the proportion of their adjacent sides]. Plato, however, maintains two infinites, the great and the small [in discontinuous number series].

---

* Plato *Timaeus* 32E; *Philebus*.

† The Pythagoreans seem to have taken the point as indivisible and patterns of even numbers as capable of division through the middle of the gaps between the points.

‡ $1+3+5 \ldots +(2n-1)=n^2$.

§ $2+4+6 \ldots +2n=n(n+1)$.

On the other hand, all those philosophers who give their attention to nature take [64h] infinity as a *character* of something different from infinity, namely, as a character *of some concrete nature* among the so-called "elements" (such as water or air or something intermediate between them); but whereas no one who regards the elements as finite in number describes them as infinite [in extent], those who treat the 20 *number* of elements as infinite, like Anaxagoras with his homogeneous parts and Democritus with his differently shaped atoms, declare the infinite [whole] to be continuous by contact [of the elements or atoms].‖ Moreover, Anaxagoras believes that any part is a mixture like that of the whole [150] because he sees anything as coming from anything. This is apparently also the reason why he says that at one time all things were together: this flesh, this bone, anything whatever, therefore all things, and all at the same time; for the separation not only of each particular thing but also of all things has, according to him, a beginning. Since whatever arises, then, arises from a body similar to it, and all things have a genesis (though not all at the same time), the genesis 30 must also have a beginning: this beginning, which is one, he calls "mind"; and "mind" in turn must begin its thinking and activity from some starting-point. Hence, all things must at some time have been together and must at some time have begun to be in movement. Democritus, however, maintains that none of the primary natures comes out of another. Still, even for him, body considered generally [92] is the source 203b of all things inasmuch as its parts differ only in size and shape.

It is clear, then, from these considerations that the theory of the infinite is a proper concern of natural scientists or philosophers. With good reason, too, all of them take the infinite to be a principle. For one thing, it cannot be altogether futile, yet the only validity [11] it can have is that of a principle: everything is either a source or derived from a source; but if the infinite were derived from a source, it would have a limit. Again, the infinite cannot be generated or destroyed, any more than can a principle: what comes into being must have [65] an [initial] terminus [100]; and all destruction has a [conclusive] ending [100a]. Hence, as we have said, the infinite does not have a beginning of its 10 own but is itself held to be a beginning of other things and to contain all things and to govern them, as those describe it who do not set up alongside the infinite some other explanatory factor such as mind or friendship; and the majority of physicists would agree with Anaximander when he says that the infinite is divine, since it is deathless and indestructible.

‖ Hence, the upshot of this view, too, is an infinite in *magnitude*.

Now, there are chiefly five lines of investigation which lead to the conviction that the infinite is something: (1) time is infinite; (2) magnitudes are divisible in such a way that mathematicians, too, work [163] with the infinite; (3) things come to be and pass away ceaselessly only
20 because the source of their generation [and destruction] is infinite; (4) the finite is always limited by something else so that it must be without any limit that something must always be limited by something else; and (5) there is, above all, the problem which everybody raises that thought has no stopping-place in dealing with numbers, with mathematical magnitudes, or with what is beyond the boundary of the universe. In particular, the infinite beyond the boundary of the universe suggests an infinite body and infinite worlds.¶ Why, in the void, would body be here rather than there? If there is mass anywhere, it must be everywhere. Even [aside from this argument] if there is void or infinite place,
30 still, there must be infinite body also, for there is no difference in the eternal between what may be [12a] and what is [23].

However, the fundamental difficulty which the theory of the infinite contains is this: either an outright denial or an outright acknowledgment of the being of the infinite leads to many impossibilities. Again, there is the question *how* the infinite is. Is it a primary being? Or is it an essential attribute of some nature? Or is it neither of these but never-
204a theless something infinite [in extent] or infinite in number?** Especially does the philosopher of nature face the problem whether there is a *sensible* infinite magnitude. First, then, we must distinguish the various meanings of the "infinite." The infinite†† is (1) what by its very nature [101c] cannot be spanned, as a voice is naturally invisible, or (2) what can be endlessly spanned, or (3) what can scarcely be spanned, or (4) what can naturally be spanned, but is not actually spanned or has a limit. (5) There is, besides, an infinity by addition, and one by subtraction, or both.‡‡

¶ According to Simplicius, Archytas insisted that, if there is nothing beyond the boundary of the universe, nothing could stop him from stretching out his hand into the beyond.

** i.2.184b21, 4.187a26-5.188a19-30, 6.189b11-20; *On Generation and Corruption* i.3.318a20.

†† The account of the infinite in *Metaphysics* xi.10 is made up of extracts from *Physics* iii.4, 5, 7.

‡‡ (1) In an accidental sense (iii.5.204a14-17), (2) strictly (a character of the subject of the process), (3) more roughly (for example, a labyrinth), (4) by analogy (iii.6.207a2-7), (5) derivatively (applied to the process). Cf. iii.7.207b1 note.

5. The Infinite as Not Actual

The infinite cannot be something that is separate from perceptible things and an independent being [4d] which is itself infinite. For if infinity is neither a magnitude nor a plurality, but is itself a primary being, 10 not an accident of it, then it will be indivisible; since the divisible is either a magnitude or a plurality. But, if indivisible, it is not infinite; except as the voice is invisible. But it is not usually so regarded, nor are we analyzing it in this sense; we regard it as incapable of being traversed. And if the infinite is an accident,* it cannot as infinite be an element of things, any more than the invisible is an element in conversation [78] because voices are invisible. And how can the infinite be by itself [2], unless number and magnitude, of which infinity is an attribute [35a], are also by themselves? Indeed, infinity would inevitably fall short of 20 number or magnitude in this respect. And it is evident that the infinite cannot be an actuality or a primary being or a source, for then any given part of it would be infinite; since "to be [88a] infinite" and "infinite" would be the same if the infinite were itself a primary being and not attributed to a primary subject [85]. Consequently, it is either indivisible or, if it has parts, it is divisible into infinites. But the same thing cannot be many infinites; for as a part of air is air, so a part of the infinite would be infinite, if the infinite were a primary being or a principle. Accordingly, the infinite must be without parts and indivisible. But an infinite in its completeness [10] cannot be indivisible; for it must be quantitative. Accordingly, subjects [82f] must be accidentally infinite. But, if so, then, as has been said, infinity cannot itself 30 be a principle, though it may be accidentally related to a principle such as air or even numbers. Hence, it is unreasonable to speak as the Pythagoreans do: they describe the infinite as a primary being and proceed to divide it.†

We seem to be inquiring into something general [concerning the infinite as an attribute], namely, whether there is among mathematical entities an [actual] infinite [extension] or among intelligibles without 204b magnitude [an actually infinite number]; but since our scientific investigation concerns sensible things, we are especially interested in asking whether or not there is among sensible things an infinitely large [actual] body. That the infinite is not among sensible things is clear from the following dialectical considerations. If the definition of a body is "what is bounded by planes," there cannot be an infinite body, whether intelli-

---

* According to another interpretation, "if the infinite is incidentally a primary being" (Simplicius), for example, by way of having air or water for its referent, it is an element only as air or water (Cod. Reg. 1947).

† iii.4.203a11.

gible or perceptible; nor a separate [or abstract] and infinite number,
since the number of what has number is numerable, and if so, then the
10 infinite can be spanned.‡ What has been said may also be clarified by
the following physical considerations. The infinite can be neither composite nor simple. It cannot be a composite body, since the elements
are quantitatively limited. For contrary elements must balance each
other, hence neither can be infinite; since if one of the two contrary
forces should fall short in power, the finite would be destroyed by the
infinite (for example, finite fire by infinite air, no matter how much
more *definitely* powerful a portion of fire than an equal portion of air).
20 And it is impossible that each should be infinite: for a body is what
has extension in every direction, and the infinite is boundlessly extended; so that, if the infinite is a body, it will be infinite in every
direction.§ Nor can an infinite body be one and simple: either, as some
say,‖ something different from the elements out of which they are supposed to be generated; or in general [105]. Some describe the infinite
in this way and not as air or water, lest an infinite element destroy the
others: the elements have contrarieties, air being cold, water being moist,
fire being hot, and if any one of them were infinite, the others would
long ago have been destroyed; as it is, these people maintain, the
30 infinite is different from them and is their source. However, the
infinite cannot be as they describe it: not because it is infinite, for this
point may be made in a general [92] argument applying to air or water,
and so forth, alike; but because there is no sensible body apart from
the socalled "elements." Everything can be resolved into that of which
it consists, and hence presumably into air and fire and earth and water
and something besides them; yet no such body besides them is ob-
205a served. Nor can the infinite be fire or any other element; for, aside
from the question how any body can be infinite, the All, even if it is
finite, cannot be or become any one of them, as Heraclitus maintains
that all things sometimes become fire. The same argument applies here
as to the One which the natural philosophers¶ present as independent
of the elements. For everything changes from one contrary to another;
for example, from hot to cold.

We must not only consider in terms of the alternative cases enumer-
ated whether or not there can be a sensible infinite body; but that this
is impossible, may also be clarified by the following general [44] consid-
10 erations. Every kind of sensible body is naturally somewhere: its whole

‡ The infinitely numerable would not be an actual infinite number.
§ It is, moreover, unnecessary and untenable to assume with Anaxagoras that the
   number of elements in the infinite body is infinite (i.4).
‖ Anaximander.
¶ Anaximander.

and its parts have their same and proper places; for example, the earth and a clod, fire and a spark. Consequently, if the infinite body is homogeneous [57a], it must be immovable or else constantly moving. But this is impossible; for why should it rather go down than up, or move anywhere? For example, if there were a clod in it, where would this move or rest? For the proper place of this homogeneous body is infinite. Will a clod in it then occupy the whole place? And how? What, then, is its rest or movement? Will it rest everywhere? Then it will not move. Or will it move everywhere? Then it cannot stop. On the other hand, if the All has unlike parts, their proper places will also be unlike: first, 20 the body of the All is not one except by contact; and, secondly, the parts will be either finite or infinite in kind. They cannot be finite, for then some will be infinite in quantity, such as fire or water, and others not, if the All is infinite; but such an infinite element would destroy those contrary to it (as we have said before). Hence, too, none of the natural philosophers identify the one infinite body with fire or earth but with water or air or something intermediate between them, since each of the former has a determinate proper place, whereas the resident place of the others is intermediate between up and down. But if the infinite parts are infinite and simple, their proper places are also 30 infinite, and the elements will be infinite in number; and, if this is impossible, and the places are finite, the All must also be finite. For places and bodies cannot exclude each other: the whole place must not be larger than so much as can hold the body, and then the body would no longer be infinite; and body must not be larger than place; otherwise, there would be either a void or a body which would naturally 205b reside nowhere.

On the other hand, Anaxagoras argues absurdly that the infinite is at rest: it stabilizes itself, he says, because it is "in itself," there being nothing else surrounding it; and where anything is, there it is naturally. But this is not true, since a thing can be somewhere not naturally but by compulsion. Even if the whole is unmoved because it stabilizes itself and is in itself (and therefore immovable), an explanation must be given as to why it is not naturally in movement: it is not enough to make the statement in question and let it go at that, since the infinite might not be in movement because it has no place to move to,°° and 10 yet there would be nothing to hinder it from being naturally in movement; thus the earth is not removed and would not be even if the earth were infinite as long as it is held in place by the center of the universe, but it would be at rest at the center not because it has no place to move to but because it naturally acts as it does, and yet it

---

°° Other versions: "since anything else might not be in movement."

could be said to stabilize itself. If, then, the earth, supposedly infinite, would be at rest where it is, not for this reason but because it has weight, and what is heavy remains at rest at the center, and the earth is at rest at the center, then the infinite would likewise be at rest not because it is infinite and stabilizes itself but for some other reason. At the same time, it is clear that any part would have to be at rest and 20 that, just as the infinite remains in itself because it stabilizes itself, so any given part would remain in itself; for the whole and its parts have similar places, for example, the earth and a clod (a lower place), fire and a spark (an upper place). Consequently, if the infinite has its resident place in itself, so will its parts. Accordingly, it will remain in itself.

In general, we evidently cannot speak of an infinite body and at the same time of a proper place for bodies if every sensible body has either weight or lightness. For a sensible body must move either towards the center (if it is heavy) or upward (if it is light), and the infinite, too, would have to do likewise. But either the whole or a half of the 30 infinite cannot do either; for how will you divide it? Or how will part of the infinite be down, and part up, or part extreme, and part central? Again, every sensible body is in a place, and the kinds or differences of place are up and down, before and behind, and right and left: these are distinguished not only relatively to ourselves and by convention, but also in the whole itself; but these cannot be in an infinite body. In gen- 206a eral [105], if there cannot be an infinite place, but every body is in a place, neither can there be an infinite body; for what is somewhere is in a place, and what is in a place is somewhere. If then the infinite cannot be quantitative, that is, a particular quantity such as two or three cubits (for this is what "quantitative" means), then also the fact that something is in a place means that it is somewhere; and this means either up or down or one of the others among the six distinguished, and each of these is a limit.††

From these considerations, then, it is evident that there is no body which is actually [9] infinite.

6. The Infinite as Potential

It is also clear that, if we deny the infinite altogether, many impossible 10 consequences would follow. Thus, time would have a starting-point and a stopping-point; there would be magnitudes not divisible into magnitudes; and numbering would not be unlimited. Since, then, neither of

---

†† Other arguments may be drawn from facts of experience to show that no nonsimple or compound bodies can be infinite (*De caelo* i.5-7).

the alternatives [72e] appears possible [that there should or that there should not be such a thing as the infinite], we must mediate between these two views and distinguish [59a] how the infinite is and how it is not.

Now, "to be" may mean to be potentially or to be actually, and there "is" an infinite by addition and one by subtraction; but, as we have shown, spatial magnitude is not actually infinite but is infinitely divisible (there being no difficulty in refuting [118] the belief in indivisible lines); consequently, we must explore the remaining alternative of a *potential* infinite. However, we must not take [65] the infinite as being potential [11c] in the ordinary meaning of a potentiality which may be completely actualized, as the bronze which is potentially a **20** statue may become an actual statue. But "to be" has many meanings, and the infinite accordingly has the kind of being which a day has or which the games have, namely, inasmuch as one after another continually comes into being: for these, too, "are" potentially or actually; thus, there "are" Olympic games both inasmuch as they may [11a] be held [116] and inasmuch as they are being held. Then, too, the infinite in time and in the generations of man clearly differs from the infinite in the division of magnitudes; although the infinite has in general the kind of being which a continually repeated process has, finite on each occasion [65], but always different. Hence, we must not take the infinite **30** as actually individual [4b], like a man or a house; rather does the infinite have the kind of being which a day has or which the games have —that is to say, the kind of being which does not belong to a concrete [4] primary being [26] that has come into being, but the kind of being which consists in continual coming to be and passing away, which is finite on each occasion, but which even so is different. But in magni- **206b** tudes, it happens [3b] that what has been taken [for example, in the division of a line] persists; whereas in time and in the generations of man, the parts taken pass away, but the supply does not fail [148a]. Moreover, within a finite magnitude, the infinite by addition is in a way the same as the infinite by division inasmuch as the addition varies inversely with the division: as we see the division going on endlessly, we observe the part marked off increasing endlessly; that is, continually taking a determinate part of a finite magnitude and inversely adding a part determined by a constant ratio (instead of keeping the parts equal) are processes which will not come to an end, whereas division **10** in an increasing ratio (to keep the parts equal) will come to an end (since every finite magnitude can be exhausted by means of any determinate quantity).

There is no actual infinite, then, but there is an infinite potentially and by division. Still, the infinite is actual in the sense in which a day is actual or the games are actual [namely, with the possibility of the next]; but potentially, the infinite does not have the independent being [2] which the finite has, but the infinite is the matter [of the actuality]. In this respect there is a potential infinite also by addition, which we have identified with the infinite by division—but only in a way: in inverse addition, it is always possible to take something beyond what has been taken, yet without exceeding the whole magnitude; in divi-
20 sion [in the same ratio], there will always be a magnitude smaller than any that may be specified [72]. Consequently, there cannot be an infinite by addition which even potentially exceeds every determinate magnitude; unless, indeed, there happens [3] to be beyond the boundary of the universe an actually infinite body whose essential being [26] natural philosophers declare to be air or something of the sort. But if there cannot be an actually infinite body of this sensible sort, then there evidently cannot be a body even potentially infinite by addition (except for the inverse addition which we have described). Although Plato maintains two infinites for this reason that it is possible both to add and to
30 divide *ad infinitum,* yet Plato does not use the two infinites he maintains: there is for him no infinite by division in numbers, where the unit is the smallest; nor is there an infinite by addition, since his number system extends only as far as the number ten.

In consequence of these considerations, the infinite is contrary to what
207a is usually described as such: there is an infinite, not when there is nothing left over and beyond, but when there is always something over and beyond! This is exemplified by the bezel-less rings which people characterize as "endless" because it is always possible to take a part beyond that taken, although this use of the term is based only on an analogy to the "infinite" in its strict meaning: the condition stated must be supplemented by the further condition that the same part is not to be taken a second time; the latter condition is not fulfilled in the case of the circle, where a part "always different" is only a neighboring part. Accordingly, *that is infinite of which it is always possible, in regard to quantity, to take a part outside what has already been taken.* On the other hand, when there is nothing left over and beyond, there is some-
10 thing "complete," that is, a "whole"; and a "whole" (for example, a whole man or a whole box) is, by definition, what has nothing wanting [23a]. It is in this same sense of an individual "whole," but with preeminent significance, that the universe is a "whole" which leaves nothing out; whereas that which has anything external whatever absent from it, is not "all." "Whole" and "complete" are either identical or akin in

meaning [101]; but nothing is "complete" which does not have an "end," and an "end" is a "limit." Hence, the view of Parmenides that the whole, being equal in every direction from the middle, is limited, is to be preferred to the view of Melissus that the whole is infinite. To connect [137c] the infinite with the "All" or the "whole" is not at all to relate commensurables. The dignity [53] and all-embracing quality of the 20 whole is only ascribed to the infinite because of a [33] superficial [4] confusion [57] of the infinite with the whole. Being the matter of the completeness belonging to a magnitude, the infinite is a whole only potentially—not actually, since division and inverse addition go on endlessly; it is a finite whole, not of itself [2], but because of something else. As infinite, the infinite does not contain but is contained; as infinite, it is therefore also unknowable, for matter is formless. Evidently, then, the infinite is to be treated [90] as an *aspect* [22] rather than a whole; for a material is an aspect of a whole, as "bronze" is an aspect of a "bronze statue." For the infinite to contain sensible things, it would also be requisite for the great and small to contain intelligible things; 30 and it would be absurd and impossible for the unknowable and indeterminate to do any containing and determining.

7. The Infinite and Its Subjects

It is reasonable for us to conclude that there is no infinite by addition in the sense of a magnitude greater than any magnitude but that there is an infinite by division. Like a material aspect, so infinity is "contained" 207b [as belonging to a determinate subject]; what "contains" it, is a "form" [20].*

With good reason, too, we hold that in counting we may proceed from a minimum to an ever greater number; but that, contrariwise, a magnitude is infinitely divisible but not infinite in prolonged dimensionality. The reason is that a unit of any sort is undivided (for example, a man is one man, not many), whereas a number is a number of units (that is, so many of them), so that numbering necessarily starts [111] from an undivided whole; so "three" or "two" or any other number- 10 name is an abstract noun, and it is always possible to think of a greater number. The infinity of numbering may be understood in terms of the continual bisection of a line, [which guarantees the endlessness of the process]. Consequently, the infinity of numbering is potential; it is never

---

* Infinity is primarily ascribed to a quantitatively determinate subject ("the continuous and sensible") which can participate in a continually repeated process where each term of the series implies the possibility of the next; secondarily, to such a continually repeated process as well as to such abstract terms as number, time, and movement or change.

actual, but consists in the fact that a number can always be found which is greater than any number suggested. But there is no separate infinite number, nor is the infinity of numbering a permanent actuality; instead, it is a process [in which one part after another continually comes into being], as in the case of time and the reckoning of time. The infinity of numbering forms a contrast to the supposed infinity of increasing magnitude: although a continuum is infinitely divisible, a magnitude is not infinitely producible. A potential extension can be only as great as the greatest possible actual extension. Consequently, since there is no

20 sensible infinite magnitude, there can be no actual infinite magnitude greater than any and every determinate magnitude, for then there would be something greater than the universe.†

Moreover, the infinite is not the same in magnitude as in movement and time, as if the infinite were some single nature; but the subsequent rests upon the prior. Thus, movement is infinite because the magnitude is infinite over which movement or alteration or increase takes place; and time is infinite because movement is so. For the present, we shall not make use of these distinctions [but shall content ourselves with calling attention to the application of the concept of infinity to such abstract terms as movement and time]; later‡ we shall also ask what this means in each case as well as why every magnitude is divisible into magnitudes.

This account does not, by denying an actual inexhaustible infinite of prolonged dimensionality, take away from the mathematicians their

30 theoretical pursuits in which they neither need nor use such an infinite. Mathematical requirements are met by a figure which may be of any size without ceasing to be determinate. Any line, however short, may be divided on the same principle [90] as any other, however large. For the purpose of their demonstrations, therefore, it will make no difference to the mathematicians that there is no actually existing infinite magnitude.

Since, among the four kinds of fundamental factors in inquiry, the infinite is evidently a fundamental factor in the sense in which a mate-

208a rial is, "being infinite" is a "privation" [not a perfection but the absence of a limit]; and its essential [2] definite subject [85] is the continuous and sensible.§ Evidently, too, all other thinkers make use of the infinite as a material factor; hence, it is absurd of them to present the infinite as a container‖ instead of something contained [as a character having a possessor].

---

† Thomas Aquinas: "non . . . est in materia prima potentia nisi ad terminatam quantitatem."

‡ vi.1, 2, 4.

§ Thomas Aquinas: ". . . ne aliquis intelligat, quod infinitum est materia, sicut materia prima. . . ."

‖ iii.4.203b11, 6.207a18-32.

8. The Infinite and the Finite

It remains for us to attack the arguments* for the view that the infinite is not merely potential but is a separate being [72d]. Some of the arguments are not at all conclusive, and others can be countered by further valid objections.

An actually infinite sensible body is not needed to provide for unfailing coming to be and passing away. The coming to be of some things 10 may be the passing away of others. And the whole may very well remain finite. Moreover, touching and being limited are different. Since everything that touches, touches something, therefore touching is relative to something else; and touching is an accident of only some of the things that are limited. But what is limited, is not limited in reference to something that surrounds it;† and not anything and everything that is limited touches anything and everything else. As for the infinite in our thinking, it would be absurd to rely on it for an actual infinite. There is not in any state of affairs, but only in the thinking referred to, something greater than the greatest or less than the least. Someone might suppose any of us to be infinitely many times his own size; but if anyone has a size greater than ours, it is not because someone thinks so, but because anyone is the size he is. The thinking is incidental. Moreover, the reason 20 why time, movement, and thought are infinite is because any part that is taken does not persist. Finally, magnitude is not, either by division or by supposed [169d] increase, actually infinite.

We have thus shown how the infinite is, how it is not, and what it is.

---

* iii.4.203b16-30.
† iii.1.200b32-201a15, 6.207a8-14.

IV. BOOK DELTA

———— ···•·◄◎||◎►·•···· ————

Place, the Void, and Time

1. Place and Its Classification

A natural scientist must inform [181a] himself not only on the infinite but also on place. Is there such a thing or not; and if so, how is it, and what is it? People generally suppose existing things to be somewhere
30 and nonexisting things to be nowhere,° for where would a goatstag or a sphinx be? Moreover, the kind of "movement" involved in all the types of change and most strictly so called is change of place or (as we are accustomed to designating it) "local motion." However, the question "What is place?" is fraught with many difficulties. All the facts [82f] appear to give rise to divergent theories [187]. Besides, nothing has come down to us from our predecessors either by way of a statement or by way of a solution of the problems concerning place.
208b Nevertheless, it clearly seems to be a fact that place "is." First, there is displacement. *Where* now there is water, *there* will be air when the water has gone (as out of a vessel); and then again some other body will occupy the same place. The place, therefore, seems to be different from all the bodies which successively displace one another. That "in" which the air is now, is that "in" which the water was before. Consequently, the place was clearly something; that is, the location was different from the bodies which, by passing into and out of it, changed places. Secondly, the motions [121] of the simple bodies (fire, earth,
10 and so forth) show not only that place is something but also that place has some kind of functional significance [11].† Unless interfered with,

---

° Plato *Timaeus* 52.
† Thomas Aquinas: "quod locus habet quamdam virtutem conservandi locata."

each of the simple bodies moves up or down to its appropriate place. The six directions, up and down and so forth [right and left, before and behind], are parts or kinds of places; and they are such not only relatively to ourselves, but they are up and down, right and left, and so forth. For us, they do not always remain the same but change as we change our position; so that often even the same thing may be both right and left, both above and below, both before and behind. But in nature each of these is distinct independently of our own position. Not any chance direction is "up," but that in which fire and light bodies 20 move; so, too, not any chance direction is "down," but that in which heavy and earthy bodies move. Thus, these distinctions of place do not only differ conventionally but also depend upon the ways in which [bodies] act [11]. This is also clarified by mathematical representations: though they do not have their being [1a] in a place, they nevertheless have distinctions of position (like "right" and "left") relative to us; they have their "position" therefore in concept [169a] only, but they do not have any position in their own nature. Thirdly, those who believe in the existence of a void acknowledge the existence of place inasmuch as a void would be a place stripped of every body.

For these reasons, then, we may regard place as something distinct [74] from bodies themselves and every sensible body as being in place. We may also credit Hesiod with rightly stressing the pre-eminence of 30 his primordial abyss. He writes:

> First of things was chaos made, and then
> Broadbreasted earth.

This judgment, that beings must first of all have location, was due to his acceptance of the popular view that everything is somewhere, that is, in a place. If such is the status of place, it must have a functional significance [11] surpassing that of the most astonishing phenomenon. If nothing else can continue in being without it whereas it remains when anything else vacates it, place must indeed rank first; for place 209a does not perish with the perishing things in it.

Yet even if we grant that place is, it is difficult to tell what place is: there is the question whether place is some sort of "bulk" of a body or is some different sort of a "nature"; and we must first ascertain the genus of place. Now, despite its length, width, and depth, which are precisely the three dimensions by which every body is defined, a place nevertheless cannot be a body: in that case, two bodies would coincide. Besides, if a body has a place or space, so will a surface or any other limit of the body, and for the same reason; since where the planes of 10 water were, there those of air will be. On the other hand, the fact that we cannot distinguish between a point and its place implies that a place, which cannot differ from a point, cannot differ from a line or a plane or a body either; therefore, a place cannot be anything different from any limit of the body. What, then, could we possibly deem place to be? Having the sort of nature described, place cannot be an element or a compound, whether corporeal or incorporeal: a place has size but has no body; but the elements of sensible things are themselves bodies, and intelligible elements cannot form a size. Again, for which of exist-

20 ing things would anyone take place to be an explanation? It does not have any of the four types of explanatory import: it is not a material of anything, since nothing is composed of it; it is not a form or definition of things; it is not their completion; and it does not move things. Then there is this difficulty: if place is itself a being among beings [4d], then place, too, must be somewhere. Zeno's problem requires attention [90]: if every being is in a place, then a place will clearly also have its place, and so on indefinitely. Finally, as every body is in a place, so every place has a body in it; what shall we say, then, about growing things? Their place would have to grow along with them if nothing has a place smaller or larger than itself.

30   Because of these difficulties, then, we must still consider our questions: What is place? And is there such a thing as place?

2. Place Not Form or Matter

We may give an account [36] of anything "by itself" [2] or with reference to something else; in particular, we distinguish "place" considered generally [92], in which all bodies are, from the specific [42] "place" in which a body is directly [17a].* Thus, you are in the cosmos because you are under the sky, which is in the cosmos; you are under the sky because you are on the earth; and you are on the earth because you

209b are in your particular place, which contains nothing in addition to yourself. If a place, then, is what directly contains any body, it would be a limit; so that the place of each body might be thought to be its form or shape, by which its magnitude or the material of its magnitude is defined, since the form is the limit of each. But whereas, so considered, the place of anything is its form, the place, considered as the extension of the magnitude, is [the latter's] material. This is different from the magnitude: it is what is contained and defined by the form, as by a limiting plane; and the material or the indeterminate† is such because,

10 when the boundary and the properties of a sphere are removed, noth-

---

* Simplicius: We learn "of what sort" anything is from the features which belong to a subject "directly," that is, "primarily."

† iii.7.207b35.

ing remains but its material. This is why Plato in the *Timaeus*‡ identifies matter and space: he identifies "what participates"§ with "space,"
although in his so-called "unwritten teachings" he gives a different account of "what participates"; at any rate, he identifies "place" and
"space." All, indeed, agree that place is something; but Plato alone
undertook to say *what* place is. In view of all this, it would seem very
difficult to find out what place is if it is either matter or form, for these   20
are hard to tell apart; so that the question requires the closest attention.

Nevertheless, we can readily see that place cannot be either form or
material. (1) Form and material cannot be dissociated from that to
which they belong, as place can: water can take the place of air, and
vice versa, as we have pointed out, and other bodies can likewise change
places; so that the place of any body is not any of its parts or any of
its states but is distinct from it. (2) Place seems to be like a receptacle,
which is a movable place but which is not itself a part of its contents.   30
Thus, as distinct from the body whose place it is, place is not the body's
form; and as the body's container, place differs from the body's material.
In general [135], what is [1] somewhere is one [15a] thing [4]; what
surrounds it, is quite [16] another [4]. Plato, by the way, ought to tell
us why forms and numbers are not in place if "what participates" is
place (whether "what participates" is the great-and-small or, as he says   210a
in the *Timaeus*, is matter). Again (3), how can anything move to its
appropriate place if place were material or form? Neither of these can
be place since there is no [local] motion to or from them and no distinction of up or down in them; it is among affairs which own these
distinctions that we must look for "place." (4) If a thing's place were
in it, as place would have to be if it were a thing's shape or material,
then place would be in place: both a thing's form and its indeterminate
aspect change and move along with the thing and do not remain in
the same place but are wherever the thing itself is; hence, place would
have a place. Again (5), when air is transformed into water, [a part of]   10
its place would vanish,‖ since the new body is not in the very same
place; but what would this disappearance be?

Thus, we have shown why place must be something and why it is
difficult to tell what [26] place is.

3. Place Not in Another Place

We must at this point distinguish different senses in which we say
that one thing is "in" another. A finger is in a hand as a part is in a

---

‡ 52.
§ 210a1.
‖ Cp. iv.1.209a27.

whole; and a whole is in its parts inasmuch as there is no whole beyond its parts. A species (for example, "man") is in a genus ("animal"); and a genus is in a species as, more generally, an [essential] aspect
20 [22] of a species is in the definition of the species. Health is in hot and cold [bodily parts] as a form is in a material; the affairs of the Greeks center in a king as events center in their primary agent; and activities culminate in what they are good for as in the perfection towards which they tend. But in the strictest sense of "in," contents are in a receptacle as a body is in a place.*

In this connection, someone may ask: Can anything be in itself?† Or can nothing be in itself; but is any one thing either nowhere or else in something else? However, the question is ambiguous. It may mean: Can anything be in itself essentially? Or it may mean: Can one aspect of a thing be in another aspect of the same thing? Now, when a whole has one of its parts in another part, the whole may be said to be "in itself." In general, it is not uncommon to attribute traits of a part to a
30 whole, for example, to characterize a man as "white" because of the color of his skin or as "scientific" because of his specialized training [176]. So there is a sense in which a "bottle of wine" is "in itself," although neither the bottle nor the wine is "in itself": since one part [the wine] is in the other part [the bottle], both are parts of the same thing [the bottle of wine]. In this way, then, a thing can be "in itself"; but not
210b strictly [17a], in the sense in which whiteness is "in" a body. The [white] skin is "in" a body, whereas disciplined intelligence [179] is "in" a way of life [154]; but because both body and soul are "parts" [or "aspects"] of a man, we also [indirectly] speak of the traits of either as being "in" a man [as a whole]. To be sure, a bottle and wine when separate are not parts of a whole; but when together they are. Thus, it is only when there are parts [in a relation of this sort] that a whole is said to be "in itself"; just as it is [indirectly] that a man is white because of his body, and his body because of its skin, whereas whiteness is "in" the skin directly. Yet a surface and its whiteness differ in kind, each having its own nature and function.‡ Accordingly, inductive investigation discloses that nothing is "in itself" in any of the meanings
10 of "in" which we have distinguished. We may also clarify this result by argument. Were it possible for anything to be [literally] "in itself," each part would have to be both: the bottle would have to be both the receptacle and its contents; and the wine would have to be both the wine and the bottle. Even if in some sense each is in the other, the

---

* Thomas Aquinas observes that what is "in time" ("the measure of motion") is reducible to what is "in place" ("the measure of what is mobile").
† iii.5.205b3; Plato *Parmenides* 138A, B5, 145B, C.
‡ The surface is analogous to a "container"; whiteness, to the "contents."

bottle contains the wine not inasmuch as the bottle but inasmuch as the wine is wine; and the wine is in the bottle not inasmuch as the wine but inasmuch as the bottle is a bottle. Clearly, then, each differs from the other in its being, for a container and its contents are differently defined. Not even accidentally§ is it possible for anything to be [literally] "in itself," for in that case two things would be in the same container: the bottle would be in itself, if what is naturally a receptacle [12] can 20 be in itself; and the contents [12] would be in the bottle also, for example, wine (if the receptacle is a wine bottle). Clearly, then, it is impossible for anything to be "in itself" in the primary sense of "in."

Moreover, it is not difficult to solve the problem raised by Zeno; namely, if place is something, it will be "in" something.‖ There is, indeed, nothing to hinder a primary place from being in something else. Yet it is not therefore in a place! It may be in something as a healthy state [33a] is in warm bodily parts¶ or as a warm temperature [35a] is in a body. Hence, the question raised does not require an infinite regress [with a consequent denial of place].

This, too, is evident: since a receptacle is no part of what it contains (inasmuch as a container in the strict sense differs from what it contains), place is neither a material nor a form but is something different 30 from both of these. Matter and form are aspects of the thing which is in a place.

Let this suffice as a statement of the difficulties concerning place.

4. Definition of Place

In order that it may become evident to us what, after all, place is, let us now go over [65] the facts [82f] which seem truly essential to place. Thus, a place surrounds [33e] that whose place it is; a place is 211a not a part of what it surrounds; a thing's primary place is neither smaller nor greater than it; a place can° be left behind by a thing and be dissociated from it; and every place is either up or down, since each of the [simple] bodies moves up or down to come to rest in its resident [55] place. These facts [85] provide the foundation on which we must construct our theory. We must, moreover, try to manage our inquiry in such a manner that it will not only yield the definition we desire, but will also solve the problems raised, bring out the factual basis of what is commonly believed about place, and uncover the reasons for the 10

§ ii.1.192b23-27. A place cannot be "in itself" even in the accidental sense in which a physician becomes his own patient.
‖ iv.1.209a23-25.
¶ iv.3.210a21. Thus, the question raised points to the pattern of spatial relations.
° MSS. FGI: place [taken simply] cannot.

perplexing questions asked about it. This may be the best way of in-
suring the demonstration of each pertinent point.

To begin with, we must keep in mind that, but for local motion, there
would be no place as a subject matter of investigation. The principal
reason why even the heavens are believed to be in place† is because
they are always in motion. However, besides local motion, "movement"
includes increase and diminution in which, too, there is change of place
or change in the lesser or greater spaciousness attained. Moreover, some
things are in movement by their own activity, but others incidentally;
20 and of the latter in turn, some can be in movement by themselves, for
example, parts of the body or a nail in a ship, whereas others can be
in movement incidentally only, for example, whiteness or science, which
change their place only in the sense that that in which they inhere
changes its place. Furthermore, when we say that we are in the "universe" as in a "place" because we are in the sublunary atmosphere, which
is in the universe, we mean that we are in the innermost surface (but
not in the whole) of the atmosphere which surrounds us; for if the
whole of the atmosphere were a place, the place of anything would
not be equal to that thing. But we think of a thing's place as equal to it;
and a thing's primary place is equal to it. Then, too, when that which
30 environs a thing is not detached from it but is continuous with it, the
thing which it environs is not in its surroundings as in a place but as a
part in a whole. But when the two are in contact only, the thing encompassed is directly in the innermost surface of what encompasses it:
this surface is not a part of the thing encompassed; but it is equal to it
in extent, since the limits at which things are in contact are coincident.
Also, a part continuous with a whole moves not "in" but *with* the latter,
for example, a pupil with an eye or a hand with a body; whereas what
211b is separable, for example, water or wine, moves *in* a container,‡ regardless of whether the container is in motion or not.

It is evident from these considerations what place *is*. Place must be
one of four things: a form, a material, an interval between the extremities of the surrounding body, or those extremities themselves if there
is no such interval apart from the extension of the surrounded body.
10 Three of these, however, place evidently cannot be. First, because it
circumscribes things, place appears to be a form: the extremities of the
circumscribing and of the circumscribed body coincide; and both form
and place are limits. Yet form and place do not limit the same thing:
form is the limit of the thing circumscribed; place is the limit of the
circumscribing body. Secondly, because a distinct content (like water)

---

† But cf. iv.5.212a31-b11.
‡ The examples given are noted in 211b1-5, which in large part repeats 211a34-b1
and which may be spurious.

may often change while the container remains the same, the interval between the extremities of a container appears to be definitely distinct from a displaced content. But there is no such [separate] interval; rather do the bodies changing their places and naturally fitting the container replace one another. If there were such an independent natural inter- 20 val abiding in the same place, there would be an infinite number of [coinciding] "places." When water and air are in the process of replacing each other, each of their parts would behave within the whole just as would all the water in the vessel: [each part would leave an independent empty interval behind]. Then, too, the "place" would change; consequently, a place would have a place, and many "places" would coincide. Yet the place within which a part moves when the whole vessel moves does not become a different place but remains the same; air and water or the parts of the water replace each other in the place in which they are, not in the place to which the vessel is moved and which is part of the place of the whole cosmos. Thirdly, matter might 30 be thought to be place if one were to view it as in a body which is at rest and which is not detached from but "continuous" with its environment. Just as we recognize the definite being of a material in a qualitative change when something black becomes white or something soft becomes hard, so we come to believe in the being of place through a similar experience [173]: we recognize matter because *what* was air is now water; and place, because *where* there was air there now is water. But as we have said before,§ a thing's material is neither dissociated 212a from it nor anything that contains it; place, however, is both. Finally, if place is none of these three things, not form or material or a distinct permanent interval independent of the extension of the displaced body, then place must be the one of the four that remains: the limit of the surrounding body (at which this body is in contact with the body it surrounds),‖ provided that the surrounded body is capable of local motion.

Now, place is regarded as something important but hard to grasp because matter and form appear with it; because a moving body is displaced in a stationary container, which apparently can have an in- 10 terval distinct from the moving bodies; and because the supposed incorporeal character of air seems to imply that place is not simply the [inner] boundaries of a receptacle but also the supposedly empty interval between them. But as a receptacle is a place which can be transported, so place is a receptacle which cannot be transported. Hence, when a body moves and changes its place within something in motion

---

§ iv.2.209b23, 31.

‖ The relative clause seems to be spurious.

(for example, a boat in a river), the [immediately] surrounding body functions as a receptacle rather than as a place; whereas place tends to be motionless (so that it is a whole river which, being motionless as
20 a whole, functions as a place). Thus, *the place of anything is the first unmoved boundary of what surrounds it.*

This is the reason why¶ the center of the cosmos and the inner surface of the rotating celestial system are conceived as functionally [55b] down and up for all men, because the former is always stationary and the latter remains coincident with itself. Since light bodies naturally tend upwards and heavy bodies downwards, the surrounding boundary towards the center of the cosmos and the center itself are "down," whereas the boundary towards the outermost part of the cosmos and the outermost part itself are "up." For this reason, too, place seems to be a sort of surface as if it were a receptacle or container. Furthermore, place
30 is in some sense coincident with the thing whose place it is, for boundaries are coincident with what they bound.

5. Ways of Being in a Place

A body is in a place if another body surrounds it; otherwise, it is not. Hence, even if water were not surrounded by anything, its parts could be in movement, since they surround one another; but the whole could be in movement only in one sense, but in another sense it could not. The cosmos does not as a whole change its place, but it can rotate;
212b and the circle [or the outer surface of the celestial system] functions as a place for its parts.* Some of its parts, then, do not move up and down but are in circular motion; but other parts also move up or down as they are rarefied or condensed.

We have previously† pointed out that some things are in a place only potentially; others, actually. The parts of a continuous homogeneous body are in a place only potentially; whereas parts which are detached yet in contact, as in a heap, are in a place actually.

Moreover, some things are essentially in a place, namely, all bodies capable of local motion or of increase; but the cosmos (as we have said) is not as a whole anywhere or in any place, since there is no body en-
10 closing it. The circle in which the cosmos moves functions as a place for the parts, which are contiguous with one another. Some things are

---

¶ 210b32-211a6.
* Alexander: "ultima sphaera nullo modo est in loco." Avicenna: "motus ultimae sphaerae non est motus in loco, sed motus in situ." Avempace: "superficies convexa sphaerae contentae, est locus primae sphaerae." Averroes: "ultima sphaera est in loco per accidens." Themistius: "ultima sphaera est in loco per suas partes." (Thomas Aquinas)
† iv.4.211a17-b5.

only incidentally in a place, for example, the soul and the cosmos. All the parts of the cosmos are somehow in a place, since one contains another on the circle. Hence, the celestial sphere moves in a circle, yet the All is not anywhere: what is somewhere both is something and must also have something else encompassing it; but beyond the All there is nothing outside the All. Thus, all things are in the cosmos, since the cosmos is the All; yet the place [of the things in the heavens] is not the heavens, but the inner surface of the heavens which is in contact with the movable body as its unmoving limit. Hence, earth is in water; **20** water, in air; air, in ether [or fire]; and the ether, in the heavens—but the heavens are not in anything else.

It is evident from these considerations that all the problems which have been raised‡ concerning place have a solution on the basis of this account. A place need not "grow" with a growing body. A point need not have a place. Two bodies need not coincide in place. There need not be a bodily interval, since what is between the boundaries of the place is not an interval in a body but any body which may happen to be there. Place, too, is somewhere, though not in the sense of being in a place but in the sense in which there is a limit in what is limited; for not everything is in a place, but only a movable body. We may also with good reason hold that each [simple] body tends to its appropriate **30** place: the elements which are in nonviolent succession and contact are akin;§ and, unlike things which form an organic unity, those in contact are capable of acting upon and of being acted upon by each other. So, too, we may with good reason hold that everything remains naturally in its resident place: a given part is in a whole place as a detached part is related to its whole; this is exemplified when we move a part of water or air. So, too, air is related to water, and the latter is like matter, **213a** and the former, like form: in a sense, water is the matter of air, and air is an actualization of water; for water is potentially air, and air is in a different way potentially water. (These distinctions will be elaborated elsewhere;‖ so that what must here be mentioned, even though obscurely stated, will there become clearer.) Thus, the same thing is matter and actuality: water is both, but is air potentially and is water actually; so that water is in a way related to air as part to whole. That is why they are in contact; whereas, when two things become actually one, they form an organic unity.

So much for the fact that place is and for an account of what place is. **10**

---

‡ iv.1.209a2-30, 4.211b7, a4-6.

§ Earth is dry and cold; water, cold and wet; air, wet and hot; fire, hot and dry.

‖ *On Generation and Corruption* i.3.

6. Arguments against and for a Void

Like the theory of place, so the theory of the "void" [or "vacuum"] must be recognized as falling within the province of the natural philosopher: the natural philosopher must investigate whether or not there is such a thing and, if so, how and what it is. Moreover, as in the theory of place, so in the theory of the void, the negative or positive convictions [183] which men hold may be traced to the different assumptions they make. Thus, those who maintain the existence of a void regard the void as a sort of place or receptacle which is full when it contains what it is capable of holding, but which is empty when its contents have been removed from it. On this view, what is "empty," what is "full," and a "place" are the same thing; although being empty, being full, and being a place are admittedly not the same. We must, therefore, at the begin-
20 ning of this inquiry try to grasp the intent [36] of those who argue for the existence of a void as well as the intent of those who argue against it and, in the third place, the commonly held opinions concerning these issues.*

Those who undertake to disprove the existence of a void, like Anaxagoras and others who argue in his manner, succeed in refuting not what people mean [177] by the "void" but only what people mistakenly say about it. Positively, those thinkers demonstrate that air is something; they do this by distending wine-skins and then by proving the resistance of the air as well as by letting air into water-siphons. But what people mean by a "void" is an interval in which there is no perceptible body:
30 thinking that every being is a body, people characterize that in which there is "nothing at all" as "empty"; consequently, what is filled with air is [according to them] a "void." Hence, what needs to be proved is not that air is something, but that there is no independent actual interval different from bodies which breaks the continuity of the bodily universe
213b (as Democritus and Leucippus and many other physical theorists say) or which may even extend beyond the continuous bodily universe [as the Pythagoreans say].

On the other hand, whereas the thinkers [who argue in the manner of Anaxagoras against the "void"] fail even to hit the door in their attack upon the problem, those who defend the existence of a "void" do much better in this respect. In the first place, the latter say, there would be no change of place, either by local motion or by expansion, since there would be no motion if there were no void and since what is full cannot admit anything more into itself. If the latter were possible, there would not only be two bodies in the same place, but any number of bodies might be together (since there is no specifiable point at

---

* The analysis of the "void" also presents striking parallels to that of the "infinite."

which this principle [36] would cease to hold). If this is so, then the 10 smallest body could admit the largest, because "many a little makes a mickle"; and therefore, if there could be many equal bodies in the same place, there could also be many unequal bodies in the same place. It is along these very lines that Melissus, by the way, argues that the All is independent of movement: if it were subject to movement, then according to him there would have to be a void which, however, is not to be found among beings. This, then, is one way in which the thinkers under discussion seek to prove that there is a definite "void." Another way is by pointing out that some things appear to contract and to be compressed. They say, for example, that a cask will hold both wine and wine-skins; this, they say, shows that a compressed body contracts into its own internal voids. Again, growth or expansion seems to all these men to depend upon there being a void: food is a body, they argue, 20 and two bodies cannot be in the same place. By way of further evidence, they represent ashes [put into a receptacle] as capable of holding an amount of water equal to that which fills the receptacle when there are no ashes in it. Finally, the Pythagoreans, too, maintained that there is a void. They taught that the void enters the universe, which inhales it [as it does the breath in living things] from the infinite air; that the void divides [72e] different "natures" by somehow separating [73] or differentiating [71] discrete [136b] things; and that the void performs this function primarily with respect to "numbers," whose "nature" it keeps discontinuous [72e].

Such and so many are the chief considerations from which men have argued for or against the [independent] existence of a void.

7. Definition and Nonbeing of a Void

In order to decide between the two kinds of views just contrasted, 30 we have to determine what the term "void" signifies. Some hold the "void" to be "a place in which there is nothing." The reason for this is because people conceive being as bodily and every body as being in a place and the "void" as therefore a place in which there is no body; hence, if in some place there is no body, they think that there is a void there. They also conceive every body as tangible, that is, as heavy or 214a light. It follows by syllogism, then, that what has nothing heavy or light in it is a void. Although this follows by syllogism (as has been said), still, it would be an absurdity for a point to be a void; in that case, a point would have to be a place in which there could be an extension [consisting] of a tangible body. However that may be, we note, at any rate, that a "void" is described in one sense as "what is not filled with

a tangible body," that is, with something heavy or light. But, we may
10 ask, what is to be said about an extension having in it color or sound:
is it a void or not? Clearly, would we not rather have to say that if such
an extension *could* hold a tangible body, then it would be a void; but
if it could not, then it would not be a void? In another sense, the "void"
is described as "that in which there is no particular primary being"
[4b], that is, in which there is no particular [4] bodily [102] being
[26]. Hence, those who define "place" in the same way° say that the
"void" is the material of a body. However, this cannot be right: the
material of things cannot be divorced from them; and what these men
are seeking is an independent "void."

Now, in our analysis of place, we have shown that the void must
be a place if it is deprived of body and also in what sense there is and
in what sense there is not such a thing as "place"; with the evident result
that in this sense there is no "void," either independently or not inde-
20 pendently [of bodies in it]. What is meant by a "void" is evidently not a
body but an interspace [111c] of body. This is the reason why the void
is held to be something; especially since place, too, is for the same reason
held to be something. It is the fact of local motion which these men
find most useful, both for the purpose of asserting that place is something independent of bodies occupying it and for the purpose of asserting that the void is something. They believe that the void is a necessary
condition [83] of movement in the sense that movement occurs in a
void; and this would be exactly the sort of thing which some declare
"place" to be. However, there is no such necessity that, if there is to be
movement, there must be a void. A void is unnecessary for any "movement" in a general sense, since (as Melissus failed to see) what is "full"
is capable of undergoing qualitative alteration. A void is even unnec-
30 essary for local motion, since bodies can take one another's place without there being any extension separate from the bodies in motion; this
is clearly so even in rotations of continuous things, as in those of liquids.
Nor is a void necessary for contraction: bodies can be "packed" through
214b expulsion of things contained in them (for example, of the air in water).
So, too, things may expand not only by something entering into them
but also by qualitative change (as when water turns into air).† Indeed,
the argument from the expansion of bodies by growth (like the argument
from the water poured on ashes) defeats itself: not every part of a growing body would grow; or [if it did], things would grow otherwise than
by addition of body; or else two bodies would be in the same place
(in which case our opponents are requiring the solution of a difficulty

---

° iv.2.209b11-16, 4.211b29-212a2.

† *On Generation and Corruption* i.5.321a9-29.

they share with others, instead of proving the existence of a void); or,
if it is in every part and by means of a void that a body grows, then
the whole body would have to be a void!‡ (The same analysis applies
to the argument about the ashes.)   10

Thus, it is evident that the arguments for the existence of a void
are readily refuted.

8. No Independent or Occupied Void

Let us continue our argument against those who ascribe independent
existence to a "void." If each of the simple bodies naturally moves in a
definite direction (fire upward, and earth downward and towards the
middle of the universe), it is clear that a void cannot explain local
motion: since the void is thought to explain local motion but not any
of those mentioned, which local motion, then, does it explain? Again, if
a "void" is a sort of "place deprived of body," where would a body
placed in a void move to? Surely not into the whole of the void! The   20
same argument applies here that was used against the conception of a
place as an independent something into which things are transported:
how would anything placed in it start moving or stop moving? The
same argument is no less applicable [56a] to "up" and "down" and to
the "void"—for those who maintain the existence of a void represent it
as a place.* Then, too, how can anything be "in" a place or a void?
This does not happen when a whole body is located in a supposedly
severed place and abiding body: any part not separately located there
would not be in a place, but in the whole [of which it is a part]!† Then,
again, if a place has no [segregated] existence, neither does a void.

Indeed, as against those who say that a void is necessary if there is
to be movement,‡ reflection shows the very opposite: were there an   30
independently existing void, there would not be a single movement!
Just as the earth is stationary (according to the arguments of some)
because of the uniformity of its medium, so, too, in a void [sharply
demarcated from bodies] there would have to be complete stability:
since there would be no differences there, neither would there be any
place for anything to move to rather than to some other place. Another 215a
reason [in support of our contention] is the difference between violent
and natural motion: being contrary [74] to natural motion, violent mo-

‡ *On Generation and Corruption* i.5.321b27.
* Besides 214b12-24, note especially 215a25-31 and 216a8-21 as well as *De caelo*
   i.2.268b20-24; ii.13.294a12-26, b3-6, 14.296a31; iii.2.301a20-b31; iv.2.308b13-28,
   309a27-b18, 4.311a16-27. Cf. Lane Cooper *Aristotle, Galileo, and the Tower of
   Pisa* (Cornell University Press, 1935).
† iv.4.211b20.
‡ Plato *Timaeus* 40, 63.

tion presupposes [18] the latter in the sense that if there is violent motion
there must also be natural motion; and therefore, if the various natural bodies do not have their natural motions, they will not be capable of violent motion either. But how can there be natural motions in
a void or in an infinite in which there are no differences? No more than
an infinite as such has any up or down or middle, does a void as such
10 have any difference of up or down: just as there are no differences in
nothing, so there can be none in nonbeing, and the void seems to be a
sort of nonbeing or privation. But natural motions are differentiated; and
their distinctions [of up and down] are natural. Either, then, nothing
has a natural motion, or else there is no void! Again, projectiles continue to move even after what has propelled them is no longer in contact with them. Some explain this phenomenon by mutual replacement.
Another explanation may be that the air which has been pushed pushes
projectiles with a motion more vigorous than their motion to their resident place.§ But none of these things can happen [82f] in a void: there,
a body can continue moving only as long as it is propelled by something
else. Again, it would be impossible to state any reason why anything
20 set in motion would stop anywhere. Why would it stop at one place
rather than at another? Hence, a body would either continue in its state
of rest or would necessarily continue in its motion indefinitely, unless
interfered with by a stronger force. Again, things are held to move into
a void because of its yielding character. However, a void would yield
equally everywhere; so that any motion in it would be in every direction.

There are further considerations which make what we have said even
more evident. We see a body of a certain weight moving at a faster
or slower rate for one of two reasons: either, on the one hand, because
of a difference in the medium, for example, in earth or water or air;
or, on the other hand, because of a difference in the moving bodies
compared such that, other things being equal, one body is heavier or
lighter than the other.

On the one hand, then, the medium explains [different velocities] be-
30 cause of its resistance; especially when the medium is moving in an
opposite direction, but also even when the medium is in a state of rest.
This is especially true of a medium which is not easily divided or which,
215b in other words, is relatively dense. Thus, a body (B) will move through
a medium such as water (W) in a given time (T_w), and through a
thinner medium such as air (A) in a different time (T_a), namely, in
proportion to the density of the hindering body (provided W and A
are equal in length): the thinner and less solid the air than the water,
the faster will B move through A than through W; one speed has

§ viii.10.266b27-267a12.

the same ratio to the other that the air has to the water. Hence, if
the air is twice as thin as the water, the body will move through W
in twice the time it takes to move through A; and the time T_w will be
twice the time T_a. Likewise, always, the less solid and resistant and the   10
more readily divided the medium, the faster will be the motion. There
is no definite ratio, however, in which a void stands to a body. So zero
is in no ratio to a number: although 4 exceeds 3 by 1, and 2 by more
than 1, and 1 by still more, there is no ratio by which 4 is greater than
0; but inasmuch as the greater has to be made up [75] of the less and
the remainder, 4 would consist of 0 and that by which 4 is greater than
0. So, too, a line is not "greater than" a point, unless a line were composed
of points. By the same token, therefore, as there is no ratio between   20
the "empty" and the "full," so there is no ratio between movement
[through the former and through the latter]; on the contrary, if a thing
moves through the thinnest medium over a certain distance in a certain
time, then it will move through the void with a velocity beyond any
ratio! Let V be a void equal in magnitude to W and to A. Then if B
is to move through V in a time T_v which is less than time T_a, the
"void" would be in that ratio to the "full." But in a time equal to T_v, B
would move through a part P of the medium A. Also, in time T_v, B
would span any body V which is thinner than air by as much as time T_a
is longer than time T_v; for if the body V is as much thinner than A as   30
T_a exceeds T_v, B would move through V (if at all) in a time inverse   216a
to the speed, that is, in a time equal to T_v. If V is empty, then, B will
pass through it still faster. But the assumption was that B passed through
V when V was empty in time T_v. Consequently, B would move through
V in an equal time regardless of whether V was full or empty. But this
is impossible. Evidently, then, if there is a time in which anything moves
through a part of a void, this impossible result will follow: a body will
move over a certain distance, full or empty, in equal times, since there
will be some body bearing to the other body the same ratio which the
one time bears to the other time. In short, the reason for this result is
clear: any motion stands to another in a certain ratio, since the motions
occur in time and there is a ratio between any finite times; but there   10
is no ratio [of density] between anything that is empty and anything
that is full.

On the other hand, having considered what follows from the differ-
ences among the media,|| let us turn to consider the difference [148]
between one moving body and another, together with its consequences.

|| Thomas Aquinas: "In hoc autem libro agitur de corpori mobili in communi.
. . . Vel potest dici, quod hic etiam procedit secundum opinionem anti-
quorum philosophorum, qui ponebant rarum et densum prima principia formalia.
. . ."

We see that, other things being equal, heavy and light bodies move with unequal velocities over an equal space in the ratio which their magnitudes have to each other. Hence, they must do so even when moving through a void. But this is impossible. Why should one of them move faster than another? To be sure, one of them does necessarily move faster in a [space] full [of matter] because the greater body divides the medium faster by reason of its force; for a body moving naturally or a projectile divides a medium as it does by reason either of
20 its shape or of its [upward or downward] tendency. But [in a void] everything would, accordingly, move with equal velocity. But this is impossible.¶

Evidently, then, the supposition of a void leads to a result which is opposed to what the defenders of a void want to establish. They think that if there is motion there would be an independently existing void. But this amounts to saying that place is an isolated something, which we have shown** to be impossible.

Then, too, if we critically examine the so-called "void" by itself, we shall find it to be truly "empty." When a cube [of wood] is placed into water, it displaces [a volume of] water equal to [the volume of the submerged part of] the cube. The same [effect is produced] in air, except that in this case the effect is not evident to sense perception. In fact,
30 any body which can be displaced [by another inserted into it] must, if it is not compressed, be displaced in the direction natural to it: downward, if it is of earth; or upward, if it is of fire; or else in both directions (regardless of the sort of thing that is inserted into it). Since a void, however, is incorporeal, this effect cannot be produced in it. Instead, the void would necessarily have penetrated the cube to the distance to
216b which the space [now filled by the cube] had extended before within the void; just as if water or air, instead of being displaced by a wooden cube, had gone all the way through it. But the size or volume of the wooden cube exactly equals the amount of the void it occupies; yet it differs in its being, although it is not cut off, from its attributes (such as hot or cold, heavy or light). Hence, if it were parted from its attributes (such as heaviness or lightness), it would occupy a portion of a place or of a void equal to itself. What, then, would be the difference
10 between the body of the cube and the void or place equal to it? And if these two coincide, why should not any number of things likewise do so? This is one absurd and impossible [consequence of the theory of a void occupied by bodies]. Again, it is evident that a cube, like all other bodies, has the same volume even if it is displaced. Therefore, if this

---

¶ Aristotle does not contend that heavy bodies would drop faster in a vacuum than light bodies but that there can be no actual independent vacuum.
** iv.4.211b18-29.

does not differ from its place, why should we ascribe to bodies a place distinct from their own volume (if their volume is without attributes)? Such an equal extension connected with the cube would be gratuitous. (Again, though a void ought to be a clear [deliverance of experience], yet it is nowhere observed in the world of moving bodies; for air is something even though it may not seem to be [so to sight]—just as there would not seem to be any water if fishes were of iron, since the tangible is discriminated by touch.)††

It is clear, then, from the reflections presented, that there is no 20 independently existing void.

9. No Void within Bodies

Some think that the existence of a void is evident from that of tenuous and dense bodies. They argue: without anything tenuous or dense, there could be no contraction or compression; and if this sort of change were impossible, then there would be no movement, or else the universe would bulge (as Xuthus maintained), or air and water would have to be transformed into equal quantities (so that transformation of a cupful of water into air would have to be balanced by transformation of an equal amount of air into a cupful of water), or [they insist] there must be a void as a necessary condition of densification and rarefaction. How- 30 ever, if they mean by a "tenuous" body one which "has many independent voids," then there evidently cannot be anything "tenuous" in this sense any more than there can be an independent void or a place having an extension all its own.° On the other hand, if they are thinking of a nonindependent void within a tenuous body, this would seem less impossible. Still, such a void would be a condition of upward motion only: a tenuous body is light; and they say that it is for this reason that fire 217a is light. Also, such a void would not be a medium conditioning [83] motion; rather would things be carried upward on it somewhat as nets float upward on inflated wine-skins. Yet how could a void move or have a place to move into? And would not such a place in the meantime be a void without a void? Again, what explanation would these men give for the downward motion of heavy bodies? Then, too, if the speed of the upward motion varies proportionately with the tenuity or emptiness of the moving body, a body altogether void would evidently move at the fastest rate of all. But it is as impossible for a void to move as it is for anything to move in a void, since the speeds [of a solid 10 and of a void] would be incommensurable.

---

†† The passage here enclosed in parentheses may be spurious.
° iv.4.211b18-29.

Now, although we deny the [independent] existence of a void, we agree that the other alternatives have been correctly stated: if there were no densification or rarefaction, there would be no movement, or else the universe would bulge, or it would always be in equal amounts that water would be transformed into air and air into water. Clearly, the air into which water is transformed is greater in volume than the water. Hence, if there were no compression, the successive parts would be pushed outward until at the last there would be a bulge; or somewhere else air would be transformed into an equal amount of water, so that the total volume of the whole would remain equal; or else no process would take place at all. Displacement would always have the result described unless, indeed, it is circular;† but, so far from being
20 always circular, motion sometimes is linear. However, in contrast to those who for the reasons stated maintain the existence of an actual [4] void, we hold a different position [36] in line with our principles [85]:‡ a single material is the subject of contraries [or of the limits within which a change occurs] such as hot and cold or any of the other natural opposites; a thing changes from what it is potentially to what it is actually; a material is not disjoined [from its changing qualities], although its being differs from theirs; and a numerically single material may happen to have a certain color and to be hot and to be cold.

Now, then, it is clearly the same material which is the subject of a larger and of a smaller bodily size. Thus, transformation of water into air is not an affair of a given material having added [65h] to itself something else which it becomes; it is an affair of the same material becoming actually what it had been potentially. This is also the sort of
30 thing that happens in the opposite transformation of air into water. But in the former transformation, there is a change from a smaller to a larger volume; in the latter, from a larger to a smaller volume. Likewise, therefore, when air contracts or expands, what happens is that the material which was potentially smaller or larger becomes [actually] smaller or larger. Analogously, it is the same cold body which, being potentially hot, becomes actually hot, and the same hot body which, being potentially cold, becomes actually cold, as by the same token
217b it is the same hot body which becomes still hotter: there is nothing in the material which, when the body was less hot, was something not hot and which thereupon becomes something hot. So when the curvature of a ring becomes the curvature of a smaller ring (whether or not it keeps its identity intact), there is nothing which in the meantime was straight or nonconvex and in which convexity then comes into being—as if a

---

† iv.1.208b2, 8.215a15.
‡ i.9.

quality had to cease to be in order to assume different degrees of more
or less. So, too, we do not find in a white-hot flame any part from which
heat and whiteness are absent. Just so, an earlier and a later degree of
heat are related [by a passage from the potential to the actual]. Accordingly, when a large or small sensible bulk is expanded, it is not
because something else has been added to the material, but because the 10
material is potentially large or small. And consequently, it is the same
body that is now dense, now tenuous; and it is the same material which
is [potentially] dense or tenuous. Moreover, what is dense is heavy,
and what is tenuous is light. But just as the contraction of a ring's curvature is not an addition of something else which is convex but is an affair
of something present being contracted, and as a fire is hot in any part
of it one may select, so what concerns us here is all an affair of the same
material contracting and expanding. There is also another type of connection between density and weight: for both what is heavy and what
is hard are dense, and both what is light and what is soft are tenuous;
yet heaviness and hardness fail to go together in lead and iron.

In view of what we have said, there evidently is no independent void 20
either absolutely [105] or within tenuous bodies; neither is there a potential void, unless one finds it desirable to speak of a "void" in a general
way as a "ground" [83] of local motion. But when the "void" is taken
in this sense, its referent would be heavy or light material as such. It is
because [39a] their material may be heavy or light [50b] that bodies,
with their varying degrees of possible density or tenuity, can [189] move
[121]; and it is because their material may be hard or soft that they
can be impervious [35d] or readily subject [35a] to changes [120a]
other than local motion.

This completes our account of the sense in which there is and the
sense in which there is not a "void."

10. Time Not a Whole or Change

After the preceding analyses, we take up the topic of time. It is 30
well for us to take off with the difficulties concerning time which are
raised in nonacademic discussions. Is time to be included among beings
or not? And what is its nature?

Certain arguments may lead us to suspect that time is a nonentity
or is at least shadowy and indistinct. Thus, infinite time as well as any 218a
selected [65] period of time comprises the past, which no longer is,
and the future, which is not yet. But if anything consists of nonbeings,
it cannot itself be held to participate in being. Then, too, when a whole
divisible into parts is present, all or at least some of its parts must be

present. But of the "parts" into which time is presumably divisible, none is present [in nature]: past events have come and gone [116]; future events are still to appear [134c]. The present is not a "part" of time: a part is a measure of the whole, whereas the present is not such a measure; and a whole must be composed of its parts, whereas time does not seem to be composed of "nows." Again, it is not easy to see whether the
10 present, which appears to divide past and future, always remains the same or is always different. Let us, on the one hand, suppose the present to be always different. But if no two "parts" of time are simultaneous (except that a shorter period of time may be comprehended in a longer one) and if what heretofore was but now is-not has necessarily at some time ceased to be, then the "nows" cannot be simultaneous with each other [any more than can the "parts" of time]; but the earlier "now" must have been always perishing. However, the earlier "now" cannot have ceased to be within itself, since at that time it was; nor can it have ceased to be in another "now," for we must recognize it to be just as impossible for one "now" to be next to another as it is for one point to be next to another point [in a line]. If, then, the earlier
20 "now" has not ceased to be in the "now" next to it but in some other "now," it would be simultaneous with the infinitely numerous "nows" between the two; and this is impossible. Let us, on the other hand, suppose the present to remain always the same. This, too, is impossible. Nothing determinate that is divisible has but one limit, whether it is continuous in one or in more than one dimension; but the "now" is a limit, and we can select a limited period of time. Also, if to be simultaneous is to be in the same "now" (neither earlier nor later) and if both earlier and later events are in this particular "now," then the events of ten thousand years ago would be simultaneous with today's events; and then no event would be earlier or later than any other.

30   Having gone over the difficulties which pertain to the facts [82f] of time, we also find that the traditional views have left no less obscure the problem: What is time? or, what is its nature? Some [like Plato]
218b identify time with celestial revolution; others [like the Pythagoreans], with the universal sphere itself. But although even a part of a celestial revolution is "a" time, it is not a revolution; and what is selected is part of a revolution, not a revolution. Moreover, if there were many worlds [as Democritus says], the course [109] of any one of them would be time just as much as the course of any other would be time; so that many worlds would imply many times at the same time. Next, the reason why time was identified with the universal sphere is because all things are in time as well as in the universal sphere. But this view is too trivial for us to consider the many impossibilities it

involves. On the other hand, we may examine the more influential view [of Plato] which identifies time with some kind of movement [109] or change [115]. But contrary to this view, a change or movement of anything is in the changing thing only or is where the moving or changing thing itself is; whereas time is everywhere and in all cases alike. Also, unlike time, change is fast or slow. We define "fast" or "slow" by reference to time: that is "fast" in which there is much going on [109a] in a short time; that is "slow" in which there is little going on in a long time. But we do not define time by reference to so much time or to such and such a kind of time. Hence, time is evidently not movement or change. (We need not, in the present context [23c], state how movement and change differ.)

11. Time as Dependent on Events

In spite of such considerations, time is not independent of change. Thus, when we have no sense [170] of change or are inattentive [205] to any change, we have no sense [175b] of the passing of time. We are in this respect like the people in the legend who, on awakening out of their long sleep in the presence of the heroes in Sardinia, link [137c] the earlier with the later time into a unified present, in disregard [118c] of the interval to which they have been insensible. As there would be no time, then, if there had been no diversified but only a single self-identical present, so we do not recognize an interval of time when we fail to note a "now" distinct from the present one. Accordingly, we are heedless of time in circumstances in which we do not discern a change because the self [154] appears to continue in a single undivided state; but we do acknowledge the passing of time in circumstances in which we sense and discriminate a change. Evidently, then, time is not independent of movement or change.

Here we have a point of departure for exploring what time is: time is evidently neither identical with nor independent of movement; so that it remains for us to determine how time is related to movement. For we experience [165] movement and time together. Even when in gross darkness we do not feel any bodily interaction [35] but only some kind of mental [154] process [109], we realize that some time has passed. Conversely, when we realize that some time has passed, we realize that some process has taken place. Time is therefore either a process or is somehow dependent upon a process; and since it is not the former, it must be the latter. Now, anything that is going on has a start and comes to a stop, and any magnitude is continuous; and so a "movement" corresponds [177] to a magnitude. Since a magnitude is continuous, so is a movement; and because of the continuity of the movement, time is also continuous, for we estimate how much time has passed by the amount of movement that has occurred. Accordingly, we distinguish "before" and "after" primarily in place; and there we distinguish them by their relative position. But movement must also have in it a distinction of "before" and "after" analogously to that in magnitude; and so must time, because of its correspondence with movement.

20 The order of "before" and "after" *which* is in process is, existentially [1d], the process; although, indeed, *what* the distinction between "before" and "after" is [88a] differs from [*what*] a process [is].* As we, then, discriminate a process with the aid of the distinction between "before" and "after," we become familiar with time; for we acknowledge the passing of time when we perceive something coming before and something coming after in a process. The distinction between "before" and "after," in turn, impresses itself upon us in the recognition that the extremes it involves are distinct both from each other and from something intermediate between them. When we, accordingly, apprehend [169a] the extremes as distinct from what intervenes between them and when we mentally [154] mark [36a] them as two "nows" (one coming earlier and the other coming later), it is then that we acknowledge and identify time. For time (let us submit) is what is limited

30 by the "now." On the one hand, then, when we experience the present as one ónly, and when we either do not experience the "now" as prior and subsequent in the process or do not even experience the "now" as having an identity related to something prior and to something subsequent, then no time seems to have passed because no process seems to have occurred. On the other hand, when we note something "before"

219b and something "after," then we acknowledge time. For this is what time is: *the number of precessions and successions in process.*

Thus, time is not sheer process but is a numerable aspect of it. This is indicated by the fact that, as we discriminate "more" and "less" by number, so we discriminate "more" and "less" movement by time. Hence, time is a kind of number. But by "number" we may mean a concrete numbered or numerable plurality or an abstract number by which we count. Time, however, is not sheer number by which we count but is something counted (which is quite different from the former). Further-

10 more, like movement, so time has a continually sequential [16b] character. To be sure, all simultaneous time is the same: the present, whatever it may then be [1e], is the same; but the respect in which it is diversified is in its being [88a] the "now" [of different events]. Yet

---

* Thomas Aquinas: "Sic igitur prius et posterius sunt idem subjecto cum motu, sed differunt ratione."

[time is not sheer "before" and "after"; but] the present determines time in its discrete character as earlier or later [and therefore also in its simultaneity].

Accordingly, the present is in one sense the same, but in another sense it is not the same. The present is different "nows" inasmuch as it inheres in one ["whiling"] and then in another, which is precisely what it meant for it to be [88a] the present; but, in being what at some time is [1d] now, the present is the same.† For a movement, as we have said, corresponds to a magnitude; and time, as we maintain, corresponds to a movement. Just so, a moving body [121], by which we recognize a process as well as what comes before and after in it, corresponds to a [mobile] point [*in the drawing of a line*]. The moving body, in being what it at some time is, is the same (a point, a stone, and so forth); but the moving body differs in the account which may be given of it [from time to time],‡ as the Sophists take "being Coriscus in the Lyceum" to differ from "being Coriscus in the market-place." The moving [point or] body differs in being now here and now there, and the present corresponds to it as time corresponds to the movement; for it is by reference to the moving body that we recognize what comes before and after in the movement, and it is in so far as the "before" and "after" can be counted [as two] that there is a present. Consequently, in what comes before and after, the present remains the same in being what-at-some-time-is-now, since the present is precisely what comes before and after in movement; yet the present is diversified in its being, namely, in the order of being successively "now" at distinguishable earlier and later stages. Moreover, it is the present with which we are most familiar. For a process is recognized by what is in process, as a local motion is recognized by what is changing its place: such a body is a particular primary being [4b], whereas a movement is not a particular primary being. Thus, like a moving [point or] body, the present is in one sense always the same; in another sense, it is not always the same.

It is evident, too, that if there were no time there would be no present, and vice versa. As a moving [point or] body and a motion go together, so do the numerable aspects of the moving body and of its motion; for time is a number belonging to a movement, whereas the present corresponds to the moving body and is a sort of unit [by which a multiplicity is differentiated]. Moreover, time depends upon the present both for its continuous and for its discrete character. In this respect,

20

30

220a

---

† Thomas Aquinas: "nunc est idem subjecto, sed alterum et alterum ratione."

‡ Thomas Aquinas: "ex ea parte, qua est quoddam ens, quodcumque sit, est idem, scilicet subjecto; sed ratione est alterum."

too, the analogy with the moving [point or] body and its motion holds: the continuity of the motion depends upon the unity of the account that may be given of the moving body (though not upon the individual unity of the body [1d], since the movement of the latter may also be discontinuous); and it is the moving body which [at a particular stage] divides a movement into earlier and later aspects. The "now" and the 10 moving body correspond in this respect to the point inasmuch as the point at once holds the line together and divides it, since the point is both the beginning [of the part of the line still to be drawn] and the end [of the part of the line already drawn]. However, [the analogy breaks down inasmuch as] taking a single point as both a beginning and an end requires a stop; whereas the "now" continually changes its character with the things in process. Consequently, time is "number," not like the dual character of the point, but rather like the "number" formed by the extremities of the line, instead of by its parts. For in distinguishing two parts of a line, the middle point must be stressed as two; and this requires a stop. And the "now" is evidently no more 20 a "part" of time than a section is part of a movement or than points are parts of a line; instead, it is two lines which are parts of one line. Accordingly, a terminal "now" is not time but an accident of it, whereas a "now" by way of "reckoning" is a number,§ for boundaries belong to that alone which they bound, whereas [abstract] number (for example, ten) belongs to "these horses" and to what-not.

It is evident, then, that time is the number of precessions and successions in process and that, since time is a number belonging to something continuous, time is therefore itself continuous.

12. Quantitative Time and Things in Time

Time is not like abstract [105] number, in which the smallest [plurality] is two; time is rather like concrete [4] number, which sometimes has a minimum and sometimes does not have a minimum. For example, the smallest number of a line as regards the plurality [6b] it has is two, 30 if not one; but the line has no minimum in magnitude, since every line is indefinitely divisible. So time [*reckoned*] has a minimum number, which is one or two; but there is no minimum length [142a] of *time*. 220b Evidently, too, we do not describe time as fast or slow but as much or little or as long or short. Continuous time is long or short; and quantitative time is [reckoned by] many or few [periods of time]. But there is no fast or slow time, any more than even the number with which we count is fast or slow. Furthermore, time is the same everywhere at

---

§ By which we count many specific durations, whatever their specific character.

any one time, but no time is the same before and after: not only do
past and future change differ from present change; but time is not a
number with which we count but is something counted, which is always 10
different before and after because the "nows" differ. Analogously, a
hundred horses and a hundred men are the same in number; but the
things counted, the horses and the men, differ. Yet the same time can
recur in the sense in which the same process can; this is true, for example, of a year, of a spring, of an autumn.

Moreover, we not only measure movement by time but we also
measure time by movement, because [measured time and movement]
determine each other. A time marks a process as its number; and so a
process marks a time. We speak of much or little time, measuring it
by the movement just as we determine the number of horses in a 20
group by using a single horse as a unit. As we know the size [6b] of a
group of horses by their number, and their number by using a single
horse as a unit, so we measure a movement by the time it takes, and
the time by means of a movement [as a unit]. There is a good reason
for this: a movement corresponds to a magnitude and time to a movement because all these are quantitative, continuous, and divisible, movement being so because magnitude is and time being so because movement is; and we measure a magnitude and a movement by each other,
saying that the way is long if the journey is long, and vice versa, and 30
therefore that the time is [much or little] if the movement is [much or
little, respectively], and vice versa. But time measures a movement, 221a
along with its being in process, by determining a movement which will
"measure" the whole movement, just as a cubit measures a length by
determining a magnitude which will "measure" the whole.

For any movement or process "to be in time," then, means that it and
its being [23] are measured by time: the movement and its being are
"timed" together; and this is what it means for it to be [88a] in time,
namely, that its being is measured. Clearly, then, this is also what it
means for other things to be in time, namely, that their being is measured by time. Now, "to be in time" may mean either to be when time 10
is or else to be in time as some things are said to be in a number-system,
namely, either if they belong to number as a part [22] or attribute [35a]
or element [4] of "number" or if they have a number. Thus, since time
is a "number," the present, the past, and so forth, are in time as elements
in it, just as a unit and the odd and even are in number as elements
in it; and events [188c] are in time in the sense of being numerable
and are therefore comprehended by time [or in a temporal structure] as
things in place are comprehended by place [or in a spatial structure].
But to be in time evidently does not mean to be *when* time is, any 20

more than to be in process or in a place means to be when a process or place is: in such a case, all things (for example, the heavens) would be in anything (for example, in a millet seed); whereas, in the case [3b] of what is "in time" (or "in process"), there is *necessarily*, not coincidentally, a time (or a process) when *it* is.

Since being in time is like being in a number-system, there is a sense in which time may be viewed as greater than anything in time; that is, everything temporal is necessarily bracketed [33e] by time [in a pattern 30 of temporal relations], just as everything local is encompassed by place [in a structure of spatial relations]. Moreover, things in time are somehow affected by time: in our way of speaking, things are ravaged and 221b aged and cast into oblivion by time, not made progressively wiser or younger or better; we attribute [83a] destruction to time because time is a number belonging to movement, and a movement is a removal [111d] of what is [82f]. Consequently, things which are continually are as such evidently not in time: they are not bracketed by time, that is, their being is not measured by time; and the fact that they are not in time is shown by the fact that they are not affected by time. On the other hand, rest is in time since time, which is a measure of movement, indirectly measures rest: what is in time need not, like what is in move- 10 ment, undergo a process; and time is not identical with movement but is a number belonging to movement. It is quite possible, therefore, even for what is at rest to be numbered by the number of movement. But as we have said before,* not everything unmoved is at rest but only what is naturally capable of the movement absent [106] from it. What is in time is measured by time as something is in a number-system if it has a number and if its being is measured by the number; and time measures the changing as changing and the enduring as enduring because it measures their movement or their rest by so much. Strictly speaking, what is in movement is not measured in its own quantitative 20 aspect by time but only in so far as movement has a quantitative aspect. Consequently, things not subject to movement or rest are not in time; for "to be in time" means "to be measurable by time," and time is the measure of movement and rest. Hence, too, it is evident that at least some of the things that are-not, are not in time; among them, things which can only "non-be," like the commensurability of a square with its side.

In general, since time measures movement directly [2] and other things indirectly [3], the things whose being time measures are clearly in motion and rest; and the things that come and go, that at one time 30 are and at another time are-not, necessarily are in time, since time in

---
* iii.2.202a4.

a sense extends beyond their being and beyond the time which measures
their being. Of the things not actual but bracketed by time, some (like
Homer) at one time were, others (like a future event) will be; and the   222a
pattern of time extends in both directions, so that some things may be
both past and future. But things in no way bracketed by time are not
past or present or future; they continually are-not, as their opposites
continually are. For example, the incommensurability of the diagonal
of the square with its side continually is; and since this will not be in
time, neither will the commensurability, but the latter forever is-not
because it is opposed to the former which forever is. On the other hand,
things whose opposites are not eternal may or may not themselves be;
and such things are subject to generation and destruction.

13. Definitions of Temporal Terms

As we have previously pointed out,* the present is time's continuity   10
since it holds past and future together and is their common boundary
as the beginning of the future and the end of the past; yet this is not
as evident as it is with a point in a line because the point is stationary.
The present is also a potential dividing of time and is as such always
different, although as unitive it is always the same. Just as mathemati-
cally a point when regarded as dividing a line is treated as if it were
two, but when regarded as one is treated as in every respect the same,
so the present functions in one way as potentially a dividing of time
and in another way as the common boundary and unity of both "parts."
And it is the same thing that at once both divides and unites, although   20
"being divisive" and "being unitive" are not the same. Besides having
this meaning, "the present" has an extended sense in which it includes
a time near the "present" in the first sense [as distinct from the more
remote past or future or both]. Thus, we say of one who will come
today or who has come today that "he will come now" or that "he has
come now"; but we do not so refer to the events narrated in the *Iliad*
or to the final cataclysm because, in spite of time's continuity, they are
remote from the present. Then, too, it is with reference to the "present"
in its first meaning that time is determinate, however indefinitely we
refer an event to "some time" in the past, for example, the capture of
Troy, or to "some time" in the future, for example, the final cataclysm.
There will be or was so much time from the present to a future or past
event. But if there were no time which is not some time, then every
period of time would be determined.

---

* iv.11.220a5.

30   Will time, then, run out? Or is it not rather the case that time will not run out since there is always some process going on? Is time, then, always different or does the same time frequently recur? Clearly, time corresponds in this regard to movement: if the same movement sometimes recurs, the time, too, will be the same; if the same movement does not recur, then time will not be the same either. Since the present is

222b the end and the beginning of time, though not of the same time, but is the end of the past and the beginning of the future, therefore time is always both at a beginning and at an end, just as a curve is both convex and concave. This is the reason why time seems to be always different, because the present is not the beginning and the end of the same time; if it were, it would at the same time and in the same respect be opposites. And because time is always at a beginning, it will never run out.

Expressions like "presently" and "just now" relate a near future or past to an indivisible [41] present, as when we reply to a question about

10   the time of a walk: "I shall go for a walk presently" or "I have just now come from a walk"; but we do not say that "Troy has just been captured," because this event occurred in a past too far removed. "Lately" also has reference to a past near the present, as in reply to a question about a recent journey; but "long ago" has reference to a distant part of the past. "Suddenly" implies a change from a former condition in a temporal interval so small as to be imperceptible. Yet all change is naturally a change from a former condition [111d]! All things that come into being and pass away do so in time. Some have therefore called time the wisest of things;† but Paron the Pythagorean more aptly called time the most stupid of things because in time we also forget. Clearly,

20   as we have said before, time is more especially a destroyer since change is a change from a former condition; only incidentally is time to be credited with what comes into being and with what is. This is sufficiently shown by the fact that nothing comes into being independently of some activity or process, whereas things can be destroyed even without undergoing a process. This is what we mean when we speak of things being "destroyed by time": not that time is a destructive agent; but even this sort of change happens to take place in time.

Such is our account of the being and definition of time, of the meaning of the "present," and of the use of such terms as "at some time," "lately," "presently," "long ago," and "suddenly."

---

† Simonides.

14. Temporal Distinctions as Relative and Objective

In the light of these distinctions, it is evident that every change or 30
movement necessarily takes place in time. Thus, every change is evidently fast or slow; and we attribute a relatively "fast" movement to a 223a
body which, moving an equal distance and along a similar path relatively to another body, arrives sooner at the terminus [85] than does the
other body (for example, if both move along the circumference of a
circle or along a straight line, and so forth). But what is earlier is in
time: both "earlier" and "later" are relative to a present, which is the
common boundary of the past and the future; so that, since any present
is in time, what is at some remove from the present, whether earlier
or later, is also in time. To be sure, there is a contrast between an earlier
and a later past and future: whereas the earlier past is farther removed 10
from the present than is the later past, it is the later future that is
farther removed from the present than is the earlier future. Since what
is "before" in time and since in every movement there is something
"before" something else, it evidently follows that every change or movement takes place in time.

However, we may well ask how in the world time is related to a living
self [154] and why time seems to be everywhere, in heaven, earth, and
sea. Undoubtedly time, as a number, is an attribute or state of a process;
heaven, earth, and sea are all subject to movement because they are 20
in place; and the things potentially or actually in time and those potentially or actually in process are, respectively, identical. Then, as to the
former question, would there be any time if there were no living self?
Without a being able to count, there would seem to be nothing count-
able, and therefore no number (since concrete "number" is what has
been or may be counted); hence, in the absence of a living or rational
being to do [101c] the counting, time would seem to be impossible. Still,
there might be time in the sense of whatever time would be in such a
case [1d].* Thus, there might be time as the order of "before" and
"after" in process in so far as these are "numerable," provided that there
can be movement in the absence of a living self.

We may also ask: Of what sort of process is time the number? Un- 30
doubtedly of every sort. Things come into being and pass away in time,
and they grow, are qualitatively altered, and move from place to place
in time. Hence, time is the number of each sort of process in so far as
it is a process; and therefore time is the number of continuous process 223b
simply, not of any one sort of process exclusively [4]. However, if each
of two simultaneously completed processes might have a number of its
own, would it not seem to follow that there would be two different

---

* In the sense of what underlies time, namely, movement.

equal times at once? Not at all! Equal and simultaneous times are the same time. Indeed, even times not simultaneous may be one in kind, just as seven dogs and seven horses are the same in number. So, too, processes having simultaneous limits take the same time even if one is fast and the other is slow and even if the one is a local motion and the other is a qualitative alteration: the time they take is the same since 10 it is equal and simultaneous; however different and mutually independent the processes, the time is everywhere the same because the amount of time which equal and simultaneous processes take is everywhere the same.

Now, processes include local motion, which in turn includes motion in a circle. Moreover, we "number" anything by means of something single which is homogeneous with it: units, by means of a single unit; horses, by means of a single horse; and therefore also a period of time, by means of some limited duration. But as we have previously said,† we measure time by movement, and movement, by time; and we can do this because it is by means of a movement determinate in time that "so much" of movement as well as of time is measured. If, then, what is primary [in any genus] is the measure of everything in the same genus, uniform circular motion is the best measure [of all 20 movement and therefore of time] because the "number" belonging to it is best known. Qualitative alteration, growth, and origination do not have the uniformity which [the circular mode of] motion has. This is also the reason why time has been identified by some with the motion of the celestial sphere: other movements are measured by means of it; and time, by means of this motion. This is also the reason why human affairs and all things that come and go in the natural course of events are commonly said to move in cycles: all affairs are judged in terms of time; and they come to a stop and then start afresh in cyclical 30 fashion. Some even look upon time itself as in some sense a cycle: time measures and is measured by circular motion; and the cyclical theory of events implies a circle of time in the sense of time measured by circular 224a motion. For, in what is measured, we observe nothing other than the measure; and the whole of it amounts to measures many times repeated.

However that may be, we may at any rate restate another point which is rightly made: if the number of a group of sheep and the number of a group of dogs are equal, the *number* of both is the same, although it is not (for example) the same *ten.* So, too, an equilateral and a scalene triangle are not the same [kind of] *triangle;* they are, however, the same [kind of] *figure,* for they are both triangles. Two things are the same [kind of] thing if they do not differ by a differentia [of their kind];

---

† iv.12.220b23.

but not if they do! Thus, triangles differing by a differentia of triangle are different triangles; they are not different figures, however, but are in the same subdivision of "figure." One kind of figure is the circle; 10 another kind, the triangle. One kind of triangle in turn is equilateral; another kind, scalene. These are the same [kind of] figures, then, since they are triangles; but they are not the same [kind of] triangles. Just so, two groups have the same number if their number does not differ by a differentia of number; but they are not therefore the same [sort of] ten, since the groups numbering ten each may differ in that the one may be a group of dogs and the other may be a group of horses.‡

So much, then, for time both as far as time itself is concerned and with respect to the considerations appropriate to the analysis of time.

---

‡ Periods of measured time are not kinds of events.

V. BOOK EPSILON

Classification of Movements

1. Types of Changes and Processes

21 Changes° are of three types: (1) accidental change [115], as when a musician turns from his music to taking a walk; (2) internal [105] change due to a change in a part, which changes the whole, as when the whole body is restored by the healing of an eye or of the chest; (3) essential change, as when a being is moved which is essentially movable. Essential change includes qualitative alteration and other kinds,
30 which in turn differ among themselves; for example, a thing may be alterable in quality by being healed or heated.

There are the same kinds of agents or movers: accidental, partial, and essential. The second is exemplified by a hand striking; the last, by a physician healing.

Any movement involves an immediate agent, a thing moved, a time
224b of change, a starting-point, and a culmination. For every single movement has a definite beginning and a definite ending; and we may distinguish in it an immediate subject (like wood) which undergoes the movement, the end to which it tends (like heat), and the condition from which it starts (like cold). Accordingly, it is clear that a movement belongs inherently to a subject:† a movement does not inhere in a form, since a form or a place or a quantity neither initiates nor undergoes a movement; but a movement has a mover or agent, a subject, and a culmination. Although a movement also has a starting-point, yet it gets its name from its end: when a perishing thing changes from being to

---

° Large parts of v.1-3 are duplicated in *Metaphysics* xi.11, 12.
† iv.11.219b29, 30.

nonbeing, it is said to "perish"; and a change from nonbeing to being 10
is called a "generation."

We have previously stated‡ what "movement" is; but the forms, the
modes, and the places into which things are carried by their movements
are themselves unmoved (for example, knowledge and heat do not
move). One might object that modes [35a] such as whiteness are movements and that therefore there are changes to movements. However,
whiteness is not a movement [or process]; the movement which ends
in whiteness is whitening. Moreover, like changes and their effective
agents, so their ends are of three types: accidental; partial, that is,
dependent on something else; and essential, that is, independent of
something else. Thus, what is growing white, becomes an object of
thought accidentally, since being thought of is incidental to color; what 20
is growing white, changes to a color, that is, to a part [or kind] of
color (just as one going to Athens may be said to be going to Europe,
of which Athens is a part); but primarily and essentially, what is
growing white, is changing to white color.

It is clear, then, in what sense a movement, its effective agent, and
its subject are such essentially and incidentally; in what sense they are
such directly, with reference to themselves, and indirectly, with reference to something else; and also that movement is not in the form of
a thing but in the thing moved, that is, in what is actually [39a] functioning [9] as a movable thing. We must, however, dismiss accidental
change from consideration, since it is present in anything, at any time,
and in respect of anything.

Nonaccidental changes are not to be found inherently in all things,
but take place between contraries or their intermediates and between
contradictories. We may convince ourselves of this by induction. A 30
change may take place from an intermediate which functions [163]
as a contrary to either extreme, since an intermediate is in a way both
of the extremes: a middle note is low compared with the highest and
high compared with the lowest; and gray is light compared with dark
and dark compared with light. Then, too, every "transformation" has a 225a
definite initial and terminal limit, as the term "transformation" itself
suggests. Hence, changes take place (1) between substantives [85], (2)
between nonsubstantives, (3) from a nonsubstantive to a substantive,
and (4) from a substantive to a nonsubstantive. By a "substantive" I
mean what is referred to in a positive assertion. Since change from one
nonsubstantive to another nonsubstantive is really not a change, involving neither contraries nor contradictories and hence no opposition [64b],
this leaves three types of change. A change from a nonsubstantive into 10

‡ iii.1.210a10, 27-29.

a substantive (its positive contradictory) is generation. Such change, when absolute [105], is the generation of a whole new subject or substantive; when partial (for example, from nonwhite to white), it is a partial [4] generation. Change from a substantive to a nonsubstantive is destruction, absolute change being absolute destruction and a partial change being partial destruction.

20   Though "nonbeing" has several senses, there is no way in which a nonbeing can be in process [109a], whether it be the nonbeing involved in [false] predication [64g] or disjunction [75], or whether it be the nonbeing of potentiality, which is opposed to complete being. To be sure, nonwhite and nongood can undergo movement incidentally, in so far as a "nonwhite" may be a moving man; but in so far as it is not a positive substantive, or "this" [4a], it cannot be moved. Hence, nonbeing cannot be moved; and, if this be so, generation cannot be movement, since it is from nonbeing that generation starts. For, even admitting that some generation is accidental, it remains true that any absolute generation must begin [82f] in nonbeing.§ Similarly, nonbeing cannot
30   come to rest. These consequences are troublesome; and so is the fact that everything that is moved is in a place, but "nonbeing" is not in a place, for then it would be somewhere. Hence, destruction is also not movement; for the contrary of a movement is another movement or rest, and of destruction the contrary is generation.

We must conclude, therefore, that since processes [109] are changes [115] and changes are of the three types mentioned, those changes which
225b are generations and destructions are not processes.‖ A process moves from one thing to its contradictory; hence, only the change from one substantive to another is a process. And these substantives are either contraries or their intermediates, for even privation is to be put down as a contrary; and privations can be expressed [59] positively, such as "naked," "bald," "black."

If the categories are divided into primary being, quality, place, relation, quantity, activity or passivity, there must be three kinds of change [109]:¶ in quality, in quantity, in place.

2. Kinds of Movement and the Immovable

10   There is no change [109] with regard to primary being, because primary being has no contrary; nor is there change in relation, since it is possible that if one correlative changes [115] what was true of the

§ i.7.190b3-5.
‖ But cp. Plato Laws x.894ff.
¶ Single "movements," distinguished into kinds according to specific beginnings and ends, include only movements in quality, quantity, and place.

other is no longer true, although this other does not itself change, so that in such cases the change [109] is accidental. Nor is there change from agent to patient, or from mover to moved; because there is no movement of movement or generation of generation or, in general, change of change.

Now there could be movement of movement in two ways: first, as change of subject matter, in the sense in which, for example, a man is a changed being when he changes from fair to dark; similarly, a change might be said to become hot or cold or to change its place or to increase or decrease. But this is impossible; for a change [115] is not literally 20 [4] a subject. Or, secondly, some other subject might change from a change into some other state of being [20]; for example, a man changes from becoming ill to recovering health. But this, too, is impossible, except incidentally. For every movement is change from something to something else; and so are generation and destruction, except that these are changes into opposites of one sort, whereas movements are changes into opposites of another sort.° A thing that changes from health to illness would, then, change at the same time from this very change into another. Accordingly, it is clear that when it has become ill it will have changed to some other change (although it is, indeed, possible for it to remain at rest), and moreover to a change that is never for- 30 tuitous; and that further change will be from something to something else. Consequently, it would be the opposite change, that of recovering health. But such change is accidental: there is a change from remembering to forgetting because that to which the process belongs undergoes a change—at one time to knowledge, at another to ignorance.

Besides, if there is to be change of change and generation of genera- tion, the process will go on to infinity. Thus, whatever a later stage is, the earlier must be also; that is, if generation itself was ever being 226a generated, then what was becoming generation itself was also at one time being generated. Consequently, it was not yet the process of gen- eration itself, but only the generation of some particular being [4], past or present. If this were true, there was a time when generation was being generated; consequently, it was not yet generated at that specific time when it was being generated. But since there is no first in the infinite the first will not be; consequently, there will not be a second either. Neither generation, then, nor movement would be possible, nor any change.

Whatever suffers a given change [109] or state of rest also suffers the contrary change or state; so what is generated is destroyed. Consequent- ly, what is being generated begins to perish when it has been in

---

° v.1.225a35-b1.

the process for some time; for it does not begin to perish either in the
10 very first stage of being generated or later, since what is perishing must
for a while be. And there must remain some material constant underneath the generation and changes. What, then, could it be? What is it
that is generated as movement or generation, as a body or a self is
which undergoes alteration? And again, what is it into which they move?
It must be the movement or generation of something [4a] from something to something. And how? For the movement of learning cannot be
learning, and the movement of generation cannot be generation. Finally,
since there are three kinds of movement, the nature persisting in the
supposed movements as well as the goals to which they tend would
have to be one of them; for example, local motion would be qualitatively altered or moved from one place to another.

20   In general, since movements are of three types, accidental, partial,
or essential, change itself would only be change accidentally, as when
a man recovering his health runs or learns. But we have already† dis-
missed accidental change from consideration.

Since there is no movement, therefore, in primary being or relation
or activity and passivity, it remains that movement is in quality and
quantity and place; for each of these can have contrariety. Let us call
a movement in quality a "qualitative alteration" [120], which refers to
both [extreme ends of the movement]. By quality [28], I do not here
mean what is in the primary being (for even the differentia is a quality),
but a passive quality [35c] with regard to which something is said to
be acted upon [35] or to be incapable of being acted upon [35d].
30 Movement in quantity has no such common designation but is named,
after one or the other of its extremes [of largeness and smallness], either
an "increase" or a "decrease," respectively. Movement in place has no
name which is derived from its extremes taken either together or
separately; but we may call it a "carrying" [121], although in the strictest
226b sense only those things are "carried" from place to place which cannot
stop or move themselves. Change to a greater or a lesser degree within
the same kind may be classified as a qualitative alteration, for a movement may be from one contrary to another in either an unqualified or
a qualified sense. A thing changing to a lesser degree of a quality may
be said to change to the "contrary," whereas a thing changing to a
greater degree of a quality may be said to change "from" the contrary
towards the quality itself; whether the change is in a qualified respect
or not, makes no difference, except that the contraries in the former
case have to be present in a qualified sense only. "Greater" or "less"
means the presence [82h] or absence of "less" or "more" of a contrary.

---

† v.1.224b26.

From these considerations, then, it is clear that there are but three kinds of "movement."

The "immovable" is (1) what is wholly incapable of being moved 10 (as voice is invisible), or (2) what is moved with difficulty after a long time or begins slowly (and is said to be "hard to move"), or (3) what can naturally be moved, but is not moved when and where and as it naturally would be. Only this last kind of immovable being I would call "being at rest"; for rest is contrary to movement, so that it must be a privation of what is capable of receiving [12] movement.

It should be evident now what movement and rest are, how many types of changes there are, and which sorts of changes are "movements."

3. Succession, Contact, Contiguity, and Related Distinctions

Let us now state what it means [87] for things to be together, apart, in contact, between, in succession, contiguous, or continuous, and to 20 what sorts of things each of these naturally belongs.

Things are "together in place" when they are in one primary place; and "apart," when they are in different places. Things "touch" when their extreme ends are "together."

That is "between" at which a thing undergoing change, if it undergoes continuous change according to its nature, arrives naturally before it arrives at the extreme [18a] towards which it is changing. Thus, "between" implies at least three terms: for the terminus [18a] of a change is a contrary; and a thing moves "continuously" when it leaves no part or hardly any part of its "path" [188c] uncovered [148a]. I do not say "of the time" (since there is nothing to prevent the time from being interrupted [148a] or to prevent a high note from being sounded im- 30 mediately after a low one), but I say "of the path" [188c] in which the movement takes place. This is evident in local changes as well as in all other kinds of change. That is "contrary in place" which is most distant in a straight line; for the shortest line, being determinate [131a], serves as a measure.

That is "in succession" which is after a beginning, as determined [72d] by position or form or in some other way, and has nothing of 227a the same kind "between" it and what it succeeds; for example, lines if it is a line, units if it is a unit, or a house if it is a house. But there is nothing to prevent something of a different kind from being between. For what is in succession is in succession to something and comes after it; for one does not succeed two, nor does the first day of the month succeed the second, but the latter succeeds the former.

That is "contiguous" which is "in succession" and "touches." Also, since all change is between opposites, and these are contraries or contradictories, and contradictories have no middle term, it is clear that what is "between" is between contraries.

10   The "continuous" is something contiguous. I call things "continuous" when their limits touch and become one and the same and are contained in each other (which is impossible when the limits are distinctly two); so that it is clear that continuity belongs to things out of whose mutual contact a unity naturally arises. And the whole is a unity in the same way in which the continuous is a unity, whether by having been nailed or glued or mixed or having grown together.

It is evident also that of these concepts it is being "in succession" that comes first: for what is in succession does not necessarily touch,
20   but what touches is in succession, which is why there is succession but no contact among things like numbers which have a logical [90] priority; and if anything is continuous, it touches, but if it touches, it is not necessarily continuous, since extremities which are together need not be one whereas extremities which are one must be together. Consequently, natural union [101f] comes last in the order of genesis: for extremities naturally united must touch, but not all things that touch are naturally united; and it is clear that there is no natural union in things in which there is no touching. Hence, if there are independently existing points and units as some say, a point is not the same as a
30   unit:° for touching would belong to points, but not to units, which are only in succession; and there is something between points (since every line is between points), but not necessarily between units (since there is nothing between one and two).†

So much for the definitions of "together," "apart," "touching," "be-
227b   tween," "in succession," "contiguous," and "continuous," and so much for the sorts of things to which each of these belongs.

4. Movements as Unified and Diversified

A movement [or process] may be "one" in various ways which correspond to the various meanings of "unity."

Movements are generically one when they are in the same category. For example, all local motions are generically one, whereas qualitative alterations differ generically from local motions.

° The Pythagoreans defined a "point" as "a unit having position" and a "unit" as "a point not having position."

† The Pythagoreans held "numbers" to be "separated from one another" by the "void."

Movements are specifically one when, besides being one in genus, they are in the same indivisible species. For example, since colors have specific differences, becoming black and becoming white differ in species; but inasmuch as whiteness is not further divided by specific 10 differences, every process of becoming white is one in species with every other. In the case of [intermediate] genera which are also species, it is clear that movements are one in species, but in a qualified sense. Among such are processes of learning; for knowledge is both a species of apprehension [174] and a genus of the various sciences. One may object: if movements are one in species when the same subject undergoes a change from the same beginning to the same end, when (for example) the same point [or particle] shifts again and again from one identical place to another, then motion in a circle and motion [back and forth] in a straight line would be the same; and so would rolling and walking. But have we not already implied [72e] that motions differing specifically in their courses differ in species; and would not motions 20 in a circle and motions in a straight line, then, differ in species?

Having shown in what senses movements are one generically and are one specifically, we may by means of further distinctions explain what sorts of movements are inherently [105] one, that is, numerically single in their being [26]. In describing a movement (or process), we specify "what" changes, and "in what respect" as well as "when." Thus, since a movement must have a subject, we distinguish "what" undergoes the movement (a man or gold); also, "that in which" it undergoes the movement (in place or in quality); and, likewise, "when" it undergoes the movement (since all movement takes time). But whereas the generic or specific unity of movements depends upon the way [188c] in which they occur, and their continuity depends upon the time, their inherent unity depends upon all three of the considerations distinguished: the 30 species of such a movement must be single and indivisible; the time of the movement must be single and uninterrupted [148a]; and the subject of the movement must be single. However, the unity of the subject must not be accidental only, as in the case of "white Coriscus" taking a walk and turning dark. Nor must the unity of the subject consist in an asso- 228a ciation [92] only, as when two people are recovering from ophthalmia at the same time and in the same way; their simultaneous recovery is not a single process but is one in species. As for the objection that Socrates may undergo the same kind [20] of qualitative alteration at different times, this movement would be unitary only if, after having ceased, it could arise again as the numerically single movement it was; if not, it is indeed the same [in kind and as regards its subject], but it is not unitary. A related question is whether our health or whether

bodily dispositions and states generally are unified in their being [26],
even though bodies themselves are apparently subject to movement and
10 flux:° thus, if the state of our health is one and the same at the dawn
of day and in this very instant, why should it not be numerically
one with the health we may have lost and after a while recovered? But
the same argument holds in this case. There is only this difference: if
the states are distinctly two, the corresponding [movements or] ac-
tivites [9] must also be distinctly two (since only a numerically single
subject can perform a numerically single activity); but if the state
is single, it does not follow that the [movement or] activity is therefore
single also (for when a man stops walking, the walking ceases until he
resumes it). If the latter were one and the same, then it would be
possible for anything, while remaining one and the same, repeatedly
to perish and again to be. But such questions are beyond the scope of
our inquiry.

20 More pertinently, since every movement, being [infinitely] divisible,
is continuous, therefore an inherently single movement must also be
continuous; and if it is continuous, it must also be single. Not that any
movement can form a continuous movement with any other movement,
any more than any chance thing can be continuous with any other
chance thing. Their extreme ends must be unified. Some things have
no extreme ends, whereas others have extreme ends differing in kind;
and how can the "end" of a line and the "end" of walking touch or
become one, even though both are called "ends"? Movements differing in
species or genus may, indeed, be consecutive; for example, one may
become feverish after running. So, too, the "legs" of a relay race are consecutive, not continuous. For, by definition, only those things are "con-
30 tinuous" whose extreme ends are one. Thus, consecutive or successive
movements are such by reason of a temporal continuity; but to be con-
228b tinuous, the movements must themselves be continuous in the sense that
the extreme end of each becomes one with that of the other. Hence,
a movement that is inherently continuous and single must be identical
in species, must have a single subject, and must take place in a time
that is single. The time must be single, lest a pause [109e] intervene
in which the movement would have come to a halt and be suspended:
movements with intervals of rest are not single but many; and therefore
any movement broken up by a standstill, as it would be in an intervening
time, is neither single nor continuous. On the other hand, even if
a specifically different movement is temporally uninterrupted, unity of
the time does not affect the specific difference of the movement; for a
10 single movement must be of a single kind, although a movement single

---

° Heraclitus.

in kind is not necessarily single in an unqualified sense. So much for the inherent "unity" of a movement.

Again, we identify a single movement as such when it has been completed [14].† This may be so with respect to its genus, or its species, or also its being [26]. It is in this way, too, that we identify any other [subject matter]; namely, as a unity which is a complete whole. Still, we sometimes ascribe unity to an uncompleted movement, provided that the movement is continuous.‡

Finally, movements have another type of unity besides those already examined; namely, when they are uniform. A uniform movement (such as movement in a straight line) is more properly characterized as one than is a nonuniform movement, which can be analyzed [into distinct movements]. But the difference seems to be one of degree [that is, of more or less irregular movements]. Moreover, any kind of movment may be uniform or nonuniform: qualitative alteration; local motion, 20 as in a circle or in a straight line; and growth and decline. Sometimes a movement is nonuniform in its path, as when its path is a broken line or a spiral or any other spatial magnitude whose parts taken at random do not fit upon [56b] other parts taken at random; for a movement cannot be uniform unless its path is so. But sometimes a movement is nonuniform, not in its subject§ or time or goal, but in its rate, which may be fast or slow: movement at a uniform rate is uniform; at a nonuniform rate, nonuniform. Hence, differences of rate, which accompany 30 [77] all the different species of movement, do not constitute species or differentiae of movement. Hence, too, earth does not, because of any difference in weight [or in rate of tending] in the same direction, differ in species from earth, or fire from fire. Even a nonuniform movement, 229a then, like a motion in a broken line, may be one if it is continuous; but it is less unified than a uniform movement, and in anything "less" there is always a mixture with a contrary. But if every unitary movement is capable [12a] of being either uniform or nonuniform, consecutive movements which differ in species cannot be one and continuous. How could a qualitative alteration and a local motion coalesce [32] in a uniform movement? Component parts ought to exhibit a harmony [56b].

---

† Pythagoreans, Platonists.

‡ Thomas Aquinas: "Et ratio hujus est: quia unum potest attendi vel secundum quantitatem, et sic sola continuitas sufficit ad unitatem rei: vel secundum formam substantialem, quae est perfectio totius, et sic perfectum et totum dicitur unum."

§ Preferable to the reading: "place."

5. Movements Contrary to Each Other

We must next determine what sorts of movements are contraries to each other and then deal in the same way with rest. (1) Are movements contrary to each other when the starting-point of one (for example,
10 health) is identical with the other movement's end? If so, then [contrary] movements would in this respect be like [5b] generation and destruction. Or (2) are movements mutually contrary when their starting-points are contraries (one movement starting from health, and the other from disease)? Or (3) when their ends are contraries (one movement ending in health, and the other in disease)? Or (4) when one movement starts from one of a pair of contraries (health) and the other ends in the other contrary (disease)? Or (5) when one movement proceeds from one to the other of a pair of contraries (from health to disease) and the other movement proceeds from the latter contrary to the former (from disease to health)? Movements must be contrary to each other either in one or in more than one of these ways, there being no other way in which they can be mutually opposed [64b].

However, a movement (4) from one of a pair of contraries (health) is not contrary to a movement to the other contrary (hence, ending in disease): these movements are the same; although their ways of being
20 [23] differ, just as changefulness [115] in health differs from changing tendencies to disease. Nor are movements contrary when (2) they start from contraries: movement *from* a contrary is incidental to movement *to* a contrary or to an intermediate (as we shall explain presently); contrariety in movements is better accounted for [83] by change *to* a contrary, which represents a gain of a contrary, than by change *from* a contrary, which represents a loss of a contrary; and a change gets its name from its culmination rather than from its starting-point (for example, "getting well" or "falling ill"). There remain the third and the fifth alternatives. However, movements which (3) end in contraries (as we were about to explain) also start from contraries; but their way
30 of being is distinct, since "to health" is not identical with "from disease," nor is "from health" identical with "to disease." Now, then, a "change" differs from a "movement" in that a movement is a change from something [4] positive [85] to something positive.* Therefore movements are contrary to each other whenever (5) one movement proceeds from
229b one to the other of a pair of contraries (from health to disease) and the other movement proceeds from the latter contrary to the former (from disease to health).

We may show inductively of what sorts the contrary terms [in contrary movements] are: falling ill is contrary to getting well, since these move-

---

* v.1.225a4, b2.

ments have contrary outcomes; being taught is contrary to being deceived by another, for it is possible to get knowledge or misinformation by one's own activity or by another's; and upward motion is contrary to downward motion in length, motion to the right is contrary to motion to the left in breadth, and forward motion is contrary to backward motion [in depth].† On the other hand, a process which is only *to* a contrary 10 (for example, "being white") inasmuch as the starting-point lacks positive determination [4], is a [partial] "change" rather than a "movement" [in sense (3)]. Where there are no contraries [but contradictories], there a "change" *from* something is contrary to a "change" *to* the same thing: thus, generation is contrary to destruction, as the loss of a thing is contrary to its acquisition; but these are [absolute] "changes," not "movements" [in sense (1)]. Where there are intermediates between the contrary terms, a movement to an intermediate is to be treated [64] as a movement to a contrary, since an intermediate operates [163] as a contrary in a movement, in whichever direction the change takes place: thus, gray operates as would black in a movement from gray to white or from white to gray, and as would white in a movement from black to gray; for a mean is in a way one of the extremes relatively to 20 the other, as we have said before.‡ In fine, two movements are contraries only when (5) one movement passes from one to the other of a pair of contraries and the other movement passes from the latter contrary to the former.

6. Movement and Rest, Natural and Violent

Not only are there contrary movements; but rest and movement, too, are contraries which call for discrimination. Strictly, a movement is contrary to a movement. But rest is also opposed to movement; for "rest" is "privation of movement," and a "privation" is in some sense a "contrary." Also, a certain kind of rest is opposed, respectively, to a certain kind of movement; for example, resting somewhere, to moving somewhere. Yet this statement is too simple. Does rest at a given place have for its opposite a movement away from or towards that place? But clearly, since a movement has two limits [85], movement from a given 30 place to another [50] has for its opposite resting in the given place, whereas movement from elsewhere [50] to the given place has for its opposite resting in the other place. Moreover, these two rests are contrary to each other; it would be an absurdity that there should be mutually contrary movements but no rests which are mutually opposed. 230a States of rest in contraries are mutually opposed: thus, (1) remaining in

† *De caelo* ii.2.284b24, 25.
‡ v.1.224b32-35.

good health is contrary to remaining in bad health; and (2) remaining in good health is also contrary to movement from good to bad health. For (2) remaining in good health cannot without incongruity be contrary to movement from bad to good health: a movement to a state of stability is rather a coming to rest, and coming to rest is at least simultaneous with movement to a state of stability; but remaining in good health must be contrary to one or to the other of these movements. However, (1) remaining white is not contrary to remaining in good health.

Where there are no contraries, there a "change" *from* something is opposed to a "change" *to* the same thing, that is, change from being is opposed to change to being; but this is not a "movement," and there-
10 fore there is also no "rest" [as the contrary] but only changelessness [115a]. If there were a positively identifiable [4] substantive [85], its changelessness in being [1] would be contrary to its changelessness in nonbeing [1b]. But if there is no identifiable nonbeing, we may ask: what is it to which changelessness in being is contrary? And is change-lessness in being, a state of rest? If so, then either not every rest is contrary to movement, or else generation and destruction are "move-ment." Clearly, then, since generation and destruction are not "move-ments," changelessness in being is not "rest" but is something like it and is contrary either to nothing or else to changelessness in nonbeing or to destruction; but it is not contrary to destruction, since destruction has changelessness in being for its starting-point whereas generation has it for its stopping-point.

There is also this objection: Why should there be natural and
20 contra-natural [74] movements and rests in local changes but not in others as well? Why is there no contrast of the natural and the contra-natural among qualitative alterations? Getting well is not more natural or contra-natural than falling ill; or becoming white, than becoming black. Why not among processes of growth and decline? Neither is thus contrasted with the other, any more than is one kind of growth with another. Why not in generation and destruction? Becoming does not differ from perishing as a natural from a contra-natural occurrence (for growing old is natural); nor do we observe generations which are either natural or contra-natural. However, if change by violence is "contra-
30 natural," would not violent destructions, being "contra-natural," be con-trary to natural destructions? Are there not accordingly also produc-tions which, so far from being inevitable, are forced and are thus
230b contrary to natural productions? Are there not processes of forced growth and decline; for example, when growth is hastened by indulgence or when the growth of plants is artificially hastened? Likewise, are there not natural and violent qualitative alterations; as when one patient re-

covers from a fever during a crisis and another gets rid of his fever at another time? But, it may be objected, will there then not be destructions which are contrary to other destructions, instead of to generations only? Yes! And why not? Is not passing away pleasantly contrary to passing away painfully; not, indeed, in the sense of an absolute contrast between one passing away and another, but in the sense of a contrast between one kind [5] of passing away and another kind?

So there are contraries among movements and states of rest gener- 10 ally [44], as we have already explained. Thus, in the category of place, upward motion is contrary to downward motion, as is rest above to rest below; and therefore the natural upward motion of fire is contrary to the natural downward motion of earth as well as to the contra-natural downward motion of fire. Likewise, there will be rests [contrary to motions]: thus, rest above is contrary to motion downward; and therefore the contra-natural rest of earth above is contrary to its natural motion downward, just as the natural upward or downward motion of 20 anything is contrary, respectively, to its contra-natural motion in the reverse direction. The question may be raised: Is every rest that is not eternal generated, as in a process of coming to a standstill? If so, a contra-natural rest (as of earth above) would be generated; so that earth, when violently carried upward, would be coming to a standstill! However, in coming to a standstill, a body moves at an ever increasing rate; whereas, under compulsion, a body moves at a decreasing rate. Therefore, the rest of such a body is not the result of a process of coming to rest. Moreover, stabilization would seem to be concomitant if not identical with a body's natural movement to its proper place.

As regards coming to a standstill,* someone may ask: Is rest opposed 231a5 to contra-natural as well as to natural motions? If not, this would seem absurd; for a body may be at rest under violence. Hence, there would be a noneternal state of rest which is not the result of a process [of coming to rest]. But it is clear that this is so; for as there is contra-natural motion, so a body can be in a contra-natural state of rest. Now, some 10 things have both a natural and a contra-natural motion; for example, fire has a natural upward motion and a contra-natural downward motion. Is it the latter, then, or is it the natural downward motion of earth, which is contrary to the former? Clearly, both are contrary to it, although in different ways: the natural downward motion of earth is contrary to the natural upward motion of fire; whereas the upward motion of fire, as natural, is contrary to the downward motion of fire, as contra-natural. Similar considerations would hold for the corresponding cases

---

* This passage (231a5-17) seems to be a different version of 230b10-28.

of rest. Yet there is perhaps a sense in which motion is opposed to a state of rest.

230b28   There is also a difficulty in the view† that rest at a given place is contrary to motion from that place. When a body is leaving a place or
30 is losing something, it appears to be still having what it is losing. Hence, if the rest at the given place is contrary to a motion away from it, the moving body will at the same time have both contraries, the rest and the motion. Or is not the situation rather such that, as long as the changing thing remains [in its place or in its initial state], it is still in some respect at rest? And, in general, is not a body in motion partly
231a at its starting-point and partly at its goal? This is why it is movement rather than rest that is contrary to movement.

So much for the unity of movement and of rest, and so much for their contrariety.

---

† 229b28-31.

VI. BOOK ZETA

Continuity of Movement

1. The Continuous as Infinitely Divisible

No continuum can be made up of indivisibles. For, in accordance with 21
our distinctions,* things are "continuous" when their ends are one; they
"touch" when their ends are together; and they are "in succession"
when they have nothing of their own kind between them. Thus, a line,
which is continuous, cannot be made up of points, which are indivisible.
For no point can have an end which could become one with that of
another, because what is indivisible cannot be divided into ends and
other parts. Nor can any point have an end which could be together
with that of another, because what is without parts cannot have any
end at all since its end would differ from that whose end it would be.
Again, in order to make up a continuum, points (like any other indivisi- 30
ble things) would either have to be continuous with one another or
would at least have to touch. But, for the reason already stated, points 231b
cannot be continuous; and a whole touches a whole, or a part touches
a part, or a part touches a whole. But since anything indivisible is without parts, therefore, in the contact of indivisibles, a whole would
have to touch a whole. However, when a whole touches a whole, this
is not a continuity; a continuity requires distinct and locally separate
parts. Again, points cannot even be in succession in such a way as to
constitute a length, any more than moments can be in succession in
such a way as to constitute a time [or a continuous duration]. Only
those things are "in succession" which have nothing of their own kind

----

* v.3.227a17-23.

between them; whereas points always have a line [or points] between them, and moments always have a time [or moments] between them.

10 Were a length or a time made up of such points or moments, it would be divisible into [successive] indivisibles; but we have found it impossible for anything continuous to be divided into any parts without parts. Nor can such points or moments even have anything of some other kind [19] between them: it would be either indivisible or else divisible, whether into indivisible parts or into parts infinitely divisible; and on the last alternative, it would be continuous [and therefore a line or a time, respectively]. Evidently, anything continuous is divisible into parts infinitely divisible; if a continuum were divisible into indivisible parts, one indivisible part would touch another (since ends of things continuous with each other are one and touch each other).

By the same argument, magnitude, time, and movement would all be 20 composed of indivisibles and would be divisible into them; or else none of them is. This will become clear in the course of the following analysis. If a magnitude consisted of indivisibles, then a motion along that magnitude would consist of an equal number of indivisible motions. Suppose [a line] *ABC* were composed of the indivisible parts *A, B,* and *C;* then [a moving point] *X* would move along *ABC* with a motion *DEF* having the corresponding parts *D, E, F,* each indivisible. If, then, a motion going on [23c] must have a definite subject and a moving subject requires an ongoing motion, a motion which is occurring would also consist of indivisible parts. Accordingly, *X* would have been moving over the distance *A* with motion *D;* over *B,* with motion *E;* and over *C,* with motion *F.* But anything moving from somewhere to somewhere 30 cannot, while in motion, *be moving* and at the same time *have completed its motion* to its destination. In walking to Thebes, for example, it is impossible to be walking to Thebes and at the same time to have 232a concluded the walk to Thebes. So *X* would have been moving along the partless part *A* with a motion *D* then going on. Consequently, if *X* was not completing its passage through *A* until after it had been passing through *A,* part *A* must be divisible; for *X,* while passing through *A,* neither was at rest nor had completed its passage but was in a stage between these extremes. But if *X* would have been passing through *A* and would at the same time have completed its passage through *A,* then it would be possible for a walk, while proceeding to its destination, to have been completed; that is, it would be possible for anything to *have* moved whither it *is* moving! On the other hand, if *X* moves over the whole of *ABC* with a motion *DEF* but could not be in motion at all but only have completed a motion over a partless part *A,* then the motion would not consist of motions but of [discrete atomic] movements [109f]. That is, *X* would have completed a movement without being in motion; for *X* would have passed through *A*, through which it 10 would never be passing. So it would be possible to have completed a walk without ever walking, for the walk would have bridged a distance without anything ever walking over it. Since, then, anything must be either at rest or in motion, *X* would be at rest at each of the parts *A*, *B*, and *C*, and therefore it would be possible for anything to be both continuously at rest and at the same time in motion; for *X* was supposedly moving over the whole of *ABC* and was supposedly resting at one or another of its parts and would consequently rest at the whole. Besides, if the indivisible parts of *DEF* were motions, it would be possible for anything, while its motion is going on, not to be in motion but to be at rest; and if they were not motions, it would be possible for movements to consist of parts which are not movements.

Thus, too, if a length and a motion were indivisible, time would likewise have to be indivisible; and so time would have to be composed of indivisible moments. For if all motion is divisible and anything mov- 20 ing at a uniform rate covers a shorter distance in a shorter time, then time will also be divisible. And if the time is divisible in which *X* is carried over the distance *A*, then *A* will also be divisible.†

2. Time and Spatial Magnitudes as Continuous

Since every magnitude, being continuous, is therefore divisible into magnitudes (for we have proved it impossible for anything continuous to be composed of indivisible parts), it follows necessarily that, of two moving bodies, the faster travels (1) a greater distance in an equal time, (2) an equal distance in a shorter time, and (3) a greater distance in a shorter time. Indeed, some define the "faster" in these very terms. Now, let *A* be faster than *B*, in the sense of changing sooner than the latter. Then (1) if *A* has changed from *C* to *D* during a time *fg*, the slower body *B* will in that time have failed to reach *D*. 30 Thus, the faster body spans a greater stretch in an equal time. But (3) even in a shorter time the faster body crosses a greater expanse. (Fig. 1.) Let *B* as the slower body arrive at *E* when *A* has come to *D*;

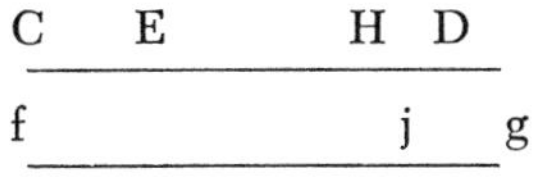

C	E		H	D
f			j	g

Fig. 1

---

† iii.7.207b21-25; iv.11.219a10-25, b15, 16, 12.220b24, 25; *Metaphysics* v.13.1020a 26-32.

232b then, since A has taken the whole of time *fg* to get to D, A' has in a shorter time *fj* passed beyond E to H. Consequently, since the distance CH bridged by A is greater than CE, and the time *fj* is shorter than the whole time *fg*, the faster body passes over a greater stretch in a shorter time. It is also evident from these proofs that (2) the faster body covers an equal distance in a shorter time. (Fig. 2.) Relatively to a

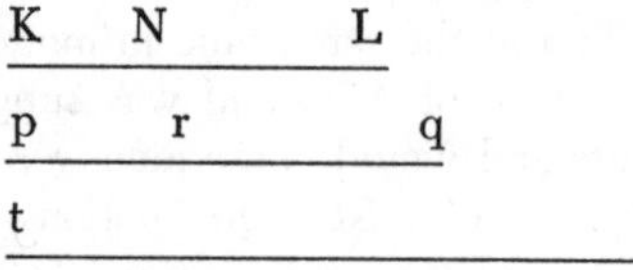

K   N   L

p   r   q

t

Fig. 2

slower body, a faster body goes a longer way in a shorter time. And taken by itself, a faster body travels a greater distance KL during a

10 time *pq* than it does a shorter distance KN during a shorter time *pr*. Hence, if the time *pq* is shorter than the time *t* which the slower body requires in passing over KL, the time *pr* is shorter still; since *pr* is less than *pq*, and anything less than what is less than a third thing is itself less than the same third thing. So the faster body moves across an equal distance in a shorter time. Again, if every moving body must move in an equal or shorter or longer time, and if a body requiring a longer time is slower whereas a body taking an equal time is equally fast and a faster body is neither equally fast nor slower, then the motion of the faster body occurs in a time neither equal nor longer but shorter. Hence, the faster body necessarily travels over an equal as well as a greater spatial magnitude in a shorter time.

20   [Now, time is continuous.] Not only does every movement take place during a time, and not only may there be movement in any period of time; but every moving thing can move relatively quickly or slowly,* and therefore there may be quicker or slower movement in any period of time. Moreover, by "continuous" I mean "divisible into parts that are infinitely divisible"; and in this sense [85] of the term, time must be "continuous." Let us follow up our demonstration that the faster body covers an equal distance in a shorter time. (Fig. 3.) Suppose that B

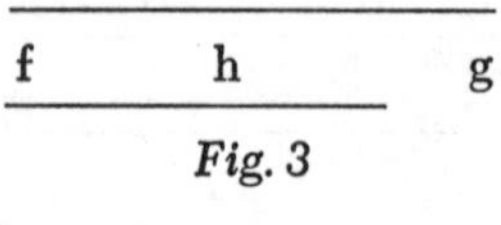

C   J   D

f   h   g

Fig. 3

---

* Thomas Aquinas: " . . . frequenter talibus propositionibus utitur in hoc sexto libro, quae sunt verae secundum considerationem communem motus, non autem secundum applicationem ad determinata mobilia."

(being slower than *A*) has finished moving across the stretch *CD* during time *fg*, with the consequence that *A* (being faster than *B*) will 30 have moved across the stretch *CD* during a shorter time *fh*; then since *A* has in the shorter time *fh* covered the whole of *CD*, the slower body *B* will in the same shorter time *fh* have covered a lesser distance *CJ*. But 233a if *B* has covered *CJ* during time *fh*, then the faster body *A* will have covered the same lesser distance *CJ* in a still shorter time; so that the time *fh* will in turn be divided. In that part [75] of the time *fh*, then, the slower body *B* will in turn have travelled over a part of *CJ*; and this part of *CJ* will be to *CJ* as the part of time *fh* is to *fh*. As we continue this procedure of substituting [65f], at each step of the demonstration, the slower for the faster body or the faster for the slower, we continue to get the same result: the faster divides the time; the slower divides the distance. Since the alternating procedure always results in this alternating division, it is evident that every time will be 10 continuous; and it is clear at the same time that every spatial magnitude is continuous, for the time and the magnitude are subject to divisions which are in the same ratio and which are equal in number.

In the customary nonacademic arguments, too, it becomes evident that, since time is continuous, spatial magnitude must also be continuous. A moving body goes a half of a distance in a half of a time or, at any rate [105], a shorter distance in a shorter time. The reason for this is because a time and a spatial magnitude are subject to the same divisions. If either is infinite, so is the other. And each is infinite in the same way as the other: if time is infinite as to its extreme ends, so is spatial magnitude; if infinite in divisibility, so is spatial magnitude; 20 and similarly, if infinite in both ways.

This is the reason why the assumption is false on which Zeno bases his [opposing] argument.† He takes it to be impossible to span infinites or to come into contact with them severally during a finite time. Now, a length or a time or anything continuous is called "infinite" in two distinct senses, according as it is infinite in divisibility or is infinite as to its extreme ends. Although it is impossible during a finite time to touch things infinite in quantity [that is, infinite in the latter sense], this is quite possible in the case of things infinitely divisible; for the time itself is also infinite in this sense. So it is in the time which is infinite, not finite, that the infinite magnitude is spanned; and the in- 30 finite parts are touched in the parts of the time which are infinite, not finite. An infinite stretch cannot be traversed in a finite time; neither can a finite stretch, in an infinite time. But if the time is infinite, so is the spatial magnitude; and if the spatial magnitude, so is the time. Let

† vi.9.239b11-13.

233b   *BA*‡ be a finite distance, and *c*, an infinite time; and let a finite part,
cd, of the time be taken. Then a moving body *X* will in time *cd* cover
a segment *BE* of the distance [*BA*]. The segment *BE* may indifferently
"measure" *BA* as an exact multiple of it or another multiple less or
greater [than *BA*]. If *X* always covers a distance equal to *BE* in an
equal time and if the number of the distances covered portions out
the whole, then the whole time in which *X* covers *BA* is finite; for the
time will be divided into as many parts as is the distance. So if *X* does
not cover the whole distance in an infinite time, but if *X* can even
during a finite time cover the segment *BE* which "measures" the whole,
10   then *X* will cover an equal distance in an equal time; and consequently,
the time [of the passage across *BA*] is finite. It is also evident that *X*
does not cross *BE* in an infinite time if we take the time [of the passage across *BA*] as limited at one end [that is, at the beginning]; for
if *X* covers a segment in a shorter time than it does the whole, then
the shorter time, being limited at one end, must be [altogether] finite.
We may in the same way disprove the supposition that the length
may be infinite and the time finite. It is evident, then, that no line or
surface or anything continuous can be indivisible [41].

This conclusion follows not only for the reasons stated but also because otherwise it would follow that the indivisible [41] would be divided. Since a body may move relatively quickly or slowly in any period
20   of time and the faster moves across a greater stretch in an equal time,
it is possible for the faster to span a stretch twice or one and a half times
as large as that spanned by the slower body; for the rates of the motions
may be so related, respectively. Suppose, then, that the faster body has
travelled one and a half times as far as has the slower during the same
time; and suppose that the greater distance has been divided into the
three indivisible [41] parts *AB*, *BC*, *CD*, and the lesser distance into
the two indivisible parts *EF*, *FG*. Then the time will also be divided
into three indivisibles, since an equal distance is covered in an equal
time. Hence, let the time be divided into the parts *jk*, *kl*, *lm*. But since
the slower body has during the same time travelled over *EFG*, the same
30   time will also be divided into two parts. Accordingly, the indivisible
[middle part] will be divisible! Also, the slower body will cover a
partless segment during a time greater than the indivisible time [in
which the faster body covers a partless segment of its path]!

It is evident, then, that no continuum is without parts.

---

‡ The text has "*AB*," but designates a segment of it "*BE*."

3. Moments, Movement, and Rest in Time

The "now" also must be indivisible: that is, the "now" in the essential and primary sense in which the "now" is inherent [82h] in all time; not, indeed, the "now" in [39a] the derived sense [16] of a period of time. For the [momentary] "now" is somehow [4] an extreme end [18a] of the past [116a], since it has nothing of the future, and somehow an 234a extreme end of the future, since it has nothing of the past; it is [functionally], we maintain, a limit of both. Accordingly, if the [momentary] "now" can be shown to be essentially of the sort described and thus to be an identity, then it will also become evident that the "now" in this sense is indivisible.

The "now" which functions as the extreme end of both past and future must, indeed, be identical. If the "now" were *different* limits [of both past and future], the one limit could not succeed the other; for the continuous cannot be made up of partless components. If the "now" were *separate* limits [of both past and future], there would be a time between them, since every continuum must have something of the same kind between its limits; and the time between them would be divisible, 10 since we have proved that all time is divisible. Accordingly, the "now" would be divisible. But if so, a dividing-point within it would mark off within it a time extending to the dividing-point and a time still to come; so that the [whole moment as] future would contain some [4] past aspect, and the [whole moment as] past would contain some future aspect. Then, too, the "now" would not be such in its essential sense of a moment; it would be a "now" in the derived sense of a period of time, for the function of dividing [occurring within it] would not belong to it essentially. In addition, part of the "now" would be past and part future, and it would not be always the same part which would be the past or the same part which would be the future; and, since the time could be divided at numerous points, the "nows" would never be the same. Hence, all these consequences being impossible, it must be the same "now" which limits either [40] the past or the future; and, 20 therefore, the "now" must obviously also be indivisible, since it would otherwise be involved in the same consequences which we have traced. Thus, it is clear that time includes something indivisible which we call a [momentary] "now."

Obviously, too, nothing can be moving during a moment; if it could, its motion could also be quicker or slower. Let a faster body travel a distance AB in a momentary "now," n; then a slower body would in the same moment travel a lesser distance AC. But since the slower body would have travelled the distance AC in the whole of the moment, the faster body in travelling the same distance [AC] would require a shorter 30 time; so that the moment would be divided. But a moment has been proved to be indivisible. Accordingly, nothing can be moving within a moment. But neither can anything be at rest during a moment; for only that is properly said to be "at rest" which naturally can move but is not moving when or where or as it naturally can. Hence, since nothing can naturally move within a moment, neither can it within a moment be at rest. Again, if the same moment functions as the limit of a past

234b   and of a future, and if a thing can be moving during the whole of one of the times and be at rest during the whole of the other time, and if the thing moving or at rest during the whole of a time would be moving or at rest during any part of the whole time within which it can naturally be moving or at rest, then (since the same moment is the limit of both past and future) it would follow that a body could be at once at rest and moving within the same moment. Again, we do not speak of a thing as being "at rest" unless both it and its parts are, in a present moment, in the same condition in which they were before; but in a present moment there is no "before" and, therefore, neither any "being-at-rest." Accordingly, anything moving is moving, as anything at rest is at rest, in time [or in a continuous duration].

4. Subjects and Kinds of Divisibility

10   Every changing thing must be divisible. Not only does every change have a beginning and an end. But also, when the changing thing has arrived at the end, it is no longer undergoing the change; and when it and all its parts are at the starting-point under consideration, it is not yet undergoing the change, since the stability of a whole and of its parts is not an instability. Accordingly, anything undergoing a change must be partly at the terminal and partly at the initial stage of the change;* for a changing thing cannot, while changing, be either at both the beginning and the end of the change or at neither of them. The "terminal" stage here meant is the one that is proximate in the change under consideration, for example, in a change from white to gray, not to black; for, in order to be changing, a changing thing need not start

20   or stop at either of two extreme opposites. It is evident, then, that every changing thing must be divisible.†

A movement is divisible both with respect to the time it takes and with respect to the movements of the parts of the moving thing. Thus, if AC as a whole is in motion, then its parts, AB and BC, will also be in motion. Let the motion of the part AB be DE, and that of the part

---

* Cp. vi.2.232a21.
† Thomas Aquinas: "Sed hoc diversimode invenitur in diversis mutationibus. . . . Aristoteles in hoc sexto libro agit de motu secundum quod est continuus."

BC be *EF;* then the motion of the whole *AC* will be *DF* [the sum
of the motion of the parts]. This must be the motion of the moving
body as a whole, because each of that body's parts has one of the two
motions which together make up the whole motion, neither part having
the motion of the other part; so that the whole motion is the motion of
the whole moving body. Again, in every motion there is something
which moves. But the whole motion *DF* cannot be the motion of one of 30
the parts; for *AB* has the motion *DE,* and *BC* has the motion *EF.*
Neither can *DF* be the motion of anything else; for its parts (*DE* and
EF) are motions of nothing other than *AB* and *BC,* respectively,
which are parts of the whole body (*AC*) having the whole motion
(*DF*), and for any single movement there never was more than a single
subject. Hence, the whole motion *DEF* is the motion of *ABC.* Again,
suppose that the whole body had another motion, *HI,* from which the 235a
motion of each of the parts could be subtracted; then the latter motions
would be equal to *DE* and to *DF,* respectively, for the reason that any
movement of a single subject is single. Hence, if the whole motion *HI*
were divided into the motions of the parts, *HI* would be equal to *DF.*
Should *DF,* however, fall short of *HI,* the remainder *JI* would not be
the motion of any subject: it could not be the motion of the whole or
of the parts, since any movement of a single subject is single; and it
could not be the motion of anything else, since continuous movement
implies continuous subjects. The same result would follow if *DF* should
exceed *HI.* Consequently, the whole motion *HI* would necessarily be
equal to *DF;* and *HI* and *DF* would, accordingly, be identical. One
way, then, in which a movement may be divided is into the movements
of the parts of the moving thing; and everything that has parts must 10
be subject to this kind of division. But a movement may also be divided
on the basis of the time it takes. Not only does every movement take
place during a time, but all time is divisible; and the shorter the time
of the movement, the more curtailed the movement. Every movement
must, therefore, be divisible in its temporal aspect.

Furthermore, since everything that is in movement "is moved" in
some respect and during some time and with a movement proper to it,
the divisions which relate to the time, to the movement [identified],
to the process of undergoing the movement, to the thing moved, and
to the respect in which the movement [constitutes a change], are the
same divisions. There is this exception, however, that division does not
relate to all the changing conditions in the same manner: place and
quantity are divisible essentially; quality, only incidentally. Let *A* be
the time during which a thing is being moved; *B,* the "movement."
Then, if the whole movement was completed in the whole of the time, 20

114   ARISTOTLE'S PHYSICS

it follows that in a half of the time less of the movement will have
occurred; in a still shorter period, still less of the movement; and so
on indefinitely. Conversely, if the movement is divisible, the time is
likewise divisible: if the whole of the movement was completed in the
whole of the time, then a half of the movement will have occurred
in a half of the time; still less of the movement, in a still shorter period
of time; and so on indefinitely. The process of undergoing the movement is divisible in the same way. Let C be "being moved." At the
half-way point of the movement, there will have been less of C than
the whole; at the half-way point of a half of the movement, there will
have been less than a half of C; and so on indefinitely. We may also
from the whole process abstract the two partial movements DC and
30 CE, in order to show that the whole process matches [39a] the whole
movement [identified], since there would otherwise have been more
than one process matching the [completed] movement. This demonstration would parallel that by which we have shown that a movement
is divisible into the movements of the parts of the moving thing. For
when the partial processes [DC and CE] are taken in their correspondence with the respective halves of the movement, it will be seen that
the whole process of "being moved" is continuous. So, too, we may go
on to demonstrate that the distance is divisible, as is any respect in
which a change occurs; except that the divisibility of some changing
circumstances is incidental to that of the thing which changes [in those
respects]. In short, if any one [of the five items we have distinguished]
235b is divisible, so are all the rest. Similarly, they are all alike finite or
infinite. The divisibility or infinity follows, for the most part, from the
divisibility or infinity which inheres immediately in the [primary] subject of change. That the latter is divisible, we have proved;‡ and we
shall later§ proceed to show that it is infinite.

5. Ends and Beginnings of Changes

Since a changing thing changes from one definite condition [4] to
another, anything which has undergone a change must, at the moment
[17a] when it has completed the change, be in the condition to which
it has changed. For a changing thing gives up or abandons its former
10 condition, and changing either means or leads to giving up and aban-
doning a former condition; and if so, then analogously completion of
the change puts the finishing touch to such abandonment. Thus, when
something has changed from a state of nonbeing to its contradictory,

‡ 234b10-20.
§ vi.7.

being, it has, at the time of the completion of the change, left the state of nonbeing behind; it is therefore at once in the state of being, since anything necessarily either is or is-not. Evidently, then, what has changed from one contradictory state to the other, is in the latter. Moreover, if such is the case in a change from one contradictory to the other, it will likewise be the case in the other kinds of change, what is true of the former in this regard being likewise true of the latter. Again, this must evidently be the case in each and every kind of change for the reason that whatever has completed a change must be somewhere 20 or in some condition; and since it cannot be in the condition from which it has departed, it must be either in the condition to which it has changed or in some other. But if what has changed to *B* were then in some other state *C*, with which *B* is presumably not contiguous [136], then, since every change is continuous, the thing would be at the same time changing from *C* back to *B;* so that what has completed a change would, at the time when it has completed the change, still be changing to the condition to which it has already changed. Since this is impossible, what has changed must be in the condition to which it has changed. So, too, it is evident that if and when something has come into being, it "is"; and if and when it has perished, it "is-not." For what we are saying holds good for the various kinds of change generally, although 30 it is especially clear in the case of a change from one contradictory to the other.

Not only is it clear that anything which has gone through a change is, at the time of the consummation of the change, in the condition to which it has changed; but the primary time in which it has completed its change must be indivisible [41]. I call that "primary" which is of a certain kind [5a] but which is not so because something else, being one of its parts, has this property.* Thus, let *AC* [as the primary time in which a change has been completed] be divisible, and let it be divided at *B;* then, if anything completes a change in time *AB* or in time *BC,* the inclusive time *AC* would not be the primary time of the completing of the change. If, on the other hand, anything was still changing in either *AB* or *BC* (in either of which it must have changed 236a or have been changing), then it must also have still been changing in the whole of *AC;* but we took it as completing its change in *AC.* The same argument applies even if in one part of the time it is changing and in one part it has completed its change, for there would then be something [*BC*] prior to the first [time *AC*]. Consequently, the time at which a change has been completed cannot be divisible; and therefore it is

* vi.6.236b20-22.

also evident that the time at which anything has finally come into being or has perished is indivisible.†

However, "the primary time in which something has changed" has two meanings: one is the primary time in which the change has come to an end [100c], that is, when it is true to say that something has changed; the other is the primary time in which something began to 10 change. The primary time which relates to the end of a change is a fact [82f] which we encounter [23]; for a change can be completed and thus have an end which we, moreover, have already shown to be indivisible because it is a *limit*.‡ On the other hand, a primary time which relates to the beginning does not exist at all: for there is no [continuous *part* of a whole change which would be an absolute] "beginning of a change"; and there is no primary part of the time in which a changing body *was first changing*. Let *AD* be such a primary time. Then *AD* would not be indivisible; for [if it were], moments would be contiguous.§ Again, let a body be at rest during the whole of [the immediately preceding time] *CA*. Then it would be at "rest" at *A* also. Therefore, if *AD* were without parts, the body would be at rest and would have 20 changed in the same time [*A* and *D* being simultaneous]; for at *A* it would be at rest, whereas at *D* it would have changed. On the other hand, since *AD* could not be without parts, it would have to be divisible, and a body would have completed a change in any part whatever of *AD*. For if *AD* were divided and if a body would have changed in another part, then it would not have changed in the whole; but if a body‖ would have changed in one of the parts, then the whole [*AD*] would not be the primary time in which it has changed. So it must have changed in any part whatever [of *AD*]. It is evident, therefore, that there is no primary time in which the body has first changed; for the divisions are infinite!

Neither does anything which has changed have a part which has changed absolutely first. Since it has been proved¶ that every changing 30 thing is divisible, let a changed body *DF* have such a primary part *DE*; and let *hi* be the time in which *DF* has changed. But if *DF* has changed in the whole of time *hi*, then only something less than *DF* could have changed in a half of the time and would therefore have changed before *DF;* and progressively something less in a progressively shorter time; and so on indefinitely. Consequently, a changing thing has no primary

---

† Against Sophistic arguments; for example, that Dion could not have died either when he was alive or when he was dead.
‡ viii.3.254b1, 8.264a4-6.
§ vi.1.231a27-29.
‖ Omitting: "had been changing in both and in the whole, but if it . . . ."
¶ vi.4.234b10-20.

part which has changed first. So there evidently is no such thing as an absolutely primary part either of a changing thing or of the time at which it has changed.

On the other hand, a different account must be given of a thing's 236b changing conditions [15a], that is, of the respects [39] in which it has changed. For there is, besides a changing thing (for example, a man) and the time in which the thing changes, the condition to which it changes (for example, white). The former two are divisible; but the last is not, except incidentally. With this restriction, all three are divisible; since that is divisible to which a quality (such as white) belongs. However, among the respects in which a thing changes, those which are not incidentally but essentially divisible (namely, magnitudes) have no first part any more than does the changing thing or the time of the change. Suppose a magnitude AB would have moved from B to a pri- 10 mary place C. Then if BC were indivisible, two things without parts would be contiguous. And if BC were divisible, AB would have moved to a place prior to C; and to another, prior to that; and so on indefinitely (because the division has no final stopping-point). Consequently, there is no absolutely first point to which anything has moved. A similar procedure is applicable to quantitative change; for this, too, is in a continuum. Therefore, it is evident that only qualitative change has any essentially indivisible part.

6. Continuity as Not Divisible into First Parts

Since any changing thing changes in time—in one sense, in the pri- 20 mary time [or the time of that change only], but also, in another sense, in the time of the change because [39a] another time is such (as when a thing changes in a certain year because it changes on a certain day of that year)—therefore a changing thing completes its change in a primary time such that the thing must be changing in any part of that time. This is clear, accordingly, from our definition of "primary time."* Our point can also be established by the following considerations. Let AB be the primary time of a movement; and since all time is divisible, let AB be divided at C. Then the movable body is either moving or not moving in AC; and likewise, in CB. But if it moves in neither, it would be at rest in the whole; for nothing can move in a given time if it is not in movement in any part of that time. And if it moves 30 in only one of the two parts of the time, then AB would not be the primary time of the movement; for the movement would relate to a different time. Therefore, the moving body must be moving in any part whatever of the primary time AB.

---

* vi.5.235b33, 34.

It follows from this proof that whatever is moving has moved before. For if in the primary time AB a certain body has moved a given distance, then a body beginning at the same time and moving at the same rate will in a half of the time have moved half that distance; therefore, the former body moving at the same rate during the same time must have travelled the same distance as has the latter; and hence, what is moving has been moving. Again, since a moment defines a time, and a time falls between moments, and if the last moment is warrant therefore for saying that a thing has been moving in the whole or in any part of the time AB, then the thing could likewise be said to have moved in the other parts of the time. But the dividing-point C is the extreme limit of a half of the time. Therefore the moving body will have moved in a half or in any part whatever of the time, since every division marks a time bounded by moments. If, then, all time is divisible, and what falls between moments is a time, every changing thing will have completed an infinite number of changes. Again, if a thing which changes continuously, which has not been destroyed, and which has not ceased changing, must at any time either be changing or have been changing, and if it cannot change in a moment, then it must at each moment of its change have been changing; and consequently, if the moments are infinite, every changing thing will have completed an infinite number of changes.

Not only must anything that is changing have changed; but what has changed must also have changed before. For whatever has changed from one [place or state] to another, has done so during a time. Suppose X has completed a change from A to B in a moment. Then the moment in which X would have changed could not be identical with that in which it is in A; for it would then be in both A and B at the same time, and we have proved† that what has completed a change is not at the time of its completion of the change in the condition from which it has changed. And if X would have changed in a different moment, there would be a period of time between the two moments; for moments are not contiguous.‡ Accordingly, X has changed in a time. But all time is divisible; so that X has in a half of the time undergone another change, and still another in a half of that time in turn, and so on indefinitely. Consequently, it must at any time have changed before. Again, what we have been saying will become even more evident from the continuity of the distance which a changing thing travels. Let X change in place from C to D. Then if CD were indivisible, two partless parts would be contiguous, which is impossible; so that there must be

† vi.5.235b6-13.
‡ vi.1.231b6-10.

a distance between them which, moreover, is infinitely divisible. Consequently, X was always previously changing in place from one to another of the infinitely numerous parts of the distance. So, then, *whatever* has changed must also have changed before. For, as to changes **237b** between two things not continuous with each other, including [qualitative] changes between contraries and [the more radical] changes between contradictories [in generations and destructions], we can use the former proof, namely, from the infinite divisibility of time.

Thus, what has changed must have been changing, and what is changing must have changed; changing is preceded by having changed, and having changed is preceded by changing; and at no stage can we lay hold of an absolute "first." The reason [83] for this is because partless parts are not contiguous. We may rather continue the process of dividing indefinitely. For example, as we continue to bisect a line, the one series of segments continues to increase in length while the other continues to decrease.§ Hence, it is also evident that anything **10** divisible and continuous which has come into being must have been coming into being before and that anything divisible and continuous which is coming into being must have come into being; although sometimes what is coming into being is a part which is something other than the whole, for example, the foundation of a house. And as with what is coming into being and what has come into being, so it is with what is perishing and what has been perishing: a continuous being which is coming into being or which is perishing has a certain infiniteness immediately inherent [82h] within itself; so that nothing is being generated without having been generated or has been generated without being generated. Just so it is with perishing and having been perishing: perishing is always preceded by having been perishing; and having been perishing, by perishing. It is evident, then, that what has been produced must previously have been in the process of being **20** produced and that what is being produced must have been produced; for all distance and all time are infinitely divisible. Consequently, at whatever [stage of a change] a thing may be found to "be," that stage would not be an absolutely first [moment or point].

7. Finite and Infinite Time and Distance

Since movement takes time and, with increasing magnitude, a longer time, a finite movement does not require infinite time unless, indeed, the same movement or part of it is continually repeated, but the whole of a finite movement does not take the whole of infinite time. Clearly,

---

§ iii.6.206b3-6.

a body moving at a uniform rate must travel a finite distance during a finite time; for it completes its movement over a whole distance in equal periods of time which, moreover, number as many as do the equal 30 parts taken of the distance. Therefore, since the parts of the distance are finite in size and number, so is the time; the latter equals the time of a part multiplied by the number of the parts. However, it makes no difference even if the body does not move at a uniform rate. Suppose 238a X has moved a finite distance AB during an infinite time cd. But then X must have travelled one part of the distance before another; for, clearly, X has in an earlier and in a later part of the time travelled different parts of the distance, because different parts of the motion are completed as the time grows longer (whether X changes at a uniform rate or not and whether the rate of motion is uniform or increases or decreases). Therefore, let AE be a part by which the distance AB is measured. Then this part of the motion would take a corresponding part of the infinite time; surely, it would not take the whole of the infinite 10 time which was assumed to be taken by the whole motion. And if I select another part equal to AE, this part of the motion would also, for the same reason, take a finite period of time. Now, if I continue to take parts in this way, there is no part by which infinite time can be measured; for an infinite cannot be made up of [a finite number of] either equal or unequal finite elements, because things finite in plurality and size can be measured by a unit (whether they are equal or not,° so long as they are finite in size). But the finite distance AB is measured by AE taken a certain number of times. Consequently, X will have travelled the distance AB during a finite time. Then, too, as with movement, so it is with coming to rest. And consequently, it is impossible for one and the same thing to be forever in process of becoming or of perishing.

20 For the same reason, there cannot be an infinite movement or coming-to-rest during a finite time, whether or not the movement is uniform. Let any part be taken by which the whole time is measured: a moving body would in that part of the time cover but a part of the distance, not the whole distance which is covered in the whole of the time; it would cover another part of the distance in another part of the time; and so on in each of the times. It would make no difference whether a part is equal or unequal to the first part taken, provided only that each is finite; for, clearly, while the time is thus being used up, the infinite is not, since only finite quantities are taken only a finite number of 30 times. Consequently, an infinite extent cannot be covered during a

---

° a4-6.

finite time; and it makes no difference to the argument whether the distance is infinite in one direction or both.

It evidently also follows from this proof that a finite magnitude cannot traverse an infinite magnitude during a finite time. The reason [83] is the same: in any part of the time, a finite magnitude covers a finite magnitude; and so in each succeeding part of the time; and therefore, in the whole of the time, it covers a finite magnitude.

Since a finite magnitude does not traverse an infinite one in a finite 238b time, it is also clear that an infinite magnitude cannot traverse a finite one during a finite time. For if the infinite traversed the finite, the finite would also traverse the infinite. Which of the two is the body in motion, would make no difference; for, in either case, the finite would traverse the infinite. Thus, when an infinite magnitude A is in motion, a part of it CD would be at a finite B; and so would one part after another; and so on indefinitely. At one and the same time, then, the infinite would have traversed the finite, and the finite would have traversed the infinite; for the infinite cannot traverse the finite without the finite 10 traversing the infinite either by way of local passage through it or by way of measuring it. Hence, since the finite cannot traverse the infinite, neither can the infinite traverse the finite.

Moreover, the infinite cannot even traverse the infinite during a finite time. If it did, it would also traverse the finite, which is inherently comprehended [82h] in the infinite. In addition, the proof from the taking of the parts of time applies here also.

Accordingly, since the finite cannot during a finite time traverse the infinite, nor the infinite the finite, nor the infinite the infinite, it is evident that there can be no infinite movement at all within a finite 20 time.† For what difference would it make whether we presented the movement or the distance as infinite? If either is infinite, the other must likewise be infinite. For every local motion takes place in a place.

8. Coming to Rest, Rest, and Stages of Change

Since everything which is naturally either in motion or at rest moves or rests when and where and as it naturally does, therefore what is coming to a stop [111] must, when doing so, be in motion. For otherwise it would be at rest and therefore not in a status [12a] of "coming to rest" [110]. It follows likewise that coming to a stop takes time. For what is in motion takes time; but what is coming to a stop has been shown to be in motion and, hence, must come to a stop in a period of time. Again, the distinction of "quicker" and "slower" presupposes 30

† vi.2.233a21-b15.

time; and stopping may be quicker or slower. Moreover, what is coming to a stop must do so in any part whatever of the primary time in which it is coming to a stop. If it were not doing so in either half of the time, it would not do so in the whole of the time and would therefore never stop moving. And if it were doing so in but one half of the time, then the whole would not be the primary time of the process but would be the "time of the process" in a dependent sense only. (We have at a previous point in our discussion° applied this proof to moving bodies.)

239a   Moreover, just as there is no first moment [17a] in which a moving body is moving, so there is none in which a body that is coming to a standstill comes to a standstill; for there is no [absolutely] first part [4] of either process. Let AB be such a primary time in which anything would begin to come to a standstill. Then AB could not be without parts: for there can be no movement in a partless time, because a moving thing must have *been moving* in a part [4] of the time; and what has been coming to a stop has been shown to be in motion. And if AB is divisible, the body is coming to a stop in any part whatever of AB; for we have just proved that what is coming to a stop must be doing so in any part whatever of the primary time in which it is coming to a stop. The primary time, then, in which anything comes to a standstill, is a [continuous] time, not a moment [41]; and since all
10   time can be divided into an infinite number of parts, there can be no first moment at which anything begins to stop moving.

Neither is there a first moment at which a body which continues to rest *began* to be at rest. It did not begin its rest in a partless moment; for no movement takes place within a moment [41], and only that can be at rest which can also be in motion. As we are accustomed to saying,† a body is, properly speaking, "at rest" when it can naturally be moved but is not being moved at a time when it naturally could. Again, we speak of a body as being "at rest" if in the present moment it is in the same condition in which it was before; so that a state of rest cannot be discriminated by reference to a single moment only, but only by reference to at least two moments. Consequently, the time in which a body continues at rest cannot be without parts. Having parts, then, it is a [continuous] time; and a body at rest will be at rest in any part whatever of the time of its rest, as we can show by means of
20   the proof used before [in similar contexts]. And consequently, there can be no first moment [at which a thing began to be at rest]. The reason [83] is because all resting or moving takes time; and since anything continuous is infinitely divisible, time can no more have an

---

° vi.6.236b25-32.
† v.2.226b12-15.

absolutely first part than can a magnitude or than can a continuum of any kind whatever.

Moreover, since every moving body moves during a time extending from the beginning to the end of the process [115], it cannot in that time (taken by itself rather than in some part) be at [39a] some primary part [of its path]. For if it (together with each of its parts) were to continue in the same [state or place] for any length of time, it would be at rest; and if it were at rest in the sense that it (together with its parts) can at one moment or another be truly said to be in the same [state or place], then a changing thing cannot in the primary 30 time of its change be completely [21] at any [identifiable part of its course]. All time being divisible, it would follow rather that the changing thing (together with its parts) would at different times be in the same [state or place]. If, on the other hand, it would at a single moment only be in the same state or place, it would not be at any [definite stage] during any time but only at a limit of the time. To be sure, it *is* at some stage at any one moment; but it could not be "at rest" 239b in a moment any more than it could be "in motion" in a moment. Surely, if a changing thing is "not moving in a moment" and if it "is at some stage," it is not therefore possible for it to be "at rest at a part" [of its passage] "during the time" [of its changing]; for it would then follow that anything in motion would be at rest.

9. Fallacies of Taking Divisibility as Prior to Movement

There is a fallacy in Zeno's way of arguing: "If anything is at rest or in motion when it is where [39a] it is [57g], and such is invariably the case with a moving body [121] at any moment, then a flying arrow cannot start moving [109e]." This is wrong because time is not made up of indivisible moments any more than any other magnitude is made up of indivisible parts.

Zeno formulated [90] four arguments about motion which worry 10 [200c] those who try to solve them. According to the first, "nothing can be in movement, since whatever is in local movement must arrive at the half-way stage before it arrives at the end."* We have commented on this contention in preceding discussions.† The second argument, known as "the Achilles,"‡ is to this effect: "The slowest runner will never be caught [65e] by the swiftest, since the pursuer must first reach the point from which the fleer has in the meantime escaped; so that the slower runner must always be some distance ahead." This

---

* Cp. vi.5.236a10-13; i.8.191b27-29.
† i.3.187a3; vi.2.233a13-34.
‡ "Achilles and the tortoise."

is the same argument as that from repeated bisection; but with this difference that ever-prolonged [65g] paths [142a] are to be analyzed
20 into segments in different relations. It follows from the argument that a slower runner can never be overtaken, but only as it follows from "bisection" [in the first argument, that there is no motion]: the conclusion of both arguments, that the limit is not reached, turns on the way in which the distance is analyzed; the second argument only enlarges on the conclusion by saying that not even the swiftest runner of traditional story achieves his goal of catching up with the slowest. Consequently, the solution must be the same in both cases. The formula [46c] that "whatever is ahead cannot be overtaken" is false. What is ahead is not overtaken as long as it is ahead, to be sure; but it is overtaken.§ One need only admit that the distance traversed is finite. These,
30 then, are the first two of Zeno's arguments. We have already reported the third: "The flying arrow continues at rest." This conclusion depends upon taking time as composed of moments; if this is not granted, the conclusion does not follow.‖

The fourth argument [known as the "Stadium"] deals with "equal bodies moving at the same rate in opposite directions in a race-course past bodies equal to themselves, one set moving in the direction away from the end [100] of the race-course and the other set moving in the direction away from the turning-point [138]." This argument is be-
240a lieved to lead to the conclusion that "one-half of the time is equal to the double." The mistake in this reasoning is rooted in the supposition [46c] that bodies moving at the same rate take the same time in moving past a moving body and past a stationary body of equal size. Let A's be stationary bodies of equal size; let B's be bodies equal to the A's in number and in size, but starting at the turning-point; and let C's be bodies equal to the A's in number and in size, but starting at the extreme end [18a] and moving at the same rate as the B's. (Fig. 4.)

A: – – – –	A: – – – –
B: 4 3 2 1	B: 4 3 2 1
C: 1 2 3 4	C: 1 2 3 4
Fig. 4.	*Fig. 5.*

10 Then, as the B's and the C's pass one another, B_1 and C_1 will be simultaneously at the [respective] end [of the A's]. (Fig. 5.) But C_1 will have passed all [the B's] while B_1 will have passed only one-half [of the A's]. Hence, since each moving body [supposedly] takes an

§ "Overtaking" is logically prior to a half of it.
‖ viii.8.263a11–b9.

equal time in passing each of the bodies it passes, the time [taken by B_1] will [supposedly] be only *one-half* [as much as the time taken by C_1]. But B_1 will at the same time have passed all the C's, for C_1 and B_1 will be at the opposite extremes at the same time.¶ The reason is because both B_1 and C_1 have taken an *equal* time in passing opposite the A's. This, then, is the argument. Its conclusion, however, rests on the false assumption which we have noted.

Similarly, we find no impossibility in change from one contra- **20** dictory to the other. "If something, in changing from nonwhite to white, is at neither extreme, it is therefore neither nonwhite or white." But the fact that it is not wholly at either extreme does not prevent it from being described as white or nonwhite: for, to be so described, it need not be wholly such, but only in most or in the most important of its parts; and "not being thus or so" is not identical with "not being thus or so at all." Likewise, in the case of a change having as its ex- tremes "being" and "nonbeing" or any other pair of contradictories: the changing thing must be in one of the opposite states, although it need not be wholly in either.

So, too, we are not confronted with an impossibility by the objection that the circle or the sphere or anything rotating on itself must be **30** at rest. Since they, together with their parts, continue for some time to be in the same place, it is concluded that they are at once both at rest and in motion. However, the parts are not in the same place for any length of time; and, besides, even the whole is always changing **240b** from one position to another. For the orbits described as starting from A or B or C, and so forth, are not the same (except incidentally, like "musician" and "man"); and none of them is ever at rest, but each is forever changing to [the position of] a different one. (Fig. 6.) Like- wise, in the case of the sphere or of anything else rotating on itself.

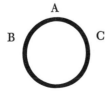

Fig. 6.

¶ The following statement appears to be a gloss: "As Zeno says, it takes as much time to pass each A as it does to pass each B."

10. The Indivisible as Motionless and Change as Definite

We may now take another step forward in our analysis by showing that what is without parts cannot be in motion, except incidentally;
10 that is, only in so far as it is an inherent part [82h] of a moving body or line [142a]. What is without parts moves, so to speak, only as anything in a boat is moving if and when the boat is in motion or as a part moves with the motion of the whole of which it is a part. By "things without parts," however, I mean such as are quantitatively indivisible. For parts of a body may have motions both essentially their own and dependent on the motion of the body as a whole. This distinction may be clearly observed in the case of a sphere and its innermost and other parts rotating with different velocities and thus having [potentially] many motions. However, to return to the point we are trying to make: what is without parts may have a motion like that of a man
20 sitting in a moving boat, but cannot have a motion essentially its own. Let a thing without parts be changing from AB to BC, whether from one magnitude to another or from one quality [20] to another or from one contradictory to the other; and let d be the primary time in which it is changing. Then the partless thing would at the time of its changing have to be in stage AB or BC or partly in the former and partly in the latter, just as would anything else that is undergoing a change. But it could not be partly in each stage, since it would on that alternative have parts; nor could it be in stage BC, since it would in that case (contrary to our assumption) have completed its change. Accordingly, it would during the time of its changing be in stage AB and,
30 since being in the same state for any length of time is being at rest, it would therefore be at rest. Consequently, what is without parts cannot move or change at all. It could have a motion of its own only if time were made up of moments [which have no parts]; for it would
241a then at any moment *have* moved or changed and would, therefore, never be moving, but would forever have finished moving. But we have proved before* that this is impossible. Time cannot be made up of moments any more than a line can be made up of points or than a motion can be made up of [atomic] movements [109f]. To attribute motion to what has no parts amounts to reducing motion to partless movements; and this would be like reducing time to moments or a line to points.

There is another way of showing that a point or anything else that is indivisible cannot move; namely, by means of the fact that no moving body can move a distance greater than itself before having moved a distance equal to or less than itself. If this is so, then it is evident

---
* vi.1.

that a point, too, would have had to move a distance less than or 10
equal to itself. Being indivisible, however, a point could not have
moved a distance less than itself. And if it would have had to move
a distance equal to itself, then a line would have to be made up of
points; for a whole line is measured by a point moving always an
equal distance. But if a line cannot be made up of points, neither can
anything indivisible move.

Still another proof is the consideration that, since all motion takes
time and no motion happens in a moment and since all time is divisible,
there must be for any moving thing whatever a period of time shorter
than the time it takes in moving a distance equal to itself. Such a
time is required for such a motion, since all motion takes time; and
we have proved† that all time is divisible. Accordingly, if a point
were to move, there would have to be a period of time shorter than 20
the time in which a point would have moved a distance equal to itself.
But this is impossible, for in a shorter time it would have to move a
lesser distance; and then the indivisible would be divisible into smaller
parts, just as the time is divisible into shorter times. In fine, what is
without parts could move if and only if movement were possible in an
atomic moment; for movement in a moment and movement of something indivisible are mutually [15a] convertible [90].

However, no single change is infinite. We have seen‡ that every
change has a definite beginning and a definite end. Moreover, these
limits are either contradictories or contraries. In changes from one contradictory to the other, each of the contradictories is a limit: in generation, for example, the final limit is being; in destruction, nonbeing. 30
And in a change from one contrary to another, each of the contraries
is a limit. Qualitative alteration, being a type of change, has such
extremes; for qualitative alteration always proceeds from one contrary
quality to another. Growth and decline, too, have such extremes: growth
has a final limit in the complete magnitude consonant with the nature 241b
proper [55] to the growing thing; and decline has a final limit in the
loss [111d] of that magnitude. As for local motion, it cannot be limited
in precisely the same way, since its termini are not always contraries.
However, "what cannot be done" has various meanings. If a thing could
not have been cut in the strict sense that it would have been an impossibility for it to be cut, then it is impossible in the same strict sense
for it to be in a process of being cut. So, in general, anything that
cannot be definitely generated cannot be in a process of being generated; and anything that cannot undergo a complete change cannot be

---

† vi.2.232b23-233a10.
‡ v.1.225a1.

in a process of changing to anything to which it cannot change completely. Now, if anything in a process of local movement is to change its location from one place to another, it must be possible for it to
10 complete such a change. Hence, its motion is not infinite and cannot be infinite, for it is impossible to traverse an infinite distance. We have shown, then, that no change can be "infinite" in the sense that it "cannot be defined by limits."

However, it remains for us to consider§ whether a movement that is single may be infinite in this sense with respect to the time of the movement. There is nothing to prevent eternal movement if the movement is not single but if, for example, local motion is followed by qualitative alteration, and the latter by growth, and this in turn by generation. In this way there can be movement through all time; but it will not be a single movement, since all these changes would not make up a unified movement. No movement which is a unitary movement can be temporally infinite, with but one exception; that exception
20 is rotation.

§ viii.7.261a17-8.265a12.

Series of Movements

1. Moved Movers and the First Mover

Anything° involved in a process [109a] is necessarily brought into 241b34
operation [109a] by some agency [4]. Obviously, if a thing subject to
change does not put [82] itself in motion, it owes its variation to
something else; in such a case, the mover or agent must be something
other than the thing acted upon. What, however, if a thing set in motion
has the beginning of its movement within itself? Take *AB* as something moved, considering *AB* by itself without reference to the movement of any of its parts. Since *AB* is presumably not given a start by
an external mover because it is itself a whole in movement, may we,
then, take *AB* as self-moved?† This would be like supposing that, when
JK both moves *KL* and is itself in movement, we would have to deny
that *JL* is moved by anything [other than *JL* itself] because we do not
see which of its parts is the mover and which is moved! Again, what is 44
[supposedly] not given a start by anything would not necessarily stop 242a35
moving because something else is in a state of rest; on the contrary,
if anything is in a state of rest because something else has stopped
moving, then the former is necessarily set going by something; and if
this is admitted, then everything movable is responsive to some agent.
Thus, since everything in movement is divisible, *AB* must be divisible.
Let *AB*, accordingly, be divided at *C*. Then if *CB* is not in movement,
neither would *AB* be in movement; if it were, then *AC* would be in

---

° Book Eta may be an independent work. The first three chapters exist in two
versions, of which this translation follows *alpha* rather than *beta*.

† Plato *Phaedrus* 245D.

movement while *CB* would be in a state of rest, and *AB* would consequently not be in movement essentially and directly as we have as-
45　sumed it to be. Hence, if *CB* is not in movement, *AB* must be in a state
of rest. But what is in a state of rest because something else is not in
movement is admittedly moved by something. Therefore, we conclude, anything involved in a process is necessarily brought into operation by some agency: for what is in movement is divisible; and if any
of its parts is not in movement, then the whole must be in a state of rest.

Anything passing through any change, then, is necessarily acted
upon by some agent, with the consequence that anything undergoing
a local motion is moved by something else in motion, the mover being
in turn moved by another thing in motion, and the latter by still another, and so on indefinitely; so that there must be a first mover, and
there cannot be an infinite regress. For let us suppose that there is no
55　first mover but that there is an infinite regress; and let *A* be moved
by *B*, *B* by *C*, *C* by *D*, and so forth. Then, since the mover supposedly
induces motion while being in motion, the motion of the thing moved
and of the mover must be simultaneous; for it is at one and the same
time that the mover induces motion and the thing moved is set in
motion. It is evident, therefore, that the motions of *A*, *B*, *C*, or of any
of the movers and of the things moved, are simultaneous. Let the motion
of *A* be *E;* that of *B*, *F;* that of *C*, *G;* that of *D*, *H*. For although each
65　thing in motion is moved by another, still, we may take each as having
a numerically single motion, since every motion has a definite [4]
starting-point and a definite stopping-point, and no motion is infinite
as to its extremes. By a "numerically single motion," I mean one proceeding from a numerically single starting-point to a numerically single
stopping-point in a numerically single period of time. For a movement
242b35　may be single generically or specifically or numerically: a movement
is generically single when it falls under a single category, such as
primary being or quality; a movement is specifically single when it
proceeds between limits which do not differ but are the same in kind
[20], for example, from white to black or from good to bad; and a movement is numerically single when it proceeds from a numerically single
starting-point to a numerically single stopping-point in a numerically
single period of time, for example, from "this" white appearance to
"that" black appearance or from "this" place to "that" place and in
"this" period of time (for a movement occurring in some other period
of time would no longer be numerically single but only specifically
single). We have discussed the unity of movement before.‡ Now, let
the time in which *A* has completed its movement be *J*. Then, since

---

‡ v.4.

the movement of *A* is finite, so is the time. On the other hand, since 45 the movers and the things moved are infinite, the movement *EFGH* (which is composed of all the particular movements) would also have to be infinite; for whether the movements of *A*, *B*, and so forth, are equal or whether they form a series of progressively greater movements, they would in either of these possible cases§ form a whole which is infinite. And since *A* and all the other things moved are moved simultaneously, the whole movement would occur in the same period of time in which that of *A* occurs and which is a finite period of time; so that there would be an infinite movement in a finite period of time. But this is impossible.‖

It might seem that this argument has proved the point we have promised [82a] to establish [namely, that there is a first mover]; but the argument does not provide the demonstration required, since it 55 fails to show conclusively any impossible consequence [of the opposite view]. For although a single subject cannot undergo an infinite process in a finite time, yet many subjects may very well [in a finite time participate in processes which together form an infinite whole]. This is precisely what happens in the case under consideration: each thing acted upon holds its own course; yet it is quite possible for many things to be acted upon at the same time. However, if anything that moves anything else by way of a local and bodily motion must touch or be continuous with the thing it moves (as observation discloses in every case), then the things so interacting must at least touch or be continuous with one another so as to form a unity. It makes no difference to the present argument whether this unity is finite or infinite: if the 65 subjects are infinite in number, then the motion which is the sum of all their motions must be infinite; and this would be so whether the particular motions are equal or progressively greater, since we may treat [65] the possible [12b] as actual [82e]. If, then, the whole made up of *ABCD*, whether finite or infinite, undergoes a movement *EFGH* in a finite time *J*, something finite or infinite would enact a finite movement in an infinite time. But neither is possible.¶ Consequently, the series must come to a stop, and there must be a definite first mover as well as something which is moved first. It makes no difference that we have derived this impossible consequence from an assumption [64h]; for we have taken the assumption as a possibility, and an 243a30 assumed possibility ought not to yield an impossible consequence.

---

§ iii.6.206b7-12.

‖ vi.7.238a32-b22.

¶ vi.7.238a32-b22.

2. Togetherness of Agents and Things Acted upon

A "proximate mover" (not the goal [96] of a movement, but its initiating [95b] factor) is correlative [137c] to what it moves, in the sense that there is nothing intermediate between them; and this "togetherness" is found in every case involving an agent and something acted upon [109a]. Moreover, since there are three kinds of "movement" (local, qualitative, and quantitative), there must also be three kinds of mover or agent: one, which transfers something from one place to another; a second, which brings about a qualitative alteration; and a third, which effects an increase or a decrease.

40
11
Let us take up local motion first, since this is the primary kind of movement. Everything undergoing a local motion is set in motion either by itself or by something else. Obviously, when things are moved by themselves, the mover and the thing moved are "together": they have their proximate mover inherent [82h] within themselves; with the consequence that there is nothing interposed between them. But when things are set in motion by something else, their motion must be produced in one of the four ways [20] in which one thing can be put in motion by another; namely, by pulling, or by pushing, or by carrying, or by rotation. All local motion may be reduced to these four kinds. Pushing onward is the kind of pushing which occurs when a mover pushes something away from itself and follows its course so as

20
243b
to keep pushing it; pushing away, when a mover does not thus follow the course of the thing it has moved; throwing, when the mover impels [34] a thing to move away from it with a motion more violent than any which the thing thrown would naturally have, so that the thing thrown continues to move as long as the motion impelled predominates over the natural motion; and pushing apart is pushing away either from the mover or from something else, as pushing together is pulling either towards the mover or towards something else. There are many varieties of the last two; among them, parting the warp threads and beating the weft threads into place. So there are other processes of combining and separating, all of which are kinds of pushing together or apart—with the exception of the processes involved in generation and

10
destruction. At the same time it is evident that combination and separation do not even constitute a [16b] distinct [4] kind [19] of movement; rather are all motions to be reduced to the four kinds already named. So, too, inhaling is a kind of pulling; and exhaling, a kind of pushing. We may similarly view spitting and other bodily actions [109], among which immissions are kinds of pulling and emissions are kinds of pushing away. Also all other kinds of local motion fall under one of the four varieties named and are accordingly to be reduced to one of them. Of

these varieties in turn, carrying and rotation are to be reduced to
pulling and pushing. Since anything carried in or on something in
motion is itself incidentally in motion and the thing which carries it is   20
being pulled or pushed or rotated, carrying is dependent upon one   244a
of these three kinds of motion and therefore belongs to all of these
three in common. And rotation is made up of pulling and pushing,
since anything which produces a rotation must push part of a thing
away and pull part of it back. Consequently, if anything which pushes
or pulls is "together" with what it pushes or pulls, it is evident that
there is also nothing interposed between anything undergoing a local
motion and its mover. This is also apparent from the definition of
"pushing" as "dislodging [109] a thing in a direction away from the
mover or from something else to some other [place]" and from the
definition of "pulling" as "dislodging a thing in a direction away from
some other [place] to the mover or to something else," namely, as long
as the motion of what does the pulling is quicker than the motion   10
tending to disjoin the two continuous things from each other (for this
is the way in which one thing is pulled onward with another). One
might, indeed, suppose that there are other forms of pulling, as in
the case of wood drawing a fire; but the only difference involved is
that a stationary body draws things to where it is, whereas a moving
body draws them to where it was. However, it is impossible in any
case to move anything away from the mover to something else or vice
versa without touching it; so that there evidently cannot be anything   244b
intermediate between anything in local motion and its mover.

Neither is there anything intermediate between a thing which is being
altered and that which is altering it. This is made plain by induction,
since a last altering factor and a first altered factor always happen to
be "together." We assume* that things undergoing qualitative alteration
are being acted upon [35] with respect [39a] to their "passive" [35c]
qualities [28]: for anything having a quality is altered inasmuch as it
is sensible, and bodies differ in their sensible traits inasmuch as one
body has more or fewer sensible traits than does another or has
the same sensible traits in a greater or lesser degree; but, in any case,
what is altered qualitatively is altered by [the action of another body
having] the same sorts of [sensible] traits. For these are modifications
[35a] of some persistent [85] quality [28]. Thus, we say that a thing is
"altered" by being heated or sweetened or condensed or dried or
whitened, whether the thing in question is a lifeless or a living being
or whether the parts of the living being in question are nonsensitive or   10

---

* This sentence represents a reconstruction of the text along lines suggested by
Simplicius. See W. D. Ross Aristotle's Physics (Oxford, 1936).

sensitive. For there is a sense in which even the senses undergo "alteration," since an actual perception is a bodily process in which a sense is acted upon in some way. A living being, then, undergoes qualitative alteration in as many ways as a lifeless being does, but not vice versa; for a lifeless being is not subject to alteration of sense organs and takes

245a no notice of action upon it as does a living being, although there is nothing to prevent even the latter from failing to take notice of a qualitative alteration which has nothing to do with the senses. Since it is by sensible objects, then, that anything is altered, it is evident that a last altering factor and a first altered factor are in every case "together": air is continuous with what alters it, and a body with air; a color is continuous with light, and the light with the eye; in hearing and smelling, the air is similarly related as proximate mover to the organ on which it acts [109a]; a flavor is likewise "together" with the

10 sense of taste; and there is an analogous "togetherness" in the case of lifeless or insensitive things. Consequently, there can be nothing intermediate between that which is being altered and that which alters it.

So, too, there is nothing intermediate between anything assuming and effecting an increase. That with which an increase starts is added to the subject in such a way that the whole forms a unity. Inversely, that with which a decrease starts becomes the occasion for the decrease when some part of it breaks loose. There must be a continuity, then, between that which starts [and that which assumes] the increase or decrease; and since there is nothing intermediate between things

245b that are continuous, it is evident that nothing intervenes between a proximate [17a] or immediate [18a] giver and a taker in a process.

3. Objects and the Senses in Qualitative Alterations

We must now proceed to show [187] that anything altered in quality is altered by sensible things and that only those things are subject [82f] to qualitative alteration which are directly [2] acted upon [35] by sensible things. One might suppose that qualitative alteration occurs in other ways also, especially by way of change in the forms [91a] or shapes [91] or in the positive states [33a] which a thing may receive or lose; but neither of these changes is a qualitative alteration.

10   When a thing has been fashioned into its final [100c] shape, we do not give it the name of its material. Thus, we do not call a statue "bronze," or a candle "wax," or a bed "wood"; rather do we describe these products, respectively, as being "of bronze," "of wax," or "wooden." On the other hand, when a thing has been acted upon so as to have been qualitatively altered, we continue to designate it by the name

of its material. Thus, we speak of "bronze" or of "wax" as being dry or liquid or hard or soft; indeed, we even speak of liquid or hot material as being bronze. In such instances, we denote a material and its property [35a] by the same term. Accordingly, if we do not distinguish 246a a product of shaping by the name of the material which has the shape, whereas we do continue to identify a thing which has undergone qualitative alteration by its material, it is evident that productions of the former sort are not qualitative alterations. Moreover, it would seem absurd to say that a man or a house or anything else has, by coming into being, been altered in quality. To be sure, when anything is generated, *something* must undergo qualitative alteration; for example, a material must be condensed or rarefied or heated or cooled. Yet it is not the things being generated that are being altered in quality; and their generation is not a qualitative alteration.

So, too, bodily or nonbodily [154] states are not qualitative altera- 10 tions. States or conditions differ in being better or worse, and neither their superiority nor their inferiority is a qualitative alteration. Rather is any excellence a kind [4] of completion; for when anything has attained its proper excellence, we say that it is "complete" in the sense that it is then most especially in its "natural" condition. For example, a circle is a "complete" figure when it is as good a circle as can be formed. On the other hand, a fault is disruptive or destructive of an excellence. Accordingly, we do not call the completion of a house an "alteration"; indeed, it would be absurd to regard the coping and the tile as an alteration of the house or to regard a house as being altered in receiv- 20 ing its coping and being tiled over, which rather marks the completion of the house. Just so it is with excellences or faults and their possession 246b or acquisition: excellences are completions, as faults are deviations from perfection; neither are qualitative alterations.*

Again, we ascribe to the being of all excellences a relational [29] status [33]. Thus, we view bodily excellences (like health and wellbeing) as dependent upon a proportionate mixing of hot and cold elements within the body or relatively to its surroundings; and beauty, strength, and other bodily excellences (as well as the lack of them) are contingent upon such relations. Positive or negative conditions like these dispose [64c] a body well or ill to those critical [55] changes [35a] by which it is naturally preserved or corrupted. Consequently, 10 since relations are not alterations and are not subject to alteration or generation or any transformation, it is evident that states or conditions and their loss or acquisition are not qualitative alterations. It may very well be so that the occurrence or nonoccurrence of a state or condition

---

* v.1.225a35, b1.

(like that of a form or shape) requires the alteration of *something*, for example, of hot or cold or of dry or wet or of other such elements on which the states or conditions in question may ultimately depend. For every excellence or fault relates to things by which its possessor is naturally altered in quality: a body in good condition is insusceptible 20 [35d] to things to which a body in poor condition is susceptible [35c]; and the former is also passive in ways in which the latter is, contrariwise, incapable of being acted upon.

247a   Likewise, nonbodily states are all dependent upon certain relations; their excellences are completions, as their faults are deviations from perfection; and their excellences dispose their possessors well, as their faults dispose them ill, to critical changes. Consequently, nonbodily states and their acquisitions and losses are not qualitative alterations, even though their occurrence requires the alteration of a sensitive organ by sensible things. For every moral virtue falls within the sphere of bodily pleasures and pains which depend, in turn, upon actions or 10 memories or hopes. The pleasures and pains arising from actions belong [39a] to the senses and must, therefore, be occasioned [109a] by sensible objects; and the pleasures of memories and hopes arise from the memory of past pleasures or from the hope of future pleasures. Hence, all such pleasure is produced by something sensible. And since virtue and vice arise in us together with the pleasure and the pain which they involve and which are qualitative alterations of something sensitive, it is evident that the acquisition and loss of moral virtues and faults requires the qualitative alteration of *something*. Accordingly, the occurrence of moral states is bound up with that of qualitative alteration; but they are not therefore themselves qualitative alterations.

247b   Like moral states, so mental [169d] states are not qualitative alterations; neither is there any generation of them. Knowing, above all, is to be understood in terms of its relation [to its objects]. And there can obviously be no generation of intellectual apprehensions as if capacity for knowledge were exercised [109a] of itself, instead of becoming an act of knowing in consequence of the occurrence [82f] of something else [to be known]; for knowledge, encountering [116] a particular [22], is somehow guided by it to universals [43].† Neither is there any generation of the use or actualization [of potential knowledge], unless one were to believe in a generation of sight and touch; for the act 10 of knowing is analogous to these [intermittent acts of perception]. Not even an initial [82a] getting [65a] of knowledge [179] is a generation or a qualitative alteration: we do not claim an intelligent [172] under-

---

† Alternate version: "for it is from the experience of a particular that we derive knowledge of a universal" (247b20, 21).

standing [179a] until our thinking has come to a stand;‡ but a process of coming to rest is not a generation, any more than any change whatever is a product of generation (as we have pointed out before).§ Besides, when one who has been intoxicated or asleep or ill has passed to the contrary condition, we do not say that he has been generated anew as a knower, even though he has been incapable of putting his knowledge to use; nor do we say this when he begins to acquire knowledge. For one cannot become intelligent or a knower unless one has passed from natural mental disturbances to tranquility of mind. This is the reason why children come short of their elders in capacity for learning or even for sense discriminations; they are at the mercy 248a of much excitement and unrest. Nature herself suffices to allay our agitation in some of our activities; in others, we must take recourse to supplementary measures; but in either case there must be some bodily alteration, as exemplified by the restoration of intellectual efficiency after one has become sober or awake.

It is evident, then, that qualitative alteration takes place in objects accessible to sense perception and in sense organs; not in anything else, except incidentally.

4. Comparability and Incomparability of Movements

The question may be raised whether every movement is or is not 10 comparable [58] with every other. If all movements are [taken to be] comparable and if things moving at equal rates [admittedly] cover equal distances in equal periods of time, [it may be objected that] curves would be equal to or longer or shorter than straight lines. Moreover, qualitative alterations and local motions completed in equal periods of time would be equal to each other; and consequently, events [35a] and lengths might be equal to each other. But this is impossible. Is it not rather the case that equal movements occurring in equal periods of time are equally fast, but that events and lengths cannot be equal to each other? It is impossible, therefore, for qualitative alterations to be equal to or less [or greater] than lengths. It follows, therefore, that not all "movements" are comparable.

How, then, shall we apply this conclusion [3b] to the example of the circle and the straight line? It would be absurd to deny that a 20 circular motion and a motion in a straight line can be co-equal [57c]—as if one were necessarily faster or slower than the other, just as if the one went downhill and the other uphill. However, it would

---

‡ Plato *Timaeus* 44.
§ v.2.225b15.

make no difference to the argument even if we were to say that the one is necessarily faster or slower than the other; for the fact that a curve would then be longer or shorter than a straight line would imply that the two can be co-equal. Thus, if in a period of time A a faster body covers a [curved] path B while a slower body covers a [straight] path C, then B is greater than C. This accords with the definition we have given* of the "faster." But the faster also covers an equal distance in a shorter period of time. Hence, there will be a part of time A in which B will trace some part of the curve equal to the line which C traces in the whole of time A. However, in spite of this, if the motions are comparable, it would follow (as we have said) that there might be a straight line which would be equal to the circle; but a circle and a straight line are not comparable, so that the motions which correspond to them will not be comparable either.

Are we to understand, then, that things not designated by a single term having a single meaning are all incomparable? Why, for example, are a pen and a type of wine and the highest note in a scale incomparable as to their "sharpness"? Obviously, because it is in different senses that they are or are not "sharp." The reason why we can compare the highest note and the note next to it is because we would call both of them "sharp" in the same sense. May not "fast," then, have different meanings as applied to motion along a curve and to motion in a straight line? And would not "fast" be much less applicable, in the same sense, to qualitative alteration and to local motion?

Or would it not be false to say, to begin with, that things unequivocally denoted by the same term are therefore comparable? Although we speak of "much" water or air in the same [quantitative] sense, we cannot compare water and air in this respect. If we do not agree to this statement, we at any rate use the term "double" to denote unequivocally a ratio of two to one; yet we cannot compare [water and air, and so forth, as respectively "double"]. Or shall we bring our previous argument to bear also upon these [relative] terms and say that "much," for example, is vague? The very definitions of some terms involve variables. Thus, even if "much" is defined as meaning "so much and more," still, "so much" varies [16b] from one case to another. The term "equal" is likewise equivocal; and so is the term "one," perhaps most directly so; hence, also the term "two." For why would some things be comparable and others incomparable if all the things in question had a single nature?

Or shall we contend that the immediate subjects [12] of an attribute are incomparable when they themselves differ? A horse and a dog are

---

* vi.2.232a25-27.

comparable in the sense that one may be "whiter" than another because the attribute of whiteness belongs primarily to the bodily exterior which both have in the same sense; and a horse and a dog are, for the same reason, comparable in size. But a body of water and a voice are not comparable [in clearness or in volume] since the aspects [of the water and of the voice] to which such attributes primarily belong differ in these cases. Or would it not, on the other hand, seem clear on this interpretation that we could then represent [34] all equivocal terms designating attributes as having a single meaning and then ascribe differences to the possessors of each attribute in different cases? Thus, 249a terms like "equal" and "sweet" and "white" would severally have the same meaning, but would become different in meaning when the respective attributes are become embodied in different subjects. Besides, not any subject is capable of receiving [12] any attribute at random; any one attribute [such as "white," and so forth] can have but a single immediate subject [such as bodily exterior, and so forth].

Shall we maintain, then, that things are comparable only on the twofold condition that they have a given attribute in the same sense and that, in addition, neither the attribute nor the thing which has the attribute can be divided into different kinds? I mean, for example, that colors can be divided into different kinds and that, consequently, things are not comparable in color, as if one of them were "more colored" than another; we compare one thing with another rather in respect of a particular color such as "whiteness."

To apply these considerations to movement: two things move at the same rate if their movements equal each other quantitatively in equal periods of time. But if one thing has been altered in quality over a 10 part of its length and another thing has been carried over a distance equal to that bodily part, would the qualitative alteration equal the local motion in speed? That would be absurd for the reason [83] that "movements" differ in kind [20]. And if we say, then, that things in *local* motion over equal distances in equal periods of time are equally fast, [we are still confronted with the difficulty that] a straight line and a curved line could equal each other. Why [83], then, [are the motions corresponding to these incomparable]? Because local motion is a genus [having different species]? Or because a line is a genus [having different species]? Now, although the time [of the motions] may be the same, still, if the paths differ [16b] in species, so do the motions differ [76] in species; for local motion is divided into kinds if its path is divided into kinds. On the other hand, although walking differs from flying as moving with the feet differs from moving with wings, yet local motion is divided into species only in accordance with the shape [91a] of its path. Thus, we conclude that things are equally fast when their movements can cover the same path [142a] in equal periods of time, 20 the same "path" being undifferentiated [76b] in species and therefore undifferentiated also in respect of the corresponding motions. In order to determine the [4] differentia [76a] of a motion, therefore, we must look [195] to its path.

Our argument shows [38] that a genus is not a unity [24] without differences [4]; but although the latter [differences] may be distinguished [74] from the genus, many of them escape notice. Thus, some differences of meaning happen to be far apart; others have some sort of similarity among them; and still others come so close to forming a genus or an analogy that they hardly seem to be cases of different meanings. When, then, is there a distinct species [of quality]? When the same attribute is found in different subjects? Or when different attributes are found in different subjects? And what is the limit [72b] defining a distinct species? On what basis are we to judge [164] sweetness or whiteness to be [in each of its instances] the same or different? Would it be different because it appears different in different subjects? Or because it is not at all the same [in different subjects]?†

30 Turning now to qualitative alteration, let us ask: how can one qualitative alteration be as fast as another? If recovery of health is a qualitative alteration and one patient gets well rapidly and another slowly 249b but some get well in the same period of time, then qualitative alterations can be equally fast; namely, when they take equal periods of time. But how shall we qualify [4] the alterations? We do not speak of "equal" qualitative alterations as we do of "equal" quantities; but qualitative alterations may be "similar." However, let things be "equally fast" [in their changes] when they undergo the *same* [kind of] change [115] in equal periods of time! Must we, then, compare with one another the subjects of the attributes [35a] or the attributes themselves? Since states of health are the same in kind, we may take cases of health recovered as accomplished [82f] changes which are without differences of degree but which are similar to one another. But when the qualities in question differ in kind, as when one subject is becoming white and another is becoming well, then the qualitative alterations are 10 not identical or equal or similar; the differences exemplified divide the qualitative alterations into different species, which cannot form a unity any more than do the local motions [which are divided according to their paths]. Consequently, we must grasp the number of species into which qualitative alterations and into which local motions are divided. If the direct rather than the indirect subjects of a change [109a]

† 248b22, 24.

liffer [76] in species, their changes will also differ in species; if the ormer differ in genus, the latter will also differ in genus; and if the ormer differ numerically, so will the latter. But in determining vhether qualitative alterations are equally fast, must we look to the ;ameness or likeness of the attributes or to the subjects altered (for ;xample, to the extent to which each has become white)? Or must ve look to both, judging the alterations to be the "same" or "different" in terms of the attributes involved or to be "equal" or "unequal" in terms of the subjects involved?

Then, too, we must ask [195] in what sense generations or destruc- 20 tions may or may not be equally fast. Two processes of generation are equally fast when two things which are identical and indivisible in species are generated in equal periods of time; for example, two men (but not two animals). One process of generation is faster than another when the things generated in equal periods of time "differ" in species. We have no [more precise] term for two "different" [primary beings in generations and destructions] corresponding to "unlike" [attributes in qualitative alterations]. To be sure, on the [Pythagorean] theory that primary being is number, [one process of generation would be faster than another when a "greater" and a "smaller" number "of the same kind" are generated in equal periods of time; since on this theory] a "greater" and a "smaller" number may be "alike in kind" [57a].‡ But we have no general [92] term [like "unlike" or "unequal" to express the relation between "different" primary beings in generations and destructions] and no pair of terms [40] corresponding to "more" [and "less"] in the case of qualities [35a] differing intensively [6c] or extensively [148] or to "greater" [and "smaller"] in the case of different quantities [27].

5. Proportions of Forces, Objects, Distances, and Durations

Any mover (or agent) not only moves (or acts upon) something definite, but does so during a definite period of time and over a definite distance. The reason is that what is moving anything must at the same time also have been moving it; so that a certain distance must 30 have been covered and a certain period of time must have been taken. Let A, accordingly, be the mover; B, the thing moved; C, the distance covered; and D, the time taken. Then (1) a force [11] equal 250a to A will in a period of time equal to D move B/2 a distance 2C; or (2) it will move B/2 the given distance C in time D/2. In this manner, we establish the proportions existing between the terms

---

‡ "Triangular" numbers, "square" numbers, and so forth.

under consideration. Also, if (3) the given force A moves the given
object B a distance C/2 in time D/2, then (4) A/2 (or E) will move
B/2 (or F) the given distance C in the given time D. Thus, the
ratio between the force and the mass remains the same; so that each
force will move a given mass the same distance in the same period
10 of time. But (5) if E moves F a distance C in time D, it does not
follow that E moves 2B a distance C/2 in time D; and hence, (6) if A
moves B a distance C in time D, A/2 (that is, E) need not therefore
move B a distance C/2 in time D or in any part of time D. For E
may not move B at all. From the fact that a given force impels a
certain amount of motion, it does not follow that half of that force will
impel a motion either in any particular amount or in any particular
period of time. If it did, one man might move a ship; for both the
force of the ship-haulers and the distance they move the ship can be
divided into as many [141] parts [as there are ship-haulers]. This is
20 the reason why we must reject as false the argument of Zeno that even
the smallest grain of millet must fall with a noise. Why should a grain
in any period of time set sound-waves in motion as would the falling
of a whole bushel? It does not even make a fraction of a noise as it might
[by a stretch of the imagination be supposed to] make if it were by
itself; for, as a part, a grain is nothing at all in a whole bushel, except
potentially. On the other hand, (7) if each of two forces moves each
of two weights a certain distance in a certain period of time, then the
two forces taken together will move the two weights taken together an
equal distance in an equal period of time; for, in these circumstances,
the proportion between the terms under consideration is preserved.

May we apply these findings also to qualitative alteration and to increase? In an increase, the effective agent, the subject acted upon,
30 the time, and the amount of the increase are all definite. So, too, in a
qualitative alteration, there is something definite which does the alter-
250b ing, something definite which is altered, a definite degree of alteration which is completed, and a definite period of time in which the
alteration has been effected. If a given period of time is doubled, the
amount of the alteration will be doubled, and vice versa. Half of an
object will be altered, or an object will be altered half as much, in
half the time; or half of an object will be altered twice as much as the
whole object in an equal period of time. But if an agent brings about
a certain degree of a qualitative alteration or a certain amount of
an increase in a given period of time, it does not follow that such an
agent will in half the time alter or increase half of the object or a
given object half as much; but as in the case of a force moving a weight,
so here, it may happen that the agent will not effect any alteration or
increase at all.

VIII. BOOK THETA

Eternity of Movement

1. Movement as Ungenerated and Indestructible

Was there ever a time when movement[*] came into being and before which it had no being? And is movement doomed to pass away, with the consequence that a time will return when nothing will be 'stirred into activity? Or is movement ungenerated and indestructible and therefore something that always was and always will be? In other words, does process have a status [82f] of being [1a] to which considerations of "death" and even of pause are irrelevant and which may, accordingly, be described as a sort of continuing "life" encompassing all of nature's performances [111i]?[†]

Now, all of those[‡] who have anything significant to say about nature acknowledge that there is movement; for they are interested in the cosmos [149b], and all their theories [187a] concern generation and destruction, which would not be possible if there were no process. But those who maintain[§] that there is an infinite number of worlds, some springing up while others vanish, say that events are always occurring, since the generations and destructions of worlds involve them; whereas those who maintain that there is a single universe, be it eternal[||] or not eternal,[¶] regard its moving system of energies as eternal or not

---

[*] Averroes: "de primo motu"; Thomas Aquinas: "universaliter de motu." Cp. viii.5.257a33.
[†] Plato *Laws* x.895.
[‡] Anaxagoras, Empedocles, Democritus.
[§] Anaximander, Leucippus, Democritus.
[||] Plato, Aristotle.
[¶] Anaximenes, Heraclitus, Diogenes of Apollonia: Cycles.

eternal in keeping with their respective theory [90] of the universe.**
As to the possibility of a time when nothing at all comes to pass, there
are only two ways in which this could be so. One way would be as set
forth by Anaxagoras; namely, that all things are together and at rest
for an infinite time until mind imports movement and distinction
[164d] into them. The other way would be in accordance with the
teaching of Empedocles; namely, that rest alternates with movement;
because there is movement when love unifies the manifold diversity
of things or when strife disrupts the unity again in all manner of ways,
and there is rest in the ensuing intervals [when love and strife are in
equilibrium]. To quote Empedocles:

30   Since unity takes its rise from plurality,
          And plurality, from a diffused unity,
251a   Things come and go and are without stability;
          But since their rotation continues forever,
          They form a cycle with eternal steadiness.

The "rotation" of which Empedocles speaks must be understood as re-
ferring to the alternating shifts from plurality to unity and from unity
to plurality. We shall find the consideration [195] of this supposed
state of affairs [33] profitable [9e] for our examination [166] of the
facts [7a] pertaining to a theory of nature as well as for our search
[198] for the first principles.

    Let us begin with distinctions which we have made before in our
10 discussions of nature.†† We define "movement" as "the functioning
[9] of the movable as movable." Accordingly, each kind of "movement"
requires the presence [82f] of actualities [188c] capable of undergoing
that kind of movement.‡‡ Even apart from the definition given of
"movement," everyone would admit that, to be in process, a thing must
be capable of that particular [40] process. Thus, to be altered, a thing
must be capable of being altered; and to be in local motion, a thing
must be capable of change in place. Thus, too, before anything is
burned, it must be capable of being burned; and before anything starts
a fire, it must be capable of starting a fire. Therefore, if there was a time
when these things did not exist, they must have been generated; if not,
they must be eternal. If, then, every movable thing was generated, its
supposed change or movement must have been preceded by another
20 in which the thing capable of giving rise or of submitting to a complete process was generated. But to hold that things existed [82g]
before there was any movement, would on first thought seem absurd;

** Anaxagoras, Archelaus, Metrodorus of Chios: beginning of movement.
†† iii.1.
‡‡ Thomas Aquinas (against Averroes): ". . . sequitur, quod productio universalis
    entis a Deo non sit motus nec mutatio, sed sit quaedam simplex emanatio. . . ."

and further reflection would show why it must be absurd. Suppose that, while there are things amenable to processes and things capable of producing them, there would be a period of time when a first mover would be [active] and a thing [acted upon] would react, but that there would be another period of time in which nothing of the sort would take place but only a continuing rest. It would follow that anything [which was at rest but is in movement] must have previously undergone a change; for its quiescent state must have had conditions [83], since coming to rest is a privation of movement [as movement, in turn, is a privation of rest]. Consequently, such a thing must have been changed before its supposed first change! To be sure, some things initiate movement in one way only, as fire heats; whereas other things initiate contrary movements, as the same science may be used in contrary ways. Even so, the way in which the former things function is not altogether unlike the way in which the latter function: thus, a cold body may become the occasion for warmth by its mere removal from the scene, just as an expert may fail to fulfill his proper role by deliberately misapplying his knowledge. However, in any case, things are not capable of acting or of being acted upon or of moving or of being moved under any and all conditions [150] but only under determinate conditions [33] and in interaction with one another. It is only as they come into close relations, then, that one thing moves another and the latter is moved by the former, provided that the former was capable of originating the process and the latter was capable of submitting to it. Clearly, then, if movement had not been always in operation, the situation would not have been such that one thing was capable of inducing movement and the other was capable of being moved; but at least one of them must have undergone a change. This would necessarily happen in the case of correlatives [29]; for example, if one thing which was not "twice" its correlative now is "twice" the latter, at least one of them (if not both) must have passed through a change. Thus, [if movement had not been always in operation] there would be a change prior to the supposed first change.

We may here interject this question: how, when there is no time, can there be any "before" and "after"; or how, when there is nothing going on, can there be time? Since time is a number belonging to a process or else is itself a sort of process,§§ then, if there always is time, movement must be eternal also. Indeed, all who have reflected about time seem, with a single exception, to agree in describing time as ungenerated. Thus, Democritus avails himself of the argument that time is ungenerated, in order to show the impossibility of the view that all

30

251b

10

§§ iv.11.219a8, 220a24, 10.218a33.

things have been generated. Plato alone presents time as generated; time, he maintains, is coeval with the heavens which, according to 20 him, have had an origin.‖‖ But if time can neither be nor be conceived without a present, and the present is a sort of "mean" in the sense of being at once the starting-point of the future and the end-point of the past, ¶¶ then there must always be time. Surely, the extreme end of any supposed last period of time must be in some present, there being no period of time immediately accessible [65] to us other than the present; and therefore, since the present is at once a beginning and an end, there must always be time in both of its directions. Accordingly, if time as an aspect [35a] of movement [is eternal], it is evident that movement must be eternal also.

To resume the main argument. This argument gives evidence that movement is [not only ungenerated but] also indestructible. Just as 30 the generation of movement would entail some change prior to a supposed first change, so the destruction of movement would have as a consequence a change subsequent to a supposed last change! A being which is no longer in movement does not therefore at the same time cease to be movable; for example, a thing which has stopped burning does not therefore cease to be combustible, for it is quite possible for things to be combustible when not actually on fire. Like-252a wise, a thing which has terminated an action [109a] does not thereby lose its capacity for action [109c]. Now, destruction being a kind of change, a destructive power which has effected a [supposedly final] destruction would still be left to be itself destroyed; and thereafter, what is capable of destroying it in turn would also have to be destroyed. But these consequences are impossible. Clearly, then, movement is eternal; and the view that there may be a time when there is some-thing "going on" as well as a time when there is "nothing doing" looks for all the world like a figment of the imagination.°°°

No less fanciful is the view which sets up a situation of this kind as a principle of the natural course of events [101c]. Empedocles, for example, suggests that, as things go, love and strife must alternate in their control of phenomena and that in the intervals they refrain 10 from all activity. No doubt, those who hold [34] with Anaxagoras to a single beginning [of movement] would go along with this opinion.

‖‖ Plato *Timaeus* 28B, 38B. Simplicius: Plato and Aristotle take "generated" and "ungenerated" in different senses.

¶¶ Thomas Aquinas (against Averroes): "Ex hoc enim quod tempus est fluens et non stans, sequitur quod unum nunc non possit bis sumi, sicut bis sumitur unum punctum. . . ."

°°° Thomas Aquinas: "Est autem ante tempus aliqua duratio, scilicet, aeternitas Dei, quae non habet extensionem, aut prius et posterius, sicut tempus, sed est tota simul. . . ." "Auctoritas autem divina praevalet etiam rationi humanae. . . ."

But nothing natural is unordered; on the contrary, nature is uniformly [150] principled [83] by order [207a]. There is no ratio, however, between an infinite [time of change] and an infinite [time of rest]; but any order consists in a ratio [90]. An infinite period of rest, followed by a movement which starts indifferently at any time—with no order to account for the beginning of the movement at a given time rather than at some previous time, is far from nature's way of working [9c]; for the natural is either invariant [105] (and not now in one way and then in another), like fire being carried upward (and not only sometimes and sometimes not), or its variability has a pattern [90]. Em- 20 pedocles and others may in a way be commended for recognizing some kind of order, at least to this extent that they speak of alternating rest and movement. But this speculative assumption, too, ought not to be illogically proclaimed as if it were an axiom; it ought to be supported by some reason [83] and be proved either by induction or by demonstration. Surely, the love and strife assumed do not explain [83] the alternation. Nor is their being [88a] an affair of alternation; rather is their being, respectively, an affair of combining and of separating. To secure the definiteness of the alternation, instances of it ought to be cited; instead, we are simply told that there is something in the universe [21] 30 like the love uniting men and like the mutual avoidance among enemies. There ought also to be a definite [4] reason [90] for the equal periods of time allotted to movement and to rest. In general, it will not do to magnify an appeal to an invariant way of being or to an invariant occurrence into an all-sufficient principle; as when Democritus reduces his natural explanations to the consideration that things have happened before as they do now. He does not deign to search for a principle ex- 252b plaining this uniformity; but though there are cases in which this attitude is justified, there are others in which it lacks justification. Thus, the interior angles of a triangle are always equal to two right angles for definite reasons which account for this eternal truth; although there are, indeed, eternal principles which do not call for further explanation.

Let these comments suffice for the statement of our conviction that there never was a time when there was no moving state of affairs and that there never will be a time when there will be none.

2. Refutation of Objections to Eternal Movement

It is not difficult to resolve the leading counter-arguments [50] envisaging a possible time when the moving tendencies [109] of existence were absent from the whole realm of being. First, it may be thought that no change [115] is eternal, since every change naturally has a 10 definite start and comes to a definite stop; so that every change must be limited by the contraries within which its course [116] is confined, and no process can continue indefinitely. Secondly, we observe even nonliving things which, though they are wholly or partially at rest and not only unmoved but without power of self-movement, are nevertheless capable of being moved and are at some time actually given a start. Hence, it is argued, if a movement which is not in being cannot be generated, then things must be either always or else never in movement. Thirdly, these considerations are said to be especially evident in the case of living beings. There are times when there is no transition

20 within ourselves but only complete repose and when nothing external moves us; but presently we find ourselves in the midst of an activity which we have ourselves set on foot. To be sure, we do not see quite the same thing happen in the case of nonliving things, which always require an external mover; but an animal, we say, "moves itself." If there is any time, then, when an animal is in a completely quiescent state, it is possible for movement to be generated in a nonmoving being by its own power without the mediation of an external mover. And if this is possible in an animal, why not in the universe? Surely, what can happen in a "microcosm" can happen in the "megacosm"; and what can happen in the cosmos can happen in the "infinite" —if the "infinite" as a whole can be in movement or at rest.°

30 In the first of these arguments, the statement is warranted that there can be no eternal passage from one opposite [13] to the other as a numerically single process remaining identical with itself. This would necessarily follow, at any rate, if one and the same thing can have a motion which, however, is admittedly not itself forever one and the same. Yet even if a given single string of a musical instrument continues in the same state and is repeatedly struck in the same way, is the sound it produces always one and the same; or is each sound

253a different? However this may be, there is nothing to prevent *some* movement which is continuous and eternal from being therefore also self-identical. This will become clearer at a later stage of our inquiry.†

The second argument, that what is unmoved comes to be moved, proposes nothing strange [61] if we consider that there are occasions when an external mover is present and occasions when an external mover is absent. Even so, we must still investigate how it is possible for the same thing to be sometimes put in motion and sometimes not by the same effective agent. This difficulty amounts to the question: why are not some beings always in a state of rest and others always in movement?‡

° iii.5.205a8-206a7.
† viii.8.
‡ viii.3.

The third argument appears to raise the most formidable problem of all. It suggests that movement can be generated in living things without being preceded by another movement in them because, it is 10 alleged, an animal may abandon its rest by walking off, apparently without the aid of any external mover. Such, however, is not the case. Observation of an animal always discloses some part [101i] in movement such that the movement is not to be attributed [83] to the animal itself but perhaps to the operation of its surroundings. We do not credit an animal with originating every one of its own movements but only its local motions. Many processes are possibly, or perhaps even necessarily, set up in an organism by the environment; and some of them arouse thought or desire which, in turn, affect the organism as a whole. This is what happens, for example, in sleep in which there is no perceptivity, but in which some process occurs in consequence of 20 which the animal awakes. However, we shall later§ clarify this issue also.

3. Rest, Movement, and Kinds of Beings

Let us begin our present line of investigation with the question which is fundamental to an examination of the second objection just noted:° Why is it that some beings are sometimes in movement and sometimes in a state of rest? We must, in this connection, consider three alternatives: either (1) all beings are always inert [110]; or (2) all beings are always transient [109a]; or (3) some beings are transient, whereas others are inert. On the third alternative, in turn, either (a) changeable beings are always in process, whereas stable beings are always in a state of repose; or (b) all beings are naturally sometimes [57] active and sometimes inactive; or (c) some beings are always independent of movement, whereas others are always subject to movement, and still others admit [65d] of both rest and movement. This 30 last possibility remains after the other alternatives have been eliminated and is therefore the one we must maintain: it holds the solution of all the problems confronting us at this point; and to establish it is, accordingly, the aim of our present enterprise [197].

Now, to hold (1) that everything reposes in a state of inertness, and to hunt for reasons [90] for this view in disregard of the testimony of the senses, is a kind of intellectual incompetence which plays fast and loose [51] with things not simply in piecemeal [22] but in wholesale [21] fashion. This is to run counter [29] not only to the natural scientist but to virtually [36b] all the sciences and to all sober opinions, 253b

§ viii.6.
° viii.2.252b12-16, 253a2-7.

which with one accord work with the fact of movement. Besides, just as a mathematician or anyone else regards assaults upon the first principles of his science [90] as beneath his notice, so a physicist need not trouble himself with an attitude which calls into question what is for him a matter of fact [64h]; namely, that nature is an originative source of movement.

Next, the view [of the Heracliteans] (2) that everything is a transient occurrence, though likewise false, is perhaps less destructive [74] of scientific exploration [198]. To be sure, we have in our writings on nature† taken nature to be a principle of movement and of rest alike. Still, nature is especially characterized [101a] by movement. Some even

10 allege that, so far from some beings being subject to and others independent of movement, all beings are in constant movement, despite our inability to perceive this because of the limitations of our senses. These men do not tell us what distinctive kind of movement they mean or whether they are talking about all kinds of movement. Yet we shall not find it difficult to reply to them. Thus, increase or decrease cannot proceed in an unbroken sequence [136a], but is interrupted by intervening states [138]. The argument in question is like that about stones being worn away by the dripping of water or being split by the growth of plants. If the dripping has moved or removed so much of the stone, it has not therefore moved or removed half as much in half the time; but just as the hauling of a ship [results from the combined labors of the haulers], so it takes a goodly number of drops to have an appreciable effect [109a] on a stone such as no portion of the number of

20 drops can have in any period of time. The amount removed is, of course, divisible into parts; but none of the parts was brought into operation separately but only in conjunction with the others. Because a decrease is infinitely divisible, then, it evidently does not follow that some part must therefore always be vanishing; on the contrary, a whole may disappear all at once. So it is also with any qualitative alteration. If the subject undergoing a qualitative alteration is infinitely divisible, the qualitative alteration is not therefore itself also infinitely divisible; it often occurs precipitately (for example, in cases of freezing). Again, an interval of time is required for one who has fallen ill to return to health; such a transformation does not occur suddenly [131] and cannot, moreover, be a change in any direction whatever, but only to health. We must conclude, therefore, that the representation of a quali-

30 tative change as continuous flies in the face [51] of obvious facts: a qualitative change is a change to a contrary state, and a stone is not [continuously] becoming harder or softer. Then, as regards local mo-

---

† ii.1.192b21.

tion, it would be very strange if the falling of a stone were beyond our powers of observation or if we could not know whether it was resting on the earth. Besides, the earth and all the other elements necessarily remain in their respective resident [55] places, from which they are removed only by violence; so that, if some of them are in their resident places, it cannot be the case that all things are in motion. 254a From these and other arguments like them, we may, accordingly, conclude with confidence that it is impossible for all things to be always transient or for all things to be always inert.

Then, too, (3) it is beyond the bounds of possibility [12b] that (a) some beings are always transient and others always inert, as if no being were sometimes at rest and sometimes in movement. Considerations like those which we have been bringing forward will also serve to show the impossibility of this view. Thus, we observe identifiable beings [15a] changing as already stated [from rest to movement and from movement to rest]. Besides, our objector's position clashes with plain matters of fact. There can be no increase [without intervening states]; and there can be no violent movement unless a being at rest can be 10 moved contrary [74] to its natural state [101]. Hence, the theory in question would abolish generation and destruction. Virtually all thinkers agree that for anything to be in process is a kind [4] of generation and destruction: whatever is changing is becoming that in which the change culminates or is coming to be in that condition; and the condition from which anything is changing is ceasing to be, or the changing thing is ceasing to be in it. Consequently, there clearly are beings sometimes in movement and sometimes at rest.

The suggestion (b) that all beings are sometimes inactive and sometimes active must be rebutted by our former arguments. Let us again take as our point of departure the alternatives we distinguished at the beginning of this analysis. Either (1) all beings are inert; or (2) all beings are transient; or (3) some beings are transient, whereas others are inert. And if some beings are inert and others are transient, then 20 either (a) [some beings are always at rest, whereas others are always in process]; or (b) all beings are sometimes inactive and sometimes active; or (c) some beings are always independent of movement, whereas others are always subject to movement, and still others are sometimes at rest and sometimes in movement. Although we have said so before, we may now repeat that (1) it is impossible for all beings to repose in a state of inertness. Even if being were in truth inert, as those thinkers‡ hold who describe being as infinite and independent of movement, this is at least not apparent to our senses; to our senses,

‡ i.2.185a32, 3.186a18.

many beings seem to be transient. Indeed, if there is any such thing as false opinion or even any opinion whatever, if there is any imagination, if anything seems different at different times, then there must 30 also be such a thing as movement since imagination and opinion are in some sense [4] held to be "movements." But to institute a protracted search for arguments in a matter of this kind where we have the good fortune of not standing in need of any arguments, shows an inability to discriminate what is better and what is worse, what is credible and what is incredible, what is fundamental [82] and what is derivative. By the same token, it is impossible (2) for all beings to be transient; and (3) it is likewise impossible (a) for some beings to be always in process, whereas others would be always in a state of repose. One 254b certitude suffices to refute all these theories: we *see* some things now in movement, now at rest.§ It being impossible, therefore, for all beings to be always inert and for all beings to be continuously transient, it is evidently no less impossible for some beings to be always in process and for others to be always in a state of repose. Thus, it remains for us to consider whether (b) all beings admit of both movement and rest or whether (c) only some beings are to be so regarded, whereas others are always independent of movement, and still others are always subject to movement. It is the latter of these alternatives which we are trying to establish [63].

4. Movements as Due to Agents

Agents or things acted upon function as such incidentally or essentially. They induce or undergo movements only incidentally in so far 10 as they happen to possess [82f] certain traits or in so far as the active or passive factor is one of their parts.* But they induce or undergo movements essentially in so far as their movements do not pertain to any of the traits or parts they happen to possess.

Now, of the things which are moved essentially, some are moved by themselves, whereas others are moved by something else; and some are moved naturally, whereas others are moved "contrary to nature" or violently. Anything moved by itself (such as any animal) is moved naturally; for when an animal is moved by itself inasmuch as it has within itself the beginning of its movement, we say that its movement is "natural." But although an animal as a whole thus moves itself naturally, its body may be moved naturally or violently; this depends 20 [76] upon the kind of movement it happens to be undergoing or upon

---

§ All movements are identified as completed movements.
* v.1.224a24, 25.

the kind of element which may enter into its composition. Anything moved by something else is moved naturally or violently. Examples of violent movement include the upward motion of earthy things and the downward motion of fire as well as such motions of animal parts as are at variance [74] with their normal [101] position [64] or manner [55c]. Violent movements afford the clearest evidence that whatever is moved is moved by an agent. Next in order, this circumstance is manifest in those natural movements which are exemplified by animals moving themselves. What may be obscure in such movements is not that they are effected by an agent, but rather how we are to distinguish [65c] in them between the mover and the thing moved. It would seem that a mover and a thing moved are to be 30 distinguished in movements of animals no less than in movements of ships or of any artifacts and that, moreover, it is only in this sense that an animal as a whole is a self-moving agent. But the distinction in question is made with the greatest difficulty in the class of things remaining according to our classification. We have noted, among things moved by something else, those which are moved violently. But there remains, among things moved by something else, a class of things opposed [64b] to those which are moved violently; namely, the class of 255a things moved naturally.

Things whose movements are not only brought about by something else, but are also natural to them, confront us with the question: By what are they moved? Consider, for example, things light and heavy. Evidently, they are moved in directions respectively alien to themselves [13] by main force; but they are naturally moved to their respective resident [55] places, that is, light things upward and heavy things downward. In their natural movements, however, it is no longer equally evident by what agency their movement is effected. Surely, they cannot set themselves in motion, any more than they can bring themselves to a stop, as animals can. If fire, for example, could of itself move up as an animal can of itself start walking, then fire could presumably also 10 of itself move downward as an animal can of itself stop walking. It would, moreover, be unreasonable to suppose that, if [elements] moved of their own accord, they would be capable of only a single kind of self-initiated "movement." Besides, how would it be possible for anything continuous and naturally unitary [101i] to bestir itself? In so far as anything is one and continuous (rather than merely contiguous), in so far it is incapable of being acted upon [35d]. Things must admit of division [73] if they are to function [101c] both actively [34] and passively [35]. Hence, nothing naturally unitary [like fire, and so forth] can be self-moving, any more than can anything else that is continuous;

but within such things, the active [109a] and the passive factors must be just as distinguishable [75] as we observe them to be in the action of living things upon nonliving things.

Now, nonliving things are [even in their natural movements] moved
20 by something. This becomes evident if we distinguish different factors [83] accounting for their movements; that is, if we divide movers, as we have divided things moved, into those whose movements are violent and those whose movements are natural to them. Thus, a lever moves [109c] a weight by main force, whereas a body which is actually [9] hot can act [109c] in accordance with this very nature upon a body which is potentially [11] hot; and so forth. Just so, a body can be acted upon [109b] in accordance with its nature if it is potentially of a certain sort or so much or somewhere; provided that the process in question can arise [82] in it, and provided that the process would not be incidental to it. (Although the same thing may be [potentially] of some character as well as of some dimension, still, the one aspect would not belong to the other essentially but would be incidental to it.) Fire and earth, then, are forcibly moved by some agent in ways at variance
30 with their nature; they are moved naturally when they are impelled to pass from their respective potential states to their corresponding activities [9].

The "potential," however, has more than one [6] meaning [36]. This is the reason [83] why the agent may escape observation [59d] in the upward motion of fire and the downward motion of earth. Consider an analogy: a learner is a potential knower at one stage; he becomes a potential knower at another stage when he has acquired knowledge which, at a given time, he is not putting to actual use [9a]. In any case, however, the interaction of something capable of acting with something capable of being acted upon is the occasion for a passage
255b [116] from a potentiality to an actuality. Thus, a learner passes from his [first] potential knowing to a second [16] potential knowing: from the knowing which is potential in the sense that the learning process has not been completed, to the differently potential knowing of one who does not turn the knowledge he possesses to actual account [187]. But one who has arrived at the latter stage puts into practice [9a] his intelligence [187], unless his activity is impeded; if he did not do so, he would really be in the opposite state of ignorance. So it is with natural bodies [or elements]: a cold body is potentially hot; when turned into fire, it burns—unless it is prevented from doing so. So, too, with heavy and light bodies: when water (the first heavy body which
10 is potentially light) is transformed into air, it becomes a light body and at once acts as such (unless hindered); for light bodies tend [9]

to rise and, when held "down," are prevented from doing so. A similar analysis may be made of quantitative and qualitative changes, with similar results.

We may, however, persist in asking: why, after all, do heavy and light bodies tend [109a] to their own respective places? The reason [83] is that they naturally [101c] tend in distinct directions. This is precisely what it means for them to be [88a] light or heavy; namely, that they tend upward or downward, respectively. But, as we have said, there is more than one way of being potentially light or potentially heavy. When a thing is water, it is at least in some sense potentially light; and when it is air, it is still only potentially light in the sense that its upward tendency may be counteracted; whereas, on the re- 20 moval of the obstacle, it becomes actually light [9a] and proceeds to rise. Analogously, what is [potentially] of a certain quality becomes [115a] actually so, as when an expert [179a] is at liberty to think [187]. Likewise, what is of a certain quantity expands when unchecked. As for anyone who removes an obstruction such as a pillar from under a roof or a weight from a wine-skin in a body of water, he in a way moves the object he has freed; that is, he moves it incidentally only, somewhat as a wall intervenes to cause a ball to rebound which a thrower has hurled against it. Clearly, then, none of the things we are talking about move themselves; yet they have a "source [82] of 30 movement." Not that they produce any movement or keep it going; but they are amenable [35] to movement.

In fine, the movements of things in their courses are natural or violent; and the latter are due to external agents, whereas the former may be enacted by self-moving agents or be due to external agents. Thus, light and heavy things have two sorts of "movers": those which 256a have produced them and made them light and heavy; and those which have freed them from the hindrances to their movements. Hence, all things in movement are acted upon by some agent.

5. The First Mover Taken as Self-Moved or Unmoved

An object is acted upon by an agent in one of two ways: its mover may itself be acted upon by another; or its mover may act of itself. A mover acting of itself may, moreover, affect an object [18a] directly [17a]; or it may do so by means of intermediate agents [6c], as when a pry which moves a stone is wielded by a hand in the control of a man who does not have another mover. We recognize as "movers" both the first and the last in the series; especially the first, which is not 10 moved by the last but moves it and which can act without the last, whereas the last cannot act without the first (a pry, for example, cannot be a "mover" without being itself "moved" by a man). Everything that is in process, then, must be acted upon by an agent, the agent being in turn acted upon by another or not being so acted upon. And a moved mover requires a first unmoved mover, whereas the latter does not require the former; for there can be no infinite series of moved movers, since an infinite series has no first term. If, then, 20 everything that is in movement is acted upon by an agent, and the first mover, though acted upon, is not acted upon by something else, then the first mover must be self-moved.

The same argument may be put in another way. Every agent acts upon an object by some means, whether by its own agency or by some other means. Thus, a man moves a stone without or with the aid of a pry; and something may be knocked down by a gust of wind or by a stone driven by a wind. But movement by any means is impossible without a mover acting by its own agency. Anything which is an agent in its own right does not require the backing of another; whereas a dependent agent requires an independent one, or else there would be an infinite regress of instrumental agents. A series of moved movers 30 must have a limit, then, if it is not to extend into infinity: if a pry which moves a stone is moved by a hand, it is the hand that moves the pry; if something else does the moving with the hand, the hand is moved by something other than itself. Accordingly, a series of instrumental agents presupposes one that is self-directing. If the latter, then, is acted upon, and there is nothing else to act upon it, such a mover must be self-256b moved. So we have another argument which shows that, if an object is not directly acted upon by a self-mover, the series of its movers can be traced to a self-mover.

The same result may be reached in still another way. Suppose that every movement is due to a moved mover. On one alternative, being moved would be an accidental trait of the movers [188c]; so that what is being acted upon would act—but not *because* it is being acted upon. On the opposite alternative, it would be essential to agents that they be acted upon.

According to the former alternative, then, being acted upon would be accidental to agents; and an agent would not be acted upon nec-essarily. If this were so, it would be possible for a time to come when 10 no being whatever would be acted upon; for what is accidental is not necessary but has the possibility of not-being. However, an assumed possibility can have no impossible consequence (although it may have a false one). But it is impossible that there should be no movement; we have proved° it necessary that there should always be movement.

° viii.1.

Then, too, it is a reasonable conclusion [that there must be an agent not acted upon by another]. For there must be a thing acted upon, an agent, and that whereby the latter acts upon the former; there must be all three of these. The thing moved must be moved, though it need not move anything. The means by which the movement is effected must be both an agent and acted upon, since it changes together with the object moved and has contact and continuity with it; as is plain from a thing which puts another in motion, where the two must have at least some contact. And the mover which is not a means 20 of this sort must be unmoved. Moreover, we see things which can be moved but cannot initiate [82] movement; and we see beings which are not moved by anything else, but only by themselves; and so it would seem probable, if not necessary, that there is also a third kind of being, namely, an unmoved mover. Hence, when Anaxagoras presented "mind" as the source of movement, he was right in describing this "mind" as impassive and unmixed. How could it be a mover unless it were itself unmoved? And if it were not aloof from the mixture of things, how could it dominate?

According to the second alternative, agents would not be acted upon accidentally but necessarily; and if an agent were not acted upon, it would not act upon anything. Hence, an agent would either have to 30 undergo the same kind [20] of process which it initiates; or else it would have to undergo a process of a different kind. I mean, for example, that either nothing could heat or heal or carry anything unless it were itself being heated or healed or carried; or else nothing could heal anything without being carried or carry anything without increasing in size. This is evidently impossible. We would have to identify 257a [36] the processes down to their smallest [41] parts [75]; so that a teacher of geometry would have to be learning precisely what he is teaching, and one who is throwing anything would have to be thrown in exactly the same manner! Or we would have to derive each kind of process from one of a different kind [19]; so that a carrier would be expanded, what expands it would be qualitatively altered, and what alters this would be subjected to some further sort of process! But such a series would soon come to a stop, since the kinds of process are limited in number. If it be suggested that the series reverts to its starting-point as if what does the altering would be carried from one place to another, this would amount to saying directly that a carrier is being carried or that a teacher is being taught; for, clearly, a thing 10 is acted upon not only by a direct agent but more especially by a prior one. But the suggestion made is impossible; it would have the consequence that a teacher would still be learning and would be ignorant of what he knows.

Then, too, there is an absurdity greater than any pointed out in the previous arguments. If everything that is moved has only a moved mover, then everything capable of moving anything would be capable of being moved. This would be like saying that every remedy is remediable and that every builder is buildable! Such an agent would be acted upon either directly or indirectly. Thus, if an agent, in directly
20 acting upon something, is acted upon by something else in a different way (as when one capable of effecting a cure would therefore be capable of being taught), still, as we have said before, the series would soon revert to the same kind of process. But the first alternative is impossible [as in the case of a teacher still having to learn what he is teaching]; and the second alternative is fanciful, for it would be absurd to suppose that what can alter anything must therefore necessarily increase in size. What is acted upon, then, is not necessarily acted upon by something else which would, moreover, be a moved mover; the series has a limit. Consequently, what is acted upon will in the first instance be acted upon by something that is at rest, or by itself.

Besides, if we had to raise the question whether the primary [82] moving [109] factor [83] is self-moved or is moved by something else,
30 anyone would reply by proposing the former. For a factor [15a] which is independent [2] ontologically [1] is always prior [17] in the order of explanation [83] to a factor having a dependent [39a] status [16].

At this point, our inquiry must make a fresh [16b] start [82]. We must ask: If anything moves itself, how in plain terms does it do so?

Everything in movement is necessarily infinitely divisible, for our
257b general reflections on nature† have shown that everything essentially in movement is continuous; so that it would be impossible for anything moving itself to move itself with all its parts. For in such an event, a self-mover, though specifically [20] one and indivisible [41], would totally and in an identical transaction place and be placed or alter and be altered and therefore teach and be taught or heal and be healed. Moreover, we have shown‡ that what is acted upon is something movable. But what is movable is not actually but only potentially in movement; what is potential is, moreover, only on the road [112] to actuality; and movement is an uncompleted actualization of what is movable. An agent, however, functions as a fully active being [9]: it
10 is a hot body which heats another body; and, in general, a formative act [116d] presupposes possession of the form [20]. Thus, [on the view we are examining], the same thing would as such be at once hot and not hot; and a similar consequence would result wherever an agent

---

† vi.4.234b10-20.
‡ viii.1.251a9-16.

would have the quality it imparts [37b].§ Accordingly, a self-mover must have one aspect according to which it initiates a process and another aspect according to which it undergoes a process.

It would evidently be impossible for anything to move itself by each of two parts moving the other. For one thing, there would then be no first mover; rather would each part, after all, be moving itself. But a process is better accounted for [83a] by a prior agent than by a dependent [136] one, by a self-mover than by a mover moved by something else, and by an agent farther from the thing acted upon but closer to the beginning of the process than by an agent relatively inter-   20
mediate. For another thing, the part moving the other part would not necessarily be moved by anything other than itself; only incidentally would it be moved by the other part in return. In the possible failure of the latter case to occur, there would be only a part that is moved and another active as the moved mover. Then, too, the eternity of process does not require that an agent be acted upon in return but only that there be some unmoved or self-moved mover. And besides, an agent [necessarily acted upon in return] would have to be acted upon in a corresponding way; for example, what gives heat would have to be receiving heat.

It would also be impossible for anything to act primarily as a self-mover by one or some of its parts moving themselves. For a self-moved whole would be moved either by one of its parts or as a whole   30
by itself as a whole. If, on the one hand, a self-mover were to operate as such only by way of some self-moved part, this part would be the primary self-mover; and this part would, accordingly, move itself inde-pendently, whereas the whole would not be moving itself in inde-pendence of the part. If, on the other hand, the whole were self-moved as such, any of its parts would be only incidentally self-moved. Let us therefore suppose that they are not moving themselves. Then a part   258a
of the whole would be an unmoved mover, and another part would be acted upon; for this is the only way in which any whole could oper-ate as a self-mover. For any whole to move itself, there would have to be a part initiating and a part undergoing the process; thus, the agent by which AB would be moved would be both AB itself and a part A.

We can show that, since a mover may be moved by something else or may be unmoved, and since a thing that is moved may or may not move something else, therefore a self-mover must have as one part an unmoved mover and as another part something that is moved but that does not necessarily move something else (but at best only by chance). Let a self-moving whole ABC have a part A, which is an unmoved

10 mover; a part *B*, which is moved by *A* and moves a part *C;* and a part *C*, which is a moved nonmover. (There could be more intermediate terms before *C*, but one suffices.) By removing *C*, we get a self-moving whole *AB*, in which *A* acts upon something and *B* is acted upon; whereas *C* would not move itself or be moved at all. Also, *BC* would not move itself without *A;* for *B* would be a mover of the kind which is moved, not by any of its own parts, but by something else. Hence, only *AB* moves itself. So a self-moving whole need have but two parts: one, an unmoved mover; and the other, something moved which does

20 not necessarily move anything else. And the two parts must be in mutual contact; or, at any rate, one part must be in contact with the other.‖ If the mover is a continuum, as the thing moved must be,¶ then the mover and the thing moved are in mutual contact. Clearly, then, it is not because some part is able to move itself that a whole moves itself, but it moves itself as a whole; and it both communicates and receives movement because it has a part which communicates and a part which receives the process. It is not the whole, but it is the part *A*, which communicates the process; as it is not the whole, but it is only the part *B*, which receives it. (For *C* to be moved by *A*, would be impossible.)

However, there is this question to be considered: if we remove a part from *A* (if the unmoved mover is a continuum) or from *B* (which

30 is acted upon), would what is left of *A* communicate any process or what is left of *B* receive any? If so, the whole *AB* would not be self-moved in a primary way; but, part of *AB* having been removed, the remaining part would form a new whole *AB* which would pre-

258b sumably be moving itself. Perhaps there is nothing to prevent each of the two parts or at least the part acted upon from being potentially divisible, yet not actually divided; whereas, after its actual division, it would no longer be in possession of the same nature [or power]. Hence, there is nothing in things potentially divisible to prevent the whole from being a self-mover, even in a primary sense.

It is evident, then, that a prime mover is unmoved. What is acted upon by something may be referred directly to the first unmoved mover or to something which is itself also acted upon but which moves and stops itself. In either case, what primarily acts upon anything that is in process is an unmoved mover.°°

‖ *On Generation and Corruption* i.6.323a21.

¶ vi.4.234b10-20; viii.5.257a34.

°° *De Anima* i.3; Plato *Laws* x.894.

6. The First Mover as Eternal, One, and Immovable

Since movement necessarily goes on eternally and without interrup- 10
tion,* there must be one or more primary movers; and the prime mover
must be unmoved. Whether each individual "unmoved mover" is eternal, has nothing to do with our present argument. We are trying to
show that there must be something which is unmoved in the sense that
it transcends all change (inherent or accidental), but which has the
function of moving [109c] other things. Suppose it possible, if anyone
insists [177], for some things to be for a period of time and then not
to be without ever passing through any gradual process of generation
and destruction; for if anything having no parts is at one time to be
and at another not to be, it would have to "be" and then "not be" 20
without gradually changing to either of these conditions. Suppose it
possible also for some "principal [82] unmoved movers" to be at one
time and then not to be. Still, this cannot be true of all of them. Clearly,
there must be something [eternal] to account [83] for self-movers which
would for a while be and then be no longer; for every *self-mover* must
have magnitude, since nothing which is without parts can be moved,
even though it is not necessary for every *mover* to have magnitude. The
continuous going on of generation and destruction, then, cannot be explained [83] by any single unmoved being which is transitory; nor can
it be explained by a succession of transitory beings bringing different
things to pass. Neither any one of them taken singly nor all of them 30
taken together can provide foundations [83] for anything eternal and
continuous: the movers must be in eternal and necessary relation [33]
to the world's eventfulness; whereas they are [supposedly] indefinite
[130] in number [150] and are not totally simultaneous beings. Clearly,
then, even if some unmoved movers and many self-movers perish and 259a
are replaced by others a thousand times and even if one unmoved
being moves another being and the latter moves still another, there
is something nevertheless which encompasses all of them, which is
distinct [74] from any of them, and to which some things owe [83] their
being and others their nonbeing and on which change depends for its
continuity; and whereas these movers induce [83] movement in other
beings, their own movements are evoked by it.

Since movement is eternal, the first mover must also be eternal if
there is but one such mover; and if there is more than one prime
mover, there must be a plurality of eternal movers. However, one is
preferable to many† and a finite number is preferable to an infinite

---

* viii.1, 2.252b17-28, 253a7-21.

† *Metaphysics* xii.10.1076a4.

10 number of them.‡ For a simpler [131a] assumption [65] throws more
light upon the same facts [3b], since it is more fitting for the determinate [131a] and the better to prevail [82f] in the natural course of
things whenever possible [12b]. Then, too, if one being is the first of
things unmoved and is eternal, that one suffices to be related to other
beings as the beginning of their movement.

We have further [4a] evidence [59d] for the unity and eternity of
a first mover. We have shown that movement must be eternal; and it
must therefore be continuous, for the eternal is continuous whereas the
successive is not. But if movement is to be continuous, it must have
unity; and to have unity, it must have a unitary mover and a unitary
subject. For a movement having one mover after another would not,
20 taken as a whole, be continuous but would be successive.

The conviction concerning a first mover may be confirmed by a re-
examination of the principles pertaining to movers. Some beings are
evidently sometimes in movement and sometimes in a state of rest.
From this, it has become clear to us that neither are all things transient,
nor are they all inert, nor are some always transient and the others
always inert. The test [63] of these alternatives is at hand in the things
which have the twofold power of being in movement and being at
rest. But this kind of beings is obvious to all. Hence, we want to show
that each of two other kinds has a distinct nature: some things are
always independent of movement; and some things are always subject
30 to movement. In proceeding to this point, we have found [64] that
everything that is in movement is acted upon by an agent, that this
agent is either unmoved or moved, and that every moved mover in a
series is acted upon either by itself or by something else; and thus we
arrived at the conclusion [65] that movements are started by self-movers
259b among the things moved but, taken in their universality, by an unmoved mover. Now, we see beings which are evidently of the kind
that move themselves; namely, living beings, especially animals. These
have given occasion for the opinion that movement may arise in things
where, before, there was no movement at all; because we seem to see
such beings pass from an unmoved state of being to a state of being
in movement. But what we need to grasp is that they move themselves
in one way only and, strictly speaking, not even in that. The crucial
factor is not in themselves. On the contrary, there are natural movements in animals, such as growth and decline and respiration, which
10 they do not produce of themselves but which they undergo even in a
state of rest when they are not moving themselves. The crucial factor
is their environment. They draw upon the environment for many things,

‡ i.6.189a15.

such as food; and the nourishment they take in accounts for some of their movements. Thus, they sleep while their food is being digested; when it is being distributed, they awake and move themselves. What primarily starts their movements, therefore, is external to them. For this reason, too, they do not move themselves continuously. It is something else that moves them; and this itself undergoes movement and change as it comes into relation with any being that moves itself. The [so-called] "prime mover or factor" in all these self-moving beings is also "self-moved," but only incidentally; for it is the body which changes its place, with the consequence that the being [supposedly] acting on the body like a lever also moves itself [incidentally].   20

Hence we may be sure that anything unmoved which incidentally moves itself cannot give rise to continuous movement; so that, since there must be movement continuously, the first mover must be immovable even incidentally. This must be so if in the world of beings there is to be, as we have said, unceasing and undying movement; and also if being itself is to continue in the same status of being. For if the first principle persists, the "All" will also persist, since the "All" is continuous with its first principle. As for anything that is incidentally moved by something else, this is different from any being incidentally moved by itself: only perishable things are incidentally moved by themselves; whereas being incidentally moved by something else belongs even to   30 some principles of celestial bodies, namely, of such as have complex orbits.

If there is an eternal mover which is eternally unmoved, then what is directly moved by it must also be eternal. This clearly follows also   260a from the fact that, without a moved mover, there would be no generation or destruction or transformation of other things. For an unmoved mover acts forever in a uniform way so as to evoke a single type of movement; its relation to its object is not subject to change. Not so with a [celestial] body which is moved by a moved mover [namely, the sphere of fixed stars], the latter being moved directly by the unmoved mover. Such a body [for example, the sun] varies in its relations to the things [188c] it influences. It therefore does not institute [83] a process which is ever the same. Being in opposite [50] positions [132] and states [20] at different times, it will affect any of   10 the things which it controls in opposite ways; so that they will sometimes be at rest and sometimes be in movement.

What we have said also answers our previous question: why is it that all things are not transient or all inert or some always transient and the others always inert, but some things are sometimes in movement and sometimes in a state of rest? The reason [83] for this is now clear.

Whereas some things are always subject to movement because they are moved by an eternal unmoved mover, other things are necessarily subject to change [or alternating movement and rest] because they are moved by a mover subject to movement and change. But, as has been said, the unmoved mover, abiding in a simple, unvarying, and identical condition, induces a single simple movement.

7. Local Motion as Primary and Continuous

20 Let us now make a fresh start, in order to shed further light on these considerations. Can any movement be continuous; and if so, which? And which kind of movement is primary [or most inclusive]? Clearly, if there must always be movement and a certain kind must be primary and continuous, this is the kind which the first mover induces; and this kind of movement is necessarily single, continuous, and primary.

Now, of the three kinds of movement, quantitative [142a], qualitative [35a], and local [132], the primary one must be local motion [121].
30 To begin with, increase presupposes [82g] qualitative alteration. A subject is increased, in one way, by its like; in another way, by its unlike. We say that a contrary is food to a contrary; but it is assimilated by becoming like to like, and this transformation to a contrary state must
260b be a qualitative alteration. And if anything is altered, there must be something which alters it; for example, something which brings about the passage from its potential to its actual hot condition. So it is clear that the mover does not have a constant relation to the thing altered, but is sometimes closer to it and sometimes farther away from it. And these interrelations cannot occur [82f] without local motion. Hence, if there must always be movement, then there must always be, first among movements, local motion; and if there is a primary and a nonprimary [18] kind of local motion, then the primary kind of local motion must be always.

Then, too, all modifications [35b] are rooted [82] in condensation and
10 rarefaction: heavy and light, soft and hard, hot and cold,° seem to be like [4] density and rarity. These, in turn, are a combination and a separation, respectively; and it is said that the generation and destruction of primary beings [26] are based [39a] on combination and separation.† But in being combined and separated, things must change from one place to another.‡ Moreover, also in their increase or diminution, extended things must undergo change in place.

° *On Generation and Corruption* ii.2.329b24-30.

† Anaximander, Anaxagoras, Empedocles, Democritus.

‡ viii.4.255a34; *Metaphysics* ix.5.

Further reflection discloses still another indication of the primacy of local motion. In movement as in other affairs, "first" has more than one meaning.§ That is "prior" on whose being that of other things depends but whose being does not depend on theirs; or which is prior in time; or which is prior in full being [26].

First, there must be movement continuously, and continuous move- 20 ment would have greater continuity of being than would successive movement; the former, too, would be better than the latter, and we invariably assume the better to be present [82f] in nature if possible; motion, moreover, can be continuous (as we shall prove later‖ and now take for granted), but only local motion can be such. Consequently, local motion must be primary. For what is subject to local motion is not necessarily subject to quantitative or to qualitative change or to generation or destruction; but none of these changes is possible without that continuous motion which the first mover induces.

Again, local motion is primary in time, since it is the only move- ment to which eternal bodies can be subject.¶ To be sure, local motion 30 is the last movement any living individual comes by; for after birth, living individuals first change in quality and grow before they achieve the competence [100d] requisite to moving locally. But prior to them 261a there must be something else, having power of local motion, to bring them into being; there must be a parent which, in producing its off-spring, is not itself being generated. Although generation may therefore be held to come first of any movements because the thing [subject to movement] must be first generated, this is true only of something generated taken individually. Generated things must be preceded by something else which is subject to movement and which, instead of having to be generated, is a being; as this being must in turn have something prior to it.°° Generation, then, cannot be primary, since all things subject to movement would in that case be perishable; and clearly, therefore, none of the subsequent movements (increase, alteration, dim- 10 inution, destruction) can be prior to local motion. All of these come after generation; so that, even if generation is not prior to local motion, neither can any of the other changes be prior to local motion.

In general, what is being generated is still undeveloped [100f] and only on its way [114g] to its prime [82]; so that what is later in the order of genesis is prior in the order of nature. Thus, in the order of generated beings, the movement which appears [82f] at the final stage of their development is local motion. That is why some living beings,

§ *Metaphysics* v.11.
‖ viii.8.
¶ ii.4.196a24-b5, 6.198a5-13.
°° ii.2.194b13.

such as plants and many kinds of animals not provided with the requisite organs, are motionless; whereas motion is in the power [82f] of animals free from such imperfection. If local motion, then, characterizes [82f] animals in the degree in which they have attained to their most fully developed nature, local motion must be prior to all other

20 forms of movement in the order of complete being [26]. Another reason is that any being deviates less from its normal state [26] in local motion than in any other kind of movement; in this movement alone its being does not become transformed in the sense in which a trait it possesses becomes transformed in qualitative alteration or its size in an increase or diminution.

Above all, local motion is clearly the chief [55b] kind of movement belonging to anything that moves itself. And among all the things that initiate and undergo movement, we single out [46] the self-mover as the foremost [17a] agent [82] acting upon the things that are in movement.

Having shown that local motion is the primary kind of movement, we must now proceed to determine which is the primary form of local motion. This analysis [198] will also bear out [59d] our present and

30 previous†† assumption that it is possible for some motion to be continuous and eternal.

It is apparent, in the light of certain facts, that no movement other than local motion can be continuous. Every other movement or change proceeds from an opposite to an opposite: in generation and destruction, the termini [72b] are being and nonbeing; in qualitative alteration, contrary qualities [35a]; in increase and diminution, a large and a small or a complete and a deficient size. Moreover, contrary changes

261b have contrary termini. But a being which is not always in motion must for some time previous to that motion be at rest. Obviously, then, what is changing [from one motion to a contrary motion] must [for some time] be at one of the contraries as at a resting-place. So, too, with opposite kinds of change: destruction and generation are mutually opposed as such [105]; and a particular [40] destruction and a particular generation are mutually opposed. Therefore, if it is impossible for anything to be transformed in opposite ways at the same time, the change will not be continuous, but a period of time must intervene between the opposite changes. Whether these opposite changes (generation and destruction) are contraries or not, makes no difference; all that matters to the argument is that it be impossible for the same thing to be

10 possessed of both at the same time. Neither is it important [163] for the argument if a thing need not be in a state of rest in one of the

†† viii.7.260b23, 3.253a29.

opposite states; nor if a change does not have as its contrary a state of rest. Undoubtedly, what is-not at all is not in a state of rest; and destruction has its terminus in nonbeing. What is important is the time-interval; for if a period of time intervenes [between opposite changes], the change cannot be continuous. So the significant [163] point in the case of the motions was not their contrariety but the impossibility of both occurring at the same time. We need not be disturbed by the consequence that one thing, motion, has more than one contrary: rest as well as motion in the contrary direction. All we need to grasp is that a motion is in some sense opposed to both (just as the equal or a measure is opposed both to the greater and to the less) 20 and that opposite motions or changes cannot belong to the same thing at the same time. Moreover, it would be a strange view of generation and destruction, or of things in general, if what has been generated must at once be in process of being destroyed without enduring for any length of time at all. We may apply this conviction to other kinds of change as well, since it would be natural for all of them to follow the same [57c] pattern [33].

8. Circular Motion as Continuous and Infinite

Let us now explain [36] which kind of motion, single and continuous, can be infinite; namely, circular motion.* Every mobile being moves in a circle or in a straight line or in a composite figure; and therefore, if either of the simple motions is not continuous, the com- 30 posite motion cannot be continuous either. Clearly, anything which moves in a finite straight line cannot move continuously: it must turn back; and the motions of what turns back in a straight line are contrary, whether up and down or forward and backward or left and right (for these are the pairs of contraries pertaining to place). We have already† defined a single and continuous movement as the move- 262a ment of a single subject in a single period of time and in a specifically single respect. We had to take into account all three: the subject, such as a man or a god; the time, such as the duration of the movement; and also the respect, that is, the place or quality or form or magnitude. But contraries, so far from being single, differ specifically; in particular, those which we have named differ locally. Thus, a motion from A to B is contrary to a motion from B to A, as is shown [38a] by the fact that, if they were to occur simultaneously, they would check and stop each other. It would be the same with motion on a circle: a motion from A to B is contrary to a motion from A to C; they would 10

* viii.2.252b7-12, 28-253a2.
† v.4.

check each other even if they were to go on continuously without having to turn back, since contraries are mutually destructive or at least obstructive. (Fig. 7.) A lateral motion, on the other hand, is not contrary to an upward motion.

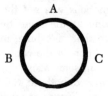

Fig. 7.

That motion in a straight line cannot be continuous, is especially apparent [59d] from the fact that turning back requires a pause. This is true regardless of whether the motion is along a straight line or a circle; for rotation is not the same as circular motion which includes, besides the continuous sort, that which turns back when it comes to its starting-point. The need of a pause in turning back may be verified [183] by sense perception as well as by argument. We may begin
20 the argument with the distinction between three points: the beginning, the middle, and the end; where the middle is a beginning relatively to the end and an end relatively to the beginning and thus, though numerically one, by definition two. We may also make use of the distinction between being potentially and being actually. In the straight line, then, any point between the extremes is a middle potentially; but it is not a middle actually unless a thing in motion divides the line at that point by stopping there and resuming its motion, and then the middle becomes the beginning of the later motion and the end of the earlier motion (as when A in its motion stops at B and resumes its motion toward C). As long as the motion is continuous, however,
30 A cannot have come to be or ceased to be at B; it is there only at a moment, not for any time—except that it is there within the period of time in which the moment constitutes a division. If someone were to insist [64] that A had come to be and ceased to be at B, then the moving thing A would always be coming to a stop: for it is impossible
262b for A to have at the same time come to be and ceased to be at B, so that it must have done so at different moments; and there would therefore be an interval of time, so that A would be in a state of rest at B and likewise at all the other points (since the same reasoning would apply to all). Thus, when A uses the middle B as both an end and a beginning, it must pause there because it treats B as two (just as

anyone might do in thinking about it). But a body which has finished
its motion and come to rest has ceased to be at the starting-point A
and come to be at C.

We are thus in a position to reply to a difficulty which may be
raised in this connection. Take [a distance] E to be equal to [a dis- 10
tance] F; take A to be in continuous motion from the starting-point of
E to C; and at the same time when A is at point B, take D to be in
uniform motion from the starting-point of F to G with the same velocity
as A. Then D would arrive at G before A arrives at C, for the reason
that "the first to start must be the first to arrive." (Fig. 8.) A would

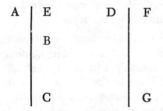

Fig. 8.

lag behind because A would not have simultaneously come and gone
at point B; had A come and gone simultaneously, A would not have
been overtaken; to be overtaken, A would have to linger. But then we
ought not to make the assumption that, when A was at B, D was
moving from the starting-point of F! For if A has "come to" B, A will 20
also have "gone from" B; and the process of going is not simultaneous
with that of coming. But A was at B, not during a period of time, but
at a [momentary] "section" of time! It is impossible, then, for us to
use the terms cited in a description of continuous motion. On the other
hand, it is necessary for us to take recourse to those terms in a description of the kind of motion which involves turning back. Take G
moving up to D and then moving back down. G thus uses the single
extreme point D in two ways, both as an end and as a beginning; and
G must therefore pause at D. But G cannot have simultaneously come
and gone at D, since it would then both be and not be at D at the
same moment. We cannot apply to this case the solution of the former
problem: we cannot say that G is at D "at a moment" and has not 30
been "coming to" D or "going from" D; for D must in this case be an
actual goal, not a merely potential one. Although the intermediate
points in the motion have potential being, the being of the terminus is
actual: it is an end from below and a beginning from above; and so 263a
it is the end of the upward motion and the beginning of the down-

ward motion. What turns back in a straight line, then, must pause; so that there cannot be any continuous eternal motion [to and fro] in a straight line.

This analysis [55c] also provides us with an answer to the objection drawn from Zeno's [first] argument.‡ Shall we suppose [46c] that half [of a distance] must be traversed [before the whole]; that the series of half-distances is infinite; and that it is impossible to traverse an infinite number of distances? Or stating the question in another form, shall we suppose that one ought to be able to count the halfway points as a body in motion comes to them; so that, when the body 10 has traversed the whole distance, one would have counted an infinite number—which is admittedly impossible?

In our former accounts of motion, we solved this problem in terms of time being infinitely divisible: there is nothing strange in traversing an infinitely divisible distance in an infinitely divisible time; the infinity inhering [82f] in the distance accords [57c] with that inhering in the time. This solution was a sufficient answer to the questioner who asked: Can an infinite number of things be traversed or counted in a finite time? But this solution does not state adequately the true [7a] state of affairs [188c]. Suppose one were to leave out the reference to distance in the question whether an infinite distance 20 can be covered in a finite period of time and one were, instead, to ask the question about the time itself, which is infinitely divisible. The solution given would then no longer be adequate; instead, the true one would have to be presented, which we have set forth in the foregoing arguments. One who divides a continuum into two halves treats a single point as two by taking it as both a beginning and an end. This is done no less by one who counts the halves than by one who bisects a line. Neither a line nor a motion, when thus divided, is continuous: a continuous motion is a motion over a continuum; and although the latter contains an infinite number of halves, it contains them potentially only, not actually. Taking the halves as actual would lead to 30 getting a motion which is not continuous, but which is arrested. This would also be the obvious result of counting the halves: a single point 263b would have to be counted as two, as the end of one half and the beginning of another; that is, if what is being counted is not a single continuous whole but is two halves. Hence, the question whether an infinite number of things can be traversed either in time or in distance, must be answered by saying: in one sense, yes; in another sense, no! If there is an actually infinite number [of moments or points], it is not possible [to pass through them]; if they are potentially only, it is possi-

‡ vi.2.233a21-31, 9.239b11-14.

ble [to pass through them]. Anything which is in continuous motion has crossed an infinite number of points incidentally, but not directly [105]; for a line can incidentally be bisected without limit, but its primary way [26] of being [23] is altogether different [16].

As for a moment or temporal "point" which keeps "before" and "after" 10 apart, any state of affairs [188c] has in any "moment" only such traits as it has afterwards; for, clearly, if we were not to regard a moment in this way, then a given state of affairs would at once both be and not be as it is and would not be so at the very time when it has come into being. True, a moment common to both the earlier and the later time is numerically single; but it is not one and the same in its definition as the end of the earlier and the beginning of the later time. And in its relation to a state of affairs, a moment always belongs to a situation in its later phase [35a]. Take a time *ACB* and an object *D* which is white at time *A* and not white at time *B;* at time *C,* then, *D* is both white and not white. For if *D* was white during the whole of time *A,* it is white in any part of *A; D* is not white in any part of *B;* and *C* is in both *A* and *B.* But in that case we ought not to agree to *D* being 20 white during the whole of *A;* we ought rather to exclude its last moment *C,* which already belongs to the later stage. If in the whole of time *A* the "nonwhite" state was coming into being and the "white" state was passing out of being, then at time *C* the "nonwhite" state has completely come into being and the "white" state has completely passed out of being. Consequently, *C* is the first moment at which *D* may truly be said to be ("white" or) "not white" [as the case may be]. If not, then when *D* has become white, it is not white, and when it has ceased to be white, it is white; or else *D* would have to be both white and not white at the same time and would indeed both have to be and not to be at all. But if anything, after not being [1b], has come [116] to be [1] and, while it is still becoming, still is-not, the time cannot be divided into atomic times! For if *D* was becoming white in time *A* and if *D* both has become and is white in another "atomic" but "contiguous" 30 time *B,* that is, if *D* was becoming white but was not white in time *A* but is white in time *B,* then there must have been between *A* and *B* a productive process [116a] and therefore also a time in which it hap- 264a pened [116]. This argument is not addressed to those who deny atomic times. For them, a state of affairs both has come into being and is at the last moment of the very time in which it was coming into being; and this moment has no other "contiguous" or "successive" to it, whereas "atomic times" are [presumably] "successive." Evidently, too, if *D* was becoming white in the whole of time *A,* the [supposed] "time" in which it both has become and was becoming such is no greater than the whole of the time in which it was becoming such.

Now, we have presented various concrete [55] arguments to convince anyone [that circular motion only can be continuous]; but the same conclusion follows also from a number of arguments based on dialectical [36] considerations [195a]. Anything which is in continuous 10 motion must, before arriving at point B, have been moving towards that point, provided that nothing deflects the moving body from its path; and it must have been moving towards B not only when approaching B but from the very beginning of its motion, for why should it have been doing so at one moment rather than at another? Such is the case also in types of movement other than local motion. But if anything were in continuous motion from A to C and back to A, it would [on the principle stated], while still moving from A to C, be moving also from C to A; in other words, it would simultaneously have contrary motions (since motions to and fro in a straight line are contraries). At the same time, the thing would be moving from C, where it is not; but 20 since this is impossible, it must come to a halt at C. Accordingly, its motion is not a single one; for no motion is single which is broken by a pause.

That this finding applies to every kind of movement, can be shown from more general considerations. We have enumerated§ all the kinds of movement to which things can be subject; and these kinds of movement have for their opposites corresponding kinds of rest. Now, what is not always subject to a given distinct kind of movement (I do not mean a given "particular" movement) must be subject to a prior opposing rest (which is the "privation" of the movement). Moreover, since motions to and fro in a straight line are contraries, and nothing can simultaneously have two such contrary motions as one 30 from A to C and one from C to A, these motions are not simultaneous; but relatively to the former, the latter motion is future. All this being so, the latter motion must be preceded by a rest at C, which we have found‖ to be the state of rest opposed to the motion from C. Clearly, 264b then, the motion [from A to C and from C to A] is not continuous. There is another argument which is even more intimately related [55] to the problem in hand. Take something as having simultaneously ceased to be nonwhite and become white. If the thing is then being altered to and fro continuously without remaining white for some time, it would have been subject to no less than three changes at the same time: having ceased to be nonwhite, having become white, and having become nonwhite. Besides, although the time [of movement to and fro] is continuous, it does not follow that the movement is therefore

§ v.2.
‖ v.6.229b28-230a7.

continuous; rather does it form a sequence [136b].¶ How can two such contraries as white and black have the same limit [18a]? [The last stage in which anything is white cannot be identical with the first stage in which it is black.]

Circular motion, however, can have unity and continuity without any 10 impossible consequence. In circular motion away from *A*, a body is forthwith in position for motion towards *A*, since it is then in motion towards the point at which it will arrive. It does not simultaneously execute contrary or opposite motions, for not every motion to a point is contrary or opposite to a motion from that point. Such motions are contrary if they are motions along a straight line, because the latter has points that are contrary in place (for example, the points at the ends of a diameter are at the greatest possible distance from each other); and such motions are opposites if they are motions along the same line. Nothing, therefore, prevents circular motion from being continuous and from going on incessantly: circular motion proceeds from a point back to the point itself, whereas motion along a straight line proceeds from the original point to a different one; and circular 20 motion does not, like motion along a straight line, proceed repeatedly between the same termini. A motion can have continuity, then, although it is ever taking place in different positions; but motions within the same limits cannot have continuity, since a moving thing would then be moving in opposite directions at once. Neither, therefore, can there be continuous motion in a semicircle or in any other arc, where a moving body has to pass over the same ground and through the same contrary changes again and again. Such motions do not join their end and beginning, as circular motion does; and the latter is therefore exclusively "finished" [100].

This distinction also shows that the other kinds of movement cannot 30 be continuous. In all of them, the same passage has to be made over and over: in qualitative alteration, through the intermediate qualities; in quantitative change, through the intermediate sizes; and similarly in generation and destruction. Whether we take the intermediate stages of a change to be few or many or whether we add any to them 265a or subtract any from them, repeated changes in any case pass through the same stages. Those physicists** are clearly in error, then, who declare all sensible things to be in constant motion: the kind of process through which these things pass must be one of those we have distinguished, particularly (according to their descriptions) qualitative alteration; for they say that things are in constant flux and decay and

---

¶ v.3.226a34, 227a10.
** Heracliteans.

even that generation and destruction is a qualitative alteration. But our argument has been general, extending to all movement and showing that no movement can be continuous except circular motion; neither,
10 then, can qualitative or quantitative change.

Let this suffice to prove that no change can be infinite or continuous, with the exception of circular motion.

9. Circular Motion as Primary; Recapitulation

Clearly, circular motion is the primary form of local motion. We have divided° local motion into the circular, rectilinear, and composite. But the simple kinds must have priority over the composite, which is made up of them; and the circular must have priority over the rectilinear, than which it is the simpler and more complete. Motion along an infinite straight line is impossible, since nothing is infinite in this way;†
and if there were an infinite of this sort, nothing could span it (since
20 the impossible does not happen, and it is impossible for the infinite to be spanned). And motion to and fro along a finite line is a compound of two motions, whereas a single motion along a finite line is incomplete and soon at an end. But the complete is prior to the incomplete, and the imperishable is prior to the perishable, in the order of nature and of definition and of time. Then, too, a motion which can be eternal has priority over one which cannot. And circular motion can be eternal, as no other motion or change can; in the latter, pauses must occur at which movement ceases.

There are good reasons [60c] for the fact [3b] that circular motion presents a contrast to the rectilinear in having unity and continuity.
30 A straight line contains a definite beginning, middle, and end; it thus has points at which a moving body starts and stops and at both of which that body is therefore in a state of rest. A circle, however, does not have such definite points: why should one point on a circle be a limit rather than another? Any point on a circle may serve as beginning or middle or end; with the consequence that a body in circular motion
265b is in some sense always and never at a beginning and at an end. Hence, a [revolving] sphere is in a sense both in motion and at rest, since it remains at the same place. The reason [83] for this is that the center is at once the primary, intermediate, and ultimate principle [of the circle or the sphere]; so that, since the center is not on the circumference, the body which is always moving round the middle, instead of to an end, has no end-point at which it can come to rest and, since the center is

---

° viii.8.261b28.
† iii.5, 8.

stationary, the whole is always somehow both at rest as well as in continuous motion. Moreover, because the measure of all other movements is rotation,‡ this must be the primary movement, for all things are 10 measured by what is primary; and conversely, because rotation is primary, it is the measure of all other movements. Then, again, circular motion alone can be uniform [in velocity]. In rectilinear motion, things do not move with uniform speed from their starting-point and towards their finishing-point; the greater the distance they put between themselves and their place of rest, the faster they move. Circular motion alone is such [101c] as to have neither its beginning nor its end within itself; but [the center which functions in both of these ways is] outside [the circumference].

All the thinkers who have given [34] attention [167] to movement uphold [199] the primacy of local motion inasmuch as they credit [46d] the things producing local motion with being the originating principles of movement. Thus, "love" and "strife" are said to "move" things 20 by combining and separating them, respectively; and combination and separation are forms of local motion. Thus, too, Anaxagoras attributes to "mind" as the "first mover" the function of separating things. Likewise, those who do not explicitly recognize [36] a moving factor [83ₗ], but have provided for motion by having recourse to "the void," ascribe local motion to nature, since this is the kind of movement which goes on in the "void" as in a place; other kinds of movement, they say, belong not to first beings but to beings which are made up of them and which increase, decline, and are altered in accordance with the combinations and separations of their constituent atoms. In much the 30 same way, those who base [149c] generation and destruction on density and rarity reduce [149a] them to combination and separation. In addition, those who regard [34] the soul as the moving factor [83] designate [46] "what moves itself" as the source of movement; but the self-induced movement of living things is local motion. Above all, in 266a speaking of things "in motion," we refer principally to things leaving one place for another; if they grow or decline or are altered while they remain in their place, we do not say that they are "in motion" but specify the kind of process through which they are passing.

To sum up, we have stated that there always was and always will be movement, what is the first principle of eternal movement, what sort of movement is primary, which kind of movement can alone be eternal, and that the first mover is immovable.

---

‡ iv.14.223b18.

10. The First Mover without Parts or Magnitude

10   Let us now proceed to show that the first mover must be without parts and without magnitude. To this end, let us first put forward [72e] certain premises [17]. One premise is that nothing finite can produce movement for an infinite time. For, besides the mover and the thing moved, we must here take into account the time of the movement; and these are either all infinite or all finite, or else one or two of them are infinite and the remaining two or one finite. Now, if a [finite] mover A moves a [finite] body B in an infinite time C, then the time F in which a part D moves a part E cannot be equal to C; for a longer time is required to move the larger body B. Consequently, time F is not infinite. But whereas, by adding to D [parts equal to D] and to E

20   [parts equal to E], we would exhaust A and B, the infinite time C cannot be exhausted by subtraction of [finite] times equal to F. Consequently, the time C in which A [presumably] moves B is finite. It is impossible, then, for a finite mover to move anything for an infinite time.°

Having proved [59d] that a finite body cannot move anything for an infinite time, we may prove [59a] the more general [44] proposition that a finite body [142a] cannot have infinite power. Let a greater power be one which produces an equal effect in less time than does another; for example, in heating or sweetening or throwing or, to put it generally, acting upon anything. Then a finite agent having infinite

30   power would have to act upon its object [35] to a greater extent than would another agent, since infinite power is greater than any other; but there could be no time in which it would so act. Take A as the time in which an agent having infinite power heated or pushed anything, and AB as the time required by [an agent having] finite power;

266b   then, by continually adding to the latter a greater finite power, we would sooner or later come to something which has completed the act [109a] in time A. For constant addition to a finite magnitude would result in a magnitude exceeding a determinate limit [72], just as constant subtraction would result in one falling short of a determinate limit. Hence, an agent with finite power would act in the same time as the agent with infinite power. But this is impossible. Consequently, a finite agent cannot have infinite power.

° Thomas Aquinas (against Avicenna): ". . . non semper per ablationem intelligenda est solutio continuitatis, quam impossibile est esse in corpore caelesti; sed ablatio intelligi potest secundum quamcumque designationem. . . . Apparet autem ex processu Aristotelis, quod hic loquitur de tempore motus, secundum quod tempus motus accipitur secundum partes mobilis, et non secundum quod accipitur secundum partes magnitudinis. . . ."

So, too, an infinite body cannot have a finite power. To be sure, a smaller body may have a greater power; but then a larger body may have a still greater power. Let AB be a finite body; and let [a part] BC move a body D in time EF. Then a part twice as great as BC 10 would move D in a time FG which, on the assumption of an inverse proportion, would be one half of the time EF. In this procedure, we would never exhaust $AB;$ but we would be always getting a smaller period of time. Hence, the power must be infinite, since it exceeds any finite power. Every finite power, however, must operate in a finite period of time; for if it moves anything in a certain period of time, a greater power will do so in a shorter period of time (according to an inverse proportion) but in a period of time which is still determinate. But a power (no less than a number or a magnitude) which ex- 20 ceeds any that is determinate must be infinite. We may also make use of another proof [63]. Take an infinite body having a [finite] power: this power would be of the same kind [19] as the power of a finite body; and the latter power would be a measure [that is, a proper fraction] of the power of an infinite body!

We have thus shown that a finite body cannot have an infinite power and that an infinite body cannot have a finite power. However, we would do well by taking up at this point a problem connected with local motion. If everything that is in motion (other than things that move themselves) is moved by an [external] agent, how is it that some things (such as projectiles) can be in continuous motion after they 30 have ceased to have contact with their mover? Suppose that what moves them also moves a medium such as air, which would then be a moved mover: it would be equally impossible for the air to continue in motion when the first mover no longer has contact with it; all the things moved would have to be in motion simultaneously, and they would have to cease to be in motion when the first mover ceases to 267a move them even if it, like a magnet, gives to what it has moved the power of moving something else. We must admit that the first mover brings into operation [34] this power [5b] of air or of water or of some other medium which is capable [101c] of being moved as well as of moving something else. But the medium does not, when it ceases to be moved, at the same time stop moving something else. Although it ceases to "be moved" as soon as its mover stops moving it, still, it continues to function as a mover: it moves something else that is con- secutive to it; and the latter may be said to do likewise. The power to move something else wears away gradually to the extent that it decreases at each successive stage. It comes to an end when a mover 10 no longer conveys the power of moving something else but only that of being moved. Moreover, the last of the movers and of the things moved must come to a stop together; and when this happens, the whole motion has run its course. This motion, accordingly, belongs inherently to things capable of both motion and rest. Also, although this motion seems continuous, it is not really so; it is rather a motion of things that are either successive or contiguous, and it has not one mover but a number of consecutive movers.† That is why motion of this kind occurs in such media as air and water. Some call this process a mutual replacement.‡ However, our own account alone can solve the difficulty in question. In mutual replacement, all the movers and things moved must function simultaneously and must cease to function simultaneously; whereas

20 what concerns us here is the apparent continuous motion of a single projectile. Since the same mover cannot continue moving the projectile, the question is: by what agency, then, is it kept in motion?

[To return to the main argument.] There must be, in the world of beings, a movement which is continuous and therefore has unity; so that it must not only be the movement of some single magnitude (without which there would be no movement at all), but it must also have, besides a single subject, a single agent (since the movement would otherwise not be continuous, but would be a succession of separate movements). Therefore, if the mover is single, it must be either a moved or an unmoved mover. As a moved being, it would have to go hand in hand with what it moves so as to be itself subject to change;

267b it would therefore itself have to be moved by an agent. Consequently, the series, to have stability, must be anchored in an unmoved mover. Such a mover need not be changed with what it changes; it can always and without effort induce movement. And the movement it induces is the only or the most uniform movement for the reason that the unmoved mover is not subject to any change. So, too, if the movement induced is to be uniform, the object in motion must not be subject to any change in its relation to its mover. The mover must therefore be either at the center§ or at the circumference of the sphere [of the fixed stars], since the center and the circumference are the originative principles of the sphere. However, since the fastest motion occurs nearest the mover and is the motion of the circumference, therefore the mover is at the circumference.

10 On the other hand, it is questionable whether something that is in motion can induce motion continuously, instead of inducing motion only successively as in successive pushing. Such a mover would have to be continually pushing or continually pulling or continually pushing and

---

† viii.6.259a18.
‡ Plato *Timaeus* 79B-80C.
§ Pythagoreans.

pulling; or else the power of moving would have to be transferred from one mover to another as in our illustration of the projectile, where motion is transmitted from one part to another of a divisible medium like air. In either case, however, the motion cannot be one; it can be successive only. It would follow, then, that only an unmoved mover, which is invariable and in an invariable and continuous relation to its correlative, can induce a motion that is continuous.

This analysis [72e] shows that the unmoved first mover cannot have any magnitude. If it had, its magnitude would have to be either finite 20 or infinite. We have proved in an earlier passage of our *Physics*‖ that there can be no infinite magnitude and have now proved that nothing finite can have an infinite power and that nothing can be moved by a finite agent for an infinite time. Since the first mover, however, induces eternal movement and therefore does so for an infinite time, it evidently follows that the first mover is indivisible and therefore without parts or magnitude.

‖ iii.5.

ANALYTICAL INDEX OF
TECHNICAL TERMS

I. BEING

1. ὄν, *(id) quod est, ens, quid est, res, existens, esse,* being, be, thing: as one, i.2, 3; viii.5.257b3; as immovable, unmoved, movable, i.2.184b26; viii.1.251b32, 33, 3.254a25, 5.256b24, 258a1, 6, 6.259a3, 13, 10.267a25; meaning of (*see also* 2–11), i.2.185a21, 3.186a25, 33,35, b1, 4, 5, 187a8; essential attributes of (*see* 12–24); as infinite, limited, i.2.185a33, b17; viii.3.254a25; itself, i.3.187a7, 8, 8.191a33; "full," i.5.188a22; knowledge of, i.6.189a13; genera of, i.6.189b24; as primary being, i.7.191a12, 13; iii.4.203a5; and becoming, destruction, i.8; vi.10.241a29; viii.7.261a7, 34; divine, i.9.192a16; blind, ii.1.193a7; not separate, ii.1.193a5; essentially, acci-dentally, ii.5.196b24; categories of (*see also* 4–6, 25–35), iii.1.200b28, 201a3, 9; in itself, iii.5.205b7; and place, iv.1.209a24, 2.209b32, 5.212b29, 10.218b12; same, iv.6.213a18, 19; as bodily, iv.6.213a29, b2, 7.213b32; vi.2.232a33; viii.6.259b19; heretofore, in time, process, iv.10.218a14, 12.221a24, 26, 27, b9, 29; vi.8.239b1; and change, v.1.224b9, 10, 225a22, 6.230a9, 11, 13; vi.5.235b14, 15, 9.240a26; continuous, vi.6.237b15; independent, viii.5.257a30; and rest, viii.7.261b1; and rest, viii.7.261b1; and not be, viii.8.263b11, 25, 26. *See also* 4c–d, 9b, 10a, 11c

1a. ὄντα, *quae sunt, entia, res, existentia,* beings: number of, i.2.184b22, 3.186a5; elements of, i.2.184b23; iii.5.204a16; many, i.2.185a29, b32, 3.187a10; come from, i.4.187a33, 36; infinite, i.4.188a4; iii.4.203a12; viii.8.263b5; natural functioning of, i.5.188a32; and simple, i.5.188b9, 7.190b2; nature of, i.6.189a27, 8.191a25; ii.1.193-a23; primary being of, i.6.189a29; v.2.225b11; principles of, i.6.189b13, 7.190b18; iii.4.203a3; natural, others, i.7.190b18; ii.1.192b8, 193a4, 10, 7.198a29; becoming, passing away, i.8.191a27; not separable, ii.2.194a1; differences of, ii.3.195a2; always, usually, by chance, ii.5.196b17, 197a19; potential, actual, iii.2.201b29; viii.8.263b5; magnitudes, iii.7.207b34; and place, iv.1.208a29, b23, 32, 209a19, 21, 22; and void, iv.6.123b14; and time, always, iv.10.217b31, 12.221a28, b4, 7, 222a7; indivisible, vi.1.232a19; and movement, rest, viii.1.250b14, 25, 251a20, 23, 2.253a7, viii.3 (inert, transient), 5.256b9, 6.259a22, b1, 5, 25, 10.267a14, 21; and prime mover, viii.6.-258b28, 31

1b. μὴ ὄν, *quod non est, non ens, nihil, quidquid non est, non esse,* what is-not, nonbeing, nonexisting: and what is, i.3.186b6, 9, 10, 187a2-6; viii.8.263b25, 26; come from, i.4.187a28, 33, 34, i.8, 9.191b36; vi.5.235b14; privation, i.8.191b15; great and small, i.9.192a7; and movement, change, generation, destruction, iii.2.201b20; v.1.224b8, 10, 225a15, 20-32, 6.230a11, 12, 17; vi.5.235b14, 15, 9.240a27, 10.241a30; viii.2.252b9, 7.261a34, b12; nowhere, iv.1.208a30; void, iv.8.215a11; and time, now, iv.10.217b31, 218a2, 14, 12.221b23, 31, 222a4; and prior, viii.7.260b18; and rest, viii.7.261b12

1c. οὐκ ὄν, *cum non sit, non ens, quod non est,* not be: something, 1.3.186b1; and what is, i.3.186b10, 11; viii.8.263b11, 12; "empty," i.5.188a23; matter, privation, i.9.192a4; moved, iii.2.201b22; viii.1.251a17, 2.252b16

1d. ὅ ποτε ὄν, *(quantum quod) id quod est, quod re ipsa est, inquantum quodcunque ens est, quatenus est id quod tandem est, quod quidem ens, re ipsa, quodcumquen ens,* what at some time is, iv.11.219a20, b14, 26, 220a8, 14.223a27

1e. ὃ ποτ' ἦν, *secundum id quod est, re ipsa,* whatever it may then be, iv.11.219b11

2. καθ' αὐτό, *secundum seipsum, per se, per seipsum,* by itself, essential, independent: privation, i.8.191b15, 9.192a5(what is-not); matter, i.9.192a26, 28; nature, not art, ii.1.192b22, 31; attributes, ii.1.192b35, 2.193b28; iii.4.203b33; constituent, ii.1.193a11; being, factor, ii.5.196b24, 26, 27, 6.198a8, 9; viii.5.257a30; and infinite, iii.4.203a4, 5.204a18, 6.206b15, 207a24, 7.208a1; place, iv.2.209a31, 4.210b33, 5.212b7, 8; vi.4.235a18 (divisible); "in itself," iv.3.210a27; movement, iv.4.211a18, 20; v.1.224a29, 33, b22, 2.226a20; vi.10.240b19; vii.1.241b38, 242a43, 44, 4.249b13; viii.4.254b8, 10, 12, 5.256b7, 257b1; void, iv.8.216a24, 26; time, now, iv.12.221b26, 13.222b20; vi.3.233b33, 234a4, 15, 16, 8.239a24; change, iv.13.222 b21; body, vi.2.232b7; indivisible, vi.5.236b8, 15; part, vii.5.250a23; not aspects, viii.4.255a28

3. κατὰ συμβεβηκός, *secundum accidens, ex accidenti,* indirectly, incidentally, by accident: infinite, i.2.185b1; iii.5.204a15, 29, 6.206b22; viii.8.263b6; come from, pass into, i.5.188a34, b4, 8.191b15, 18, 24; not explain, not source, i.7.190-b18, 26; matter, what is-not, i.9.192a4; tendency, i.9.192a24, 25; not inhere, i.9.-192a32; ii.1.192b23; factor, ii.1.192b31, 3.195a5, b4, 5.196b25, 26, 28, 197a5, 15, 24, 33, 6.198a5-9, 8.199b23; arrangement, ii.1.193a15; come about, ii.5.196b23, 197a12, 8.199b22; not "in itself," iv.3.210b18; movement, change, iv.4.211a18, 19, 21; v.1.224a21, 27, 31, b16, 19, 23, 26, 28, 225b13, 23, 226a19, 20, 22; vi.10.240b9; vii.2.243b19, 4.249b13; viii.4.254b7, 8, 255a26, b27, 5.256b5, 7, 28, 257b21, 33, 6.258b15, 259b18, 21, 28; place, iv.5.212b11; time, iv.12.221b8, 26, 13, 222b22; one, v.4.227b31, 33; divisible, vi.4.235a18, 36, 5.236b6, 9; alteration, vii.3.248a9; immovable, viii.6.259b24

3a. συμβεβηκός, *accidens,* attribute, happen to be, incidental, accidental: and subject, i.3.186a34, b17, 18, 33, 34; chance, i.5.188a35; privation, i.7.190b27; in science, ii.2.193b27, 32, 194a3; factors, ii.3.195a33, 36, b1, 9, 14, 15, 5.197a13; occurrence, ii.8.199b25; infinite, iii.4.203a5, b33, 5.204a10; not necessary, viii.5.-256b10

3b. συμβαίνειν, *contingere, proficisci, evenire, accidere, sequi, incidere,* have conditions, follow, happen, occur, take place, be incidental to, be the case: scientific knowledge, i.1.184a10; conclusions, i.2.185a12, 3.186a10, 4.187a36; iii.4.-203b16; iv.7.214a2, 4; viii.8.264a9; raise physical problems, i.2.185a19; consequences, results, i.2.185b21, 8.191a32, 9, 192a19; iii.4.203b32, 6.206a10; iv.8.214b25, 28, 216a5, 8, 12, 21, 9.216b34, 217a18; v.1.225a30; vi.2.233b18, 3.234a22, 5.236-a16, 7.238a8, 8.239b4, 9.239b21, 22, 31, 35, 240a9, 10, 13, 18, 30; vii.1.242b73, 4.248b5; viii.1.251a23, b30, 5.256b4, 11, 14, 257a13, 15, 8.264b10; "being," i.3.186-a35, b4, 7; and definition, i.3.186b14-35; fact, i.5.188a31 (and logic); viii.6.-259a9; pattern, i.5.188b11; to consist, ii.1.192b19; physician, musician, ii.1.192b25; v.1.224a23; vi.9.240b4; properties, attributes, aspect, ii.2.193b33, 5.196b29; iii.5.-204a31; vi.5.236b7; viii.4.255a28; factor, ii.3.195a34; by chance, ii.4.196b4, 5.196b35, 6.197b19; growth, ii.8.198b21, 23, 29; iii.8.208a19; to define, iii.1.200b18; movement, change, rest, iii.1.201a6, 2.202a8; v.2.225b31, 4.229a2, 5.229a21, 28, 6.230a5, b27; vi.2.233a29, 3.234b3; vii.1.242b57, 69; viii.1.250b24, 251b8, 2.252b26, 253a9, 18,

4.255a18, 6.259b5, 8.263a31, 264b30, 265a2, 9.265a27, b8; numbers, iii.4.203a13; infinite, iii.6.206b33, 7.207a33; viii.8.263a10; touch, iii.8.208a13; time, iv.11.218b30, 220a18, 22, 12.220b9, 24, 221a24, 13.222b26, 14.223b24; thought, v.1.224b19; alteration, vii.2.244b4; circle, line, vii.4.248a19; viii.8.263b7, 9.265b3; mover, viii.5.258b8

4. τί, *quoddam, quiddam, aliquod, aliquid, quidpiam,* some, fact, something, particular, a sort of: whole, part, i.1.184a25, b11, 4.187b12, 15, 20, 35; 5.188b21, 6.189a26, b1, 7.191a16, 21, 22; ii.2.194a20; iii.6.206b10, 18, 207a1, 3, 8; iv.9.217-b14; v.1.224a24, 27, 32; vi. 1, 2, 4-6; vii.1.241b38; viii.3.253a34, b21, 4.254b11, 10.266a17; number, i.2.184b20; iii.5.204b18; iv.12.221a13, 16, b14; principle of, i.2.185a4, 5; quality, quantity, number, i.2.185a26, 33, b5, 4.187b9; iii.5; iv.11; subject, i.3.186a34, b18, 6.189a30, i.7; v.2.225b20, 21; nonbeing, being, i.3.186b1, 3, 9, 4.188a8, 8,191a31, b2; iii.4.203b15, 6.206a32; v.6.230a12; elements, principle, beginning, i.4.187b12, 6.189a26, b1, i.7, 9.192a29; ii.1.192b21, 8.199b7, 16; iii.4-7; iv.1.208b4; intermediate, i.5.188b2, 8; v.3.227a30; product, i.5-9; ii.1-4, 6; v.1.225-a17, 2.226a2; reason, i.6.189a21, b18; viii.1.252a32, b3; nature, i.6.189b2, 21, 9.192-a10; difference, ii.2.193b23, 6.198a2; vii.4.249a21; good, bad, ii.5.197a26; mover, ii.7.198b1; iii.1.201a27; vii.1.242b72; viii.1.251a24, 2.252b22, viii.4, 6, 10.266b31, 267a4; moved, iii.3.202b7; vii.4.249b27; viii.2.253a12, 10.267a7; infinite, iii.5.204-a17, 205a2, 3, 7.207b20; vi.6.237b15; place, iii.5; iv.1-8; touch, iii.8.208a12, 13; beyond universe, iv.6.213b1; time, iv.8, 10-14; vi.4.235a14, 8.239a5, 26, 9.240a31; vii.5.249b27, 28; viii.8.264b4, 10.266b17; change, movement, iv.10.218b10, 11.219a5, 14.223b1; v.1, 4, 5; vi.7; viii.1.251b9, 13, 30, 252a3, 2.253a1, 20, 3.254a30, 7.260a22, 261a29, 8.264a23, 265a5; start, stop, iv.11.219a10, 11; v.1.224b1, 225a1, 5.229b11; vi.4.234b11, 5.235b6, 6.237a19, 20, 8.239a23, 10.241a27; viii.1.242a66; viii.2.252b10; in succession to, v.3.227a4; unity, v.3.227a15; vii.4.249a22; altered, vii.3.246a7, 247a17; completion, vii.3.246a13; life, viii.1.250b14; dense, rare; viii.7.260b10; *et passim*

4a. τόδε, τοδί, *hoc, illud,* this, anything, fact, primary being: body of water, i.3.196a17; materials, i.5.188b19; ii.9.200b1-4; iii.4.203a26; question, i.6.189a28; in process, i.7.190a27, 28, 8.191a36; iii.3.202b8, 21, 22; v.2.226a13, 14; explain, ii.3; heavens, universe, ii.4.196a25, 6.198a12; factor, ii.7.198b5-7; affairs, ii.8.198b14; function, end, ii.9.200a10, 11, b2-4; as category, iii.1.200b35, 201a4, 2.201b26, 202a9; argument, iii.5.204b11, 205a7, 9; iv.10.217b33; vi.6.236b25, 10.241a6; vii.-3.245b5; viii.5.257b15, 6.259a14, 8.264a8, 21, b2, 10.266a26; place, iv.2.209a35; v.4.227b16, 17, 6.230b29; consequences, iv. 8.216a13; present, iv.10.218a27; things counted, iv.11.220a23; time, iv.13.222a27; vi.3.234a1, 2; vii.1.242b40, 5.250a4; and movement, vii.1.242b39, 40, 4.249a9; viii.6.258b28, 29, 259a2, 3, 7.260a24, 261a 32, b1, 8.264a25; beginning, viii.8.262a19; difficulty, viii.8.262b9

4b. τόδε τι *hoc aliquid,* primary being: product, i7.190a21, 32, 191a12, 13; material, i.7.190b25; actual, potential, iii.1.200b27; not the infinite, iii.6.206a30; and the void, iv.7.214a12; body, iv.11.219b30

4c. ὄν τι *ens aliquod, ens quiddam,* an existing something: and being, i.3.186b2; not an opposite, i.3.187a6; and place, iv.4.211b17

4d. τι ὄν, *aliquid quod, quid ens,* any being: not attribute, 1.4.188a8; come into being, i.8.191b24; not the infinite, iii.5.204a9; τιτῶν ὄντων *aliquid, eorum quai sunt, aliquod ens,* a being among beings: not place, iv.1.209a23 *See also* 1, 26a-b

5. τοιόνδε, τοιονδί, *tale, hujusmodi, ejusmodi,* of such a sort, qualitative: come into being, i.4.187a30; ii.4.196a32, 33, 9.200b1; way, ii.8.198b13, 9.200b6; material, ii.9.200a10, b6; actual, potential, iii.1.200b27; as a category, iii.2.201b26, 202a10; figure, iv.14.224a10, 11; passing away, v.6.230b27; arguments, viii.2.252b8

5a. τοιοῦτος, *talis, hujusmodi,* such, of a sort: quality, i.2.185a26; part, whole, i.4.187b15, 19, 22; vi.9.240a24; viii.5.258a23, 6.258b20; argument, i.6.189b17; ii.8.-198b34; iii.5.204b4; iv.6.213b28; viii.1.252a22, 3.254a3, 8.264a8; natural, ii.1.192b11, 193a4, 2.194a14, 8.199a6; genus, ii.1.192b16; beginning, ii.1.192b33; iii.4.203a29; undiscriminating, ii.1.193a8; element, ii.1.193a23, 3.195a17, 8.198b14; iii.6.206b-24; iv.1.208b9; viii.4.255a32; limits, ii.2.193b33; factor, ii.4.196a19, 30, 34, 8.199a29; viii.9.265b23; events, ii.5.196b16, 22; mover, ii.7.198b1; iii.1.201a24; viii.5.256a16, b3, 6.259b32, 10.267a4; animals, ii.8.199a23; being, iii.1.201a11, 16; viii.6.259a27, b2; actuality, iii.2.202a1; movable, iii.2.202a6; potential, iii.3.202b27; infinite, finite, iii.4.202b33, 5.204a12, b30, 205a24; philosophy, iii.4.203a2; body, iii.5.204b32; iv.7.-214a1, 11.219b19; direction, iv.1.208b24; place, iv.1.208b34, 2.209b28, 210a5, 4.211-a28, 5.212a32; nature, iv.1.209a15; material, iv.2.209b9, 9.217b23; experience, iv.4.211b34; void, iv.8.215a23; dense, iv.8.215a31; coincide, iv.8.216b10, 16; magnitude, iv.12.220b27; temporal, iv.12.221a24; nonbeing, iv.12.222a4; present, iv.13.-222a14; vi.3.233b35, 234a3; movement, iv.14.223b30; v.4.227b22; viii.2.252b17, 4.255a24, 9.265b19, 16.267a16, b8; continuous, vi.3.234a9; primary, vi.5.235b34; pleasure, vii.3.247a13; beings, viii.3.254b4; beginning, viii.4.255a26

5b. οἷον, *ut, ut puta veluti, possibile, posse, exempli causa, tamquam,* for example (*passim*), be possible: not one, i.3.186a19, 5.189a12; not belong, i.3.186b8; not opposite, i.3.187a4; not tend to itself, i.9.192a20; chance, not invariable or typical, ii.5.197a31, 8.199a4; not be futile, iii.4.203b5; and infinite, finite, iii.5.204a9, 205-a23, 206a3, b21, 25; vi.2.233a31, 7.238a9, 21; viii.10.266a12, 22; not place, iv.1.-209a15; not void, iv.8.215b20, 216b17, 9.217a4; contraction, iv.9.216b24; not perish, become, iv.10.218a17; vi.7.237a19; not continuous, v.3.277a13; not point, unit, moment, time, v.3.227a28; vi.1.231b12; viii.8.263b27; movement, rest, v.5.229a10; viii.3.254a23, 7.260b25, 8.261b30, 10.267a2, 3, b15; not increase, decrease, viii.3.-253b14; self-moved, viii.5.258a2; come, cease to be, viii.8.262a29. *See also* 11, 12a

6. πολλαχῶς, πολλαχῇ, *multipliciter, multis modis,* various in many ways, πλεοναχῶς, *multipliciter, multifariam, multifarie, multis modis,* in more ways: meanings, i.2.185a21, b6, 3.186a25; iii.6.206a21 (being); i.2.185b6; v.4.227b2 (unity); i.7.190a31 (becoming); ii.3.195a4, 29 (factor); v.1.225a20 (nonbeing); v.4.227b2 (unity of movement); viii.4.255a31, b17 (potential); viii.7.260b17 (primary); divisions, vi.3.234a18

6a. ποσαχῶς,*quot modis,* in how many senses: nature, ii.1.193b22; infinite, iii.4.204a2; in, iv.3.210a14; present, iv.13.222b28

6b. πλῆθος, *multitudo, numerus,* number, plurality: in mixture, i.4.187b4; and infinite, finite, i.4.187b8, 10, 34; iii.4.204a1, 5.204a10, 7.207b3, 13; vi.7.238a14; viii.10.266b19; varieties of factors, ii.3.195b12, 5.197a17, 6.198a5; divisible, iii.5.-204a12; elements, iii.5.204b13; and time, iv.12.220a29, b21; of parts, vi.7.237b33

6c. πολλά, *multa,* many, πλείων, plus, major, plurality, more, wider, greater, πλεῖστον, *plus, plurimum,* most, majority: parts, i.1.184a26; vi.9.240a24; viii.-3.253b20; principles, i.2, 6.189a26, 7.190b35, 191a15; and one (*see* 24); beings (*see also* 1a), i.3.186b12, 8.191a32; ii.1.193a4; constituents, components, elements, i.4.187b6, 23; iii.5.204b13; philosophers, i.5.188b27, 37; contrarieties, contraries, i.6.189b23; viii.7.261b16; differ, i.9.192a3; vii.2.244b5; factors, ii.3.195b32, 4.196a11, b4; parts of animals, ii.4.196a24; automatism, ii.6.197a36, b14, 198a12; ways, iii.-1.201a23; physicists, iii.4.203b15; iv.6.213b1; impossibilities, iii.4.203b32, 6.206a9; infinites, iii.5.204a25; number, iii.7.207b3, 7, 10; vii.4.249b24; difficulties, iv.1.208-a32; view, iv.1.208b33; places, iv.4.211b24; bodies, iv.6.213b10-12; voids, iv.9.216-b30; air, iv.9.217a14, 31; dimensions, iv.10.218a24; worlds, times, iv.10.218b4, 5; going on, iv.10.218b16; time, iv.10.218b16, 11.219b4, 5, 12.220b1, 3, 29-32, 221a27,

184   ARISTOTLE'S PHYSICS

b30; v.2.226b11; vi.7.237b24, 238a4; viii.8.264a5, 10.266a18; measures, iv.14.224a1; distant, v.3.226b33; viii.8.264b16; movement, v.4.228b4; vi.2, 4.235a31; vii.1.242b57, 58; viii.2.253a16, 3.254a26; qualities, vii.4.249b26; intermediated, viii.5.256a6, 257-a19, 8.264b33; movers, viii.6.259a1; orbits, viii.6.259b31; power, viii.10.266a26, 3, b7, 8; πολλάκις *multoties, saepe, plerumque, saepenumero,* often, i.9.192a14; ii.7.198a25, 8.199a1; iii.1.200b19; iv.1.208b17, 4.211b15, 13, 222a31; v.4.228a19; viii.3.253b25, 4.254b23, 8.264b20, 23, 25, 31, 265a2

6d. τοσόνδε, τοσονδί, τοσοῦτον *tantum, (usque) hoc, hucusque, quantum hoc,* so far, so many, quantitative, so much: go along, i.5.188b26; ignorance, ii.8.191b11, 31; tendency, ii.1.19b20; elements, ii.1.193a24; factors, ii.3.195-a3, 27, b17, 7.198a14, 15, 21; actual, primary, iii.1.200b27; kinds of movement, change, iii.1.201a9; as category, iii.2.202a10; extension, iii.7.207b18; arguments, iv.6.213b28, 10.218a30; viii.1.252b6, 8.265a12; and movement, iv.8.215b5, 26, 30, 216a1; v.1.224b6; vii.2.243b2, 4.249a9, 5.250a5, 6, 22, 26, b4; viii.3.253b16, 18; displacement, change, iv.8.216a28, b3, 9.216b28; time, iv.11.219a14; vi.7.237b29, 32; vii.5.250a26, b4; difference, v.4.228a12; varies, vii.4.248b18; altered, vii.4.249b17; power, viii.10.266b17

7. ἀληθές, *verum,* true, valid: assume, i.3.186b10, 4.187a28, 6.189a35; not, i.3.187a4; iii.5.205b5; vii.4.248b12, 5.250a20; pattern, i.5.188b21; objections, iii.8.-208a7; generation, v.1.225a28; result, vi.2.233a9; change, vi.5.236a8; rest, movement, vi.8.239a28, 32, b2; infinite divisibility, viii.8.263a22, b18, 23; ἀληθῶς, *vere, revera,* truly: chance, ii.4.196a7; place, iv.4.2, b33; void, iv.8.216a27, 9.217a11

7a. ἀλήθεια, *veritas, (revera),* truth: contraries, i.5.188b30; nature, i.8.191a25; viii.1.251a6; unmoved, viii.3.254a24; infinite divisibility, viii.8.263a18

7b. ἀληθεύεσθαι, *verum esse, vere dici,* be true, v.2.225b12

8. ψεῦδος, *falsum, mendacium,* false: premises, i.2.185a9, 3.186a7, 24; vi.2.233a22; not in abstraction, ii.2.193b35; time, vi.9.239b8, 240a4, 18; no mover, viii.2.253a11; and movement, viii.3.253b7, 254a27; consequence, viii.5.256b12

8a. ψεύδεσθαι, *mentiri,* derive erroneously: in demonstration, i.2.185a15

9. ἐνέργεια, *actus, actio,* actuality, actual being, operation, activity, exercise, actual functioning: what is not, i.8.191b28; movement, process, iii.1.201b9, 10, 13, 2.201b29, 31, 32, 34, 202a1, 2; viii.1.251a9; quantity, iii.2.201b30; of mover and moved, iii.3.202a15, 18, 22, 33, 36, b1, 6, 7, 22; v.1.224b26; viii.5.257b9; and infinite, iii.5.205b7, 6.206a16, 21, 7.207b12, 18, 29, 8.208a9; and potentiality, iii.6.-206a24; iv.5.213a3, 9, 9.217a29; viii.4.255a23, 30, 34-b31, 7.260b2, 8.262a22, 23; moved by own, iv.4.211a18; place, iv.5.212b4, 6; and time, iv.14.223a21; being, v.1.225a22; one, v.4.228a14, 16; and sense perception, vii.2.244b11; of knowledge, vii.3.247b7, 248a5

9a. ἐνεργεῖν, *operari, agere,* operate, be realized, function, act, use: factor, ii.3.195b5, 16, 17, 28; movable, iii.1.201a28; power, iii.1.201b8; on movable, iii.2.-202a6, 3.202a17, b10; knowledge, vii.3.247b9; viii.4.255a34, b4; vs. hindrance, viii.4.255b10, 21

9b. ἐνεργείᾳ ὄν, *actu ens, actu existens, quod est actu,* actual being: not infinite, iii.5.204a21; not interval, iv.6.213a33; change to, iv.9.217a23; viii.4.255b22; goal, viii.8.262b31, 32. *See also* 1

9c. ἔργον, *opus, munus,* task, use, action, function, result: of philosophy, i.192a36; ii.2.194b14; in art, ii.2.194b8; as means, ii.3.195a3; by means, ii.9.200a13, b5; of functioning, iii.3.202a24; and nature, viii.1.252a16

9d. ἐργάζεσθαι, *operari,* act, *operari, perficere,* carry to a finish, ii.8.199a16, 22; iii.4.203a31

9e. πρὸ ἔργου, *praeopere, conferre,* profitable, viii.1.251a5

9f. περίεργον, *otiosum, supervacuum,* superfluous, i.6.189b22

9g. εὐεργόν, *operose, ad opus idoneum,* useful, ii.2.194a34

10. ἐντελέχεια, *endelechia, actus,* actuality, actualization: one, i.2.186a3; being, ii.1.193b7; iii.1.200b26, 27, 201a10, 17, 6.206a14; and movement, iii.1.201-a11, b5, 2.202a7; viii.5.257b6-8; and potentiality, iii.1.201a21, 28, 30, 33, b6, 2.202-a11, 6.206a14; iv.5.213a7, 8; viii.5.257b7, 258b2; of mover and moved, iii.3.202a14, 16; infinite, iii.6.206b13, 22, 25, 207a22; viii.8.263a29, b5

10a. ἐντελεχείᾳ ὄν, *esse actu,* complete being: and infinite, iii.5.204a28. *See also* 1

11. δύναμις, *potentia, potestas, vis,* potentiality, import, power, validity, functional significance, force: many, i.2.186a3; being, nonbeing, i.8.191b28; ii.1.193b8; iii.1.200b26, 201a10, 6.206a14; v.1.225a22; subject, i.9.192a2; matter, i.9.192a27; ii.1.193a13 nature, art, ii.1.193a34, 36; factor, ii.3.195b16, 20, 27; and actuality, iii.1.201a22, 30, 32, 2.201b28, 34, 3.202b26, 6.206a14, 23; iv.5.213a3, 4, 7, 9.217a29, 32, 34, b9; viii.4.255a23, 25, 35b31, 5.257b7, 7.260b2, 8.262a22, 23; infinite, iii.4.-203b6, 6.206a18, b13, 14, 16, 21, 26, 207a22, 7.207b11, 18, 8.208a6; viii.8.263a29, b6, 10.266a24, 267b22; element, iii.5.204b15; place, iv.1.208b11, 22, 34, 5.212b4, 5; and nature, iv.3.210b7; and void, iv.9.217b21; and time, iv.13.222a14, 18, 14.223-a20; knowing, vii.3.247b4; and movement, rest, vii.5, 6.259a26; part, vii.5.250a24; meanings of, viii.4.255a30-b31; divisible, viii.5.258a32, b3; and first mover, viii.-10.266a24-b24, 267b22

11a. δύνασθαι, *posse,* be able: to generate, i.6.189b21; to discriminate, ii.1.193a5; factor, ii.3.195b4; and contraries, iii.1.201a35; to classify, iii.2.201b19; to function, move, be moved, iii.3.202a17; v.2.226b13; viii.1.251a11, 13, b6, 5.256-b21, 10.267b3; to happen, iii.6.206a24; be, not be, iv.12.222a8; to learn, vii.3.247-b18; to stop, viii.4.255a7

11b. δυνατόν, *possibile, posse,* potential, capable: factor, ii.3.195b28; not, ii.8.199b6; iv.10.218a22; realization of, iii.1.201b4, 5, 32; viii.4.255a35; quantity, iii.2.201b30; span, iii.5.204b10; vi.7.238b9; to be, iii.6.206a19; change, move, v.2.-225b23; vi.2.232b21, 4.234b17, 10.241a24, b9; viii.1.251a20, b1, 2, 2.252b13, 25, 32, 7.260b23; assumption, viii.5.256b11, 7.260b23; movers, viii.6.258b23

11c. δυνάμει ὄν, *quod potentia est, id quod est potestate,* potential being, iii.1.201a11, 27, 2.202a4, 12, 3.202b10, 6.206a19; iv.9.217a23; viii.4.255a30, b1, 8.262b31, 32. *See also* 1

11d. ἀδυνατεῖν, *nequire,* be unable to, ii.8.199a16

11e. ἀδύνατος, *non posse,* incapable, ἀδύνατον, *impossibile,* impossible: "all things one," i.2.186a27, 3.186a4, 187a10; and primary being, i.2.185a30, 3.186-b18, 30, 4.188a9, 10; come from, being, from what is-not, i.4.187a34, 8.191a29, b17; infinite, finite, i.4.187b11; iii.4.203b32, iii.5, 6.206a9, 207a31, vi.7.237b23, 238a12; vi.10.241b10; vii.1.242b53, 71; viii.5.256a17, 23, 8.263a6, 11, 9.265a20, 10.266b5, 267b22, 23; size, i.4.187b16, 20, 34; and contrariety, i.6.189b23, 7.190b33; objection to natural ends, ii.8.198b34; end without means, ii.9.200a11; movement, change, rest, iii.1.200b21, 3.202a36; v.1.225a25, 2.225b20, 226b10, 4.228b22; vi.1.231b31, 4.235a8, 5.235b26, 6.236b30, 8.239a25, 9.240a20, 10.241a2, 8, 11, 14, b6-8; vii.1.242-b59, 2.244a14; viii.1.250b17, 3.254a1, 5, 33, b2, 4.255a6, 5.256a25, b12, 34, 257a12, b2, 6.259b22, 7.261b5, 9, 8.262a13, 264a19, b10, 10.266b32, 267b18, span, iii.4.-204a3; place, iv.1.209a6, 2.209b21, 210a3, 3.210b10, 22, 6.213b20; full, void, iv.6.213b6, 8.216a4, 5, 17, 21, 26, 34, b12, 9.216b34, 217a8; time, iv.10.218a2, 18, 21, b8, 14.223a22, 23, 26; vi.3.234b19, 10.241a21; viii.1.251b19, 8.262a32; continuous, contiguous, vi.1.231a24, 2.232a24, 6.237a32; viii.8.262b21; cut, vi.10.241b3, 6;

meanings of, vi.10.241b5; and proof, vii.1.242b55, 72, 243a31; viii.5.256b11; knowledge, vii.3.247b15; went, length, vii.4.248a15, 11; generation, destruction, viii.251-b16, 252a3, 7.261a8, 9.265a19; increase, viii.7.260a29; solve, viii.10.267a19

12. δεδεγμένος, *susceptibilis*, subject, i.3.186a29; δεκτικός, *receptivus*, *suscipiendi vim habens*, *susceptivus*, *capax*, *receptaculum*, container, capable of holding, capable of receiving, subject: receptacle, iv.3.210b14, 20, 21, 6.213a17; movement, v.2.226b16; incomparable, vii.4.248b21, 249a2. *See also* 66

12a. ἐνδέχεσθαι, *contingere*, *posse*, *convenire*, be possible, be able: infinite, i.2.185b1; iii.5, 6.206a13, 7.207b17-19, 8.208a9; vi.2.233a22, 26, 27, b8, 10.241b13; viii.2.252b28, 8.263a16, 19, b3, 5, 6, 10.266a23, 24, b5, 7, 25, 267b21; one, many, i.2.186a2; accidental, essential, i.3.186b19, 28; ii.5.196b25; viii.5.256b10; size, i.4.-187b13, 14, 27; derivation, i.6.189a14; potential, actual, i.8.191b28; iii.4.203b30; viii.1.251b33, 4.255b19, 8.262b29; failing, ii.1.193a6; uses, ii.5.196b21; happen, ii.5.197a34; v.4.228a5; conduct, ii.6.197b6; mistakes, ii.8.199a35; to function, iii.1.-201b7; movement, change, iii.2.201b35, 202a2, iv.4.211a19, 21, 7.214a28, 29, 32, 8.214b30, 14.223a27; v.1.225a21, 23, 2.225b11, 16, 226a21, 4.229a3; vi.1.232a16, 2.232b21, 233b21, 3.234a25, 35, 5.236a11, b18, 7.238a33, vi.10, vii.1.242b47, 48, 50, 55, 66, 4.249a12; viii.1.250b23, 2.252b8, 3.253a28, 254a4, 4.254b18, 250b23, 2.252b8, 28, 3.253a28, 254a4, 4.254b18, 255a12, 5.256b9, 257b22, 6.259b3, viii.7, 9, 265a24, 25, 266a8, 10.267a12, b9; place, iv.2.209b24, 4.211b10, 212a10, 6.213b7, 9, 11, 7.214b6; number, iii.5.204b9; "in itself", iv.3.210a25, 33, 34, b12, 18, 20; not void, iv.9.216b29, 31; not perish, be, iv.10.218a18; v.4.228a18; recurrence, iv.-12.220b13; rest, iv.12.221b11; v.2.225b29; vi.8.238b25, 239a30, b3; viii.3.253a28, 254a4; not otherwise, iv.12.221b24; between, v.3.227a30; time, vi.8.239a4; viii.10.-266a31; and proof, vii.1.243a1; be, not be, viii.6.258b17, 22; determinate, viii.6.-259a11; uniform, viii.9.265b11. *See also* 5b

13. ἀντικείμενον *oppositum*, opposed, opposite: one, many, i.2.186a2; stages, i.5.188b4, 15; not, i.7.190a18; from, i.7.190a26; and persistent being, i.7.190b13, 15; increase, decrease, iii.1.201a13; and movement, change, rest, iii.2.201b24; v.2.-225b26, 30, 3.227a8, 6.229b25, 29, 230a1, 8, 231a6, 17; vi.9.240a28; viii.2.252b30, 7.261a32, 33, b6, 18, 21, 8.264a23, 27, 33, b13, 14, 16, 24; being, nonbeing, iv.12.-222a4; v.1.225a19, 23; places, viii.4.255a3; question, destruction, viii.7.261b4. *See also* 49, 50.

14. τέλειον, *perfectum*, complete, finished, τελειότης, *perfectio*, completeness: vs. incomplete, iii.1.201a6; vs. infinite, iii.6.207a9, 13, 14, 21; size, v.2.226a31; vi.10.241b1; viii.7.261a36; movement, v.4.228b12, 13; viii.8.264b28, 9.265a17; excellence, vii.3.246a14, 15; prior, viii.9.265a23. *See also* 100a

15. τὸ αὐτό, ταὐτό, *ipsum*, *idem*, same: and intelligible, i.1.184a18; relation, i.1.184b10; definition, meaning, 1.2.185b8, 7.190a16, 17; iii.3.202b16-22; vii.4.248-b10-13; not good, bad, i.2.185b21-25, 27; one, many, i.2.185b33, 186a2; argument, reason, proof, i.3; iii.1; iv.1, 7-9; v.4, 6; vi.1, 2, 5-10; vii.4; viii.1, 7, 8, 10; contraries, i.5; iii.1; v.5; viii.7; science, ii.2.194a18, 22; factor, ii.3.195a5, 11; not chance, automatism, ii.4; and chance, thought, ii.5.197a7, 6.197b13; happiness, ii.6.197b4; terminus, ii.8.199b17, 18; viii.8.262a17, 264b20; interval, iii.3.202a18, b19; infinite, iii.5.204a23, 25, 6.206b13, 17, 7.207b21; place, iii.5; iv.1, 2, 4-8; vi.8, 9; viii.9.265b2; ratio, iii.6.206b8; iv.8.215b6; part, whole, iii.6.206b8, 10, 207a5, 13; vi.3.234a4; something, iv.2.209b32; limit, iv.4.211b13; v.3.227a11; viii.8.264b8; being, iv.6.213-a18, 19; v.5.229a29; vi.7.238a19; viii.8.263b11; other things, iv.8.215a28; material, iv.9; present, time, iv.10-14; vi.3.6; vii.1; viii.8; point, iv.11.220a13, 15; v.3.227a29; change, movement, v.4.227b15-18, 228a18; vi.4.235a8, 31, 5.238b9, 7.237b25, 10.-241b13; vii.1.242a67-b36; viii.2.252b29-35, 6.260a4, 7.260a26, 8.262a15, b30; divisions, vi.2.233a11, 16, 4.235a15; path, vi9.240b2; vii.4.249a19, 20; judge, vii.4.249a27, 29; and aspects, viii.4.255a27; results, viii.5.256b3; nature, viii.5.258b3; facts, viii.6.259a9; velocity, viii.8.262b13; power, viii.10.266b22; *et passim*

15a. αὐτό, *idem*, itself: unity, i.2.185b6; part, i.2.185b12, 15; being, i.3.186b2, 6, 187a7, 8, 8.191a33; iii.4.203a5, 5.204a8, 10, 17; vi.9.240a25; whole, i.4.187b14; iii.5.205b34; vi.8.239a27, 29, 32; not tend to, i.9.192a20; nature, i.9.192a30; subject, ii.2.194a3; iii.1.201a16; good, ii.3.195a26; agent, ii.5.197a2; vii.2.243a13; viii.5; moved, move, iii.1.201a25; v.1.224a28; viii.4.254b32, viii.5; realized, iii.1.201a28; infinite, iii.5.205b; quality, v.2.226b5; mover, factor, viii.2.253a4, 5.257b2; *et passim*

16. ἕτερον, *alterum, diversum*, other, different: science, i.2.185a2; ii.2.193b26; species, i.3.186a21; iv.3.210b6; vii.4.249a26, b2; not separate, 186a31; iii.4.203a5; being, i.3.186a35, b13, 33, 4.188a8; iv.8.216b5, 9.218a24, 11.219a21; viii.5.25a31, 7.266a6, 7, 8.263b8; contraries, opposites, i.5.188b36, 37, 189a3, 6.189a18, b22, 7.191a1, 6, 9.192a14; iii.1.201b1, 2.201b25, 5.204b15; v.2.225b21; vi.9.240a28; nature, i.6.189b20; iii.4.203a17, 33, 34; terms of change, i.7.189b33; vi.4.234b15; matter, privation, i.9.192a3; triad, i.9.192a8; meaning, ii.3.195a7; factor, ii.4.196a31; viii.1.-252b4; means, end, ii.8.199a15; moved, mover, iii.2, 3; vii.1; viii.1, 2, 5-7; infinite, finite, iii.4.202b34, 203b22, 5.204b29, 6.206a29, 33; part, iii.6.207a7; iv.5.212b11; vi.1.231a28, 5.235b34, 7.238a2, 3, 9; touching, being limited, iii.8.208a11; place, iv.1-4; v.3.226b23; vi.9.240b5; time, iv.8, 11-14; vi.2, 6, 8; viii.8; present, iv.10-13; vi.3; movement, iv.14; v.1.4; vi.7, 10; iii.2, 8, 10; vs. same, iv.14.224a8; vii.4.249-a28; alteration, vii.4.249a30; potential, viii.4.255b1; *et passim*

16a. ἑτερότης, *alteritas, diversitas*, otherness: movement, iii.2.201b20; and unlikeness, vii.4.249b23

16b. ἄλλο, *aliud, ceterum*, other: categories, i.2.185a31, 7.190a33, 36, b1; ii.1.193a25, 27; iii.1.200b28, 201a1, 2.201b27; being, i.3.186a28, b1, 5, 4.187a15, 23, 8.191b12; viii.6.259a13, 7.260b18; subject, i.3.186b18, 31; iii.3.202b6; vii.4.248b21, 25, 249a3, 26, 28; element, body, i.4-6; ii.1; iii.5, 6; iv.1, 2, 8; change, process, movement, i.5; iii.1, 3; iv.11, 14; v.1, 3, 6; vi.5; viii.1, 6, 7; product, result, i.5, 7; ii.1, 6, 9; vi.6; contraries, terms of change, i.6.189a17, 24, 7.189b33, 190b34, 191a16; iv.9.217a22; v.2.225b24; viii.8.264b19; beginning, i.9.192b4; iii.4.203b11; viii.5.257-a32, 7.260a20; factors, ii.1.192b8, 3.194b26, 195a24, 25, 30, b1, 4.196a18, 197a24, 8.198b15; iii.4.203b12, 5.205b17; agent, mover, ii.1.3; iii.3; vii.2; viii.4-6, 10; science, ii.2.194a18, 26; matter, form, ii.2.194b8, 8.199a32; iii.4.203a14; place, iii.5; iv.1-5; one after another, iii.6.20622, 28; vi.7.238b7; viii.6.259a19, 8.264b21; "in", iv.3.210-a14, 26, b24; equal, iv.8.215a28, 216a14; time, iv.10, 11, 13; v.4; vi.6, 8; viii.8; so much, vii.4.248b18; path, vii.4.249a15; potential, actual, viii.8.262a21; *et passim*

16c. ἄλλως, *aliter, alioqui*, in other ways: natural science, astronomy, ii.2.193b28; move, ii.4.196a23; strange, ii.4.196b1; chance, ii.6.197b13; define, iii.-2.201b17; iv.2.209b20; infinite, iii.6.206a25, b12; vi.7.238a10; void, iv.9.216b29; one, v.4.228b15; opposed, v.5.229a16; pulling, vii.2.244a12; nature, viii.1.247a18; potential, viii.4.255b33; not, viii.6.260a2; relation, viii.6.260a7; ask, viii.8.263a7; solve, viii.10.267a17

16d. ἀπαλλάττεσθαι, *evadere dimitti, aliud esse*, let go at that, v.5.229a24; ἀπαλλαγή, *discessio, liberatio*, loss, v.5.229a24

17. πρότερον *prius, antea*, before, previous, prior, preceding, earlier, reference, i.5, 6; ii.8; iii.3, 7; iv.4, 7, 8, 10, 12, 13; v.1, 5; vi.1, 3, 4, 8-10; vii.1, 3; viii.1, 3, 5, 7-10; contraries, i.5.188b31, 6.189a18, b24; subject, i.6.189a31, 34; thinkers, i.8.-191b32; not claim, ii.3.194b18; factors, ii.3.195a30, b24, 6.198a8, 9, 11; viii.5.257-a30; and succeeding, ii.8.199a9, 20; iv.1.208b6, 209a9, 4.211a16, 8.216a35, 9.217a7; v.3.227a20; vii.3.247b16; viii.1.251b10, 3.254a10; and movement, change, rest, iii.-

1.201b7; v.1.225a2, 2.225b35, 3.226b23; vi.2.232a28, 6.236b34, 237a18, 28, 34, 35, b5, 7.238a2, 9.239b12, 10.241a8, 11; viii.1.250b11, 18, 251a28, b9, 30, 2.253a9, 10, 5.256a33, 257a12, b16, 7.261a10, 12, b2, 8.263a8, 264a26, 32, 9.266a15, 22, 24; place, iv.1.208b34; vi.5.236b13; and time, iv.10, 218a11-30, iv.11, 12.220b6, 9, 221a14, 13.222a25, 14.222b23, 223a4-15, 28; vi.3.234b7, 5.236a32, 7.238a3, 8.239-a15; viii.1.252a16, 3.253b17, 7.260b18, 8.263b10, 13, 26, 264a13: and primary, vi.5.236a4; viii.7.260b18, viii.9; becoming, perishing, vi.6.237b10, 20, b18; viii.1.-252a35; be capable, viii.1.251a15; be, viii.7.261a1, 6, 7, b2; in nature, viii.7.261a14; and mover, viii.10.266a11, 267a10

17a. πρῶτος, *primus*, fundamental, first, ultimate, direct, immediate, proximate: facts, factors, i.1.184a13; ii.3.194b20; principles, beginning, i.1.184a14, 2.184b18; viii.1.251a8, 6.259b13; in procedure, i.1.184a15, 21, b13, 7.189b30, 31; ii.8.198b10; iii.1.200b25; iv.1.209a4, 4.211a12, 10.217b30; v.5.229b8; viii.1.251a8; elements, i.2.184b23, 7.190b18, 9.192a29; iii.4.203a34; contraries, i.5.188a29, 6.189b23; chance, thing, i.5.188b3; thinkers, i.8.191a25; persistent being, i.9.192a31; philosophy, i.9.-192a35; ii.2.194b15; nature, ii.1.192b22; constituent, ii.1.193a10, 29; agent, mover, ii.3.194b30, 7.198a19, 26, 33, 35; iv.3.210a22; v.1.224a33, 34; vii.1.242a53, b59, 72, 2.243a32, 14, 245a8, b1, 245a8, 12, b1; viii.1.251a24, viii.5, 6, 7.260a25, 629, 9.265-b23, 266a9, 10.266b32, 267a1, 3, b18, 24; being, iii.7.198b3; viii.9.265b27; seed, ii.8.199b8, 9; infinite, iii.1.200b17; abyss, iv.1.208b30, 31, 32; place, iv.1.209a1, 2.209a33, b1, 3.210b24, 4.210b34, 211a2, 28, 32, 212a20, 11.219a15; v.3.226b22; "in itself", iv.3.210a33, b22; container, iv.3.210b29; numbers, iv.6.213b26; measure, iv.14.223b18; viii.9.265b10; moved, v.1.224a28, b1, 24; vii.1.242a43, 44, b72; viii.5.-257a27, 6.259b33; end, v.1.224b17; not in infinite, v.2.226a5; successive, v.3.227a18; present, vi.3.233b34; terminal stage, vi.4.234b17; part of time, or change, vi.5.-235b7, 31-236a35, b8-16, vi.6; time, vi.6, 8.238b31-239a10, 10.240b22; viii.b.263b23; movement, vi.10.241a10; vii.2.243a39; viii.7, 8.262a26, 9.265a13, b9, 10, 17, 266a7; elements, vii.3.246b17; subject, vii.4.248b21, 23, 249a3; change, viii.1.251a27, b10, 30, 7.260b32; potential, viii.8.263a11; problem, viii.10.226b28

18. ὕστερον, *posterius, deinde, postea*, later, more recent, posterior: in procedure, i.1.184a22, b14; iii.1.200b24; thinkers, i.2.185b26, 6.189b15; contraries, i.5.-188b31, 6.189b24; reference, i.9; ii.1, 5; iii.7; iv.5; v.5; viii.2, 7; factors, ii.3.195a30, 6.198a9; terms in series, ii.8.199a19; v.3.227a4; movement, change, iii.1.201b7; iv.8.215a4, 9.217b8; v.1.225a2, 2.225b35, 226a9; vi.1.232a2; viii.1.251b31, 7.260b7, 31, 261a11, 8.262a26, b19; and time, iv.10.218a26, 27, 29, iv.11, 12.220b6, 9, 14.223-a4-15, 28; vi.7.238a3; viii.1.251b11, 8.263b10, 13, 15, 21; destroyed, viii.1.252a2; in genesis, viii.7.261a14

18a. ἔσχατος, *ultimus, extremus, extremitos*, last, extreme, ending, ultimate, extremity: matter, i.9.192a33; process, end, ii.2.194a30, 32; v.3.226b24, 26; viii.5.-256b21; explanation, ii.7.198a16, 18; and infinite, iii.5.205b31; vi.2.233a18, 19, 25; vii.1.242a66; and place, iv.4.211a26, 32, 34, b8, 11, 212a21, 24, 27, 28, 5.212b19; part, iv.9.217a16; and time, iv.11.220a16; vi.3.233b35, 234a5, b5; vi.6.237a5, 7; viii.1.251b23, 8.264a3, b8; two, one, v.3.227a13, 4.228a24, 26, 30, b1; vi.1.231a22, 26-29, b18; and race-course, vi.9.240a7, 9, 15; altering factor, vii.2.244b4, 245a4, b1; object, viii.5.256a6; and circle, viii.9.265b7

18b. ἄκρος, *ultimus, extremus*, close, extreme, iv.2.209b20, 11.219a27; v.1.224b32, 3.226b23, 227a22, 25, 5.229b20; vi.4.234b19, 10.241a31; viii.8.262a23, b11, 12, 24

19. γένος, *genus*, genus, (including) kind: principles, i.2.184b21; of contraries, i.5.188a24, 6.189a14, b25, 26; category, i.6.189a14, b24; iii.1.201a10; art, ii.2.192b16; factors, ii.3.194b26, 27, 195a33, b14, 25; movement, change, iii.2.201b19; v.4.227-

b4-7, 12, 14, 20, 27, 228b12; vii.1.242a69, b35, 2.243b10, 4.249b14; viii.5.257a4; of place, iv.1.209a4; and species, iv.3.210a18, 19; v.4.227b12; vii.4.249a14, 22, 24; and "succession", v.3.227a1, 5.228b27; vi.1.231b12; of living beings, viii.6.259b3, 7.261a17; power, viii.10.266b22

19a. συγγενής, *cognatus, proximus, ejusdem generis, sui generis,* homogeneous, akin, of the same kind, iii.5.205a16; iv.5.212b31, 14.223b13, 18; vi.1.231a23

20. εἶδος, *species, forma,* species, (included) kind, form, state: of principles, i.2.184b21; differ in, i.3.186a19, 21; iii.3.202b1; iv.3.210b6; v.4.228a25; vii.4.239a16, 26, b12, 13; viii.8.262a2, 5; unity, i.4.187a18; contraries, i.4.187a20; infinite, finite, i.4.187b9, 10; iii.5.205a22; and definition, i.7.190a16; formal aspect, principle, i.7.190b24, 9.192a34; one, i.7.190b28; viii.5.257b4; and essential being, i.7.191a19; cannot tend to itself, i.9.192a21; natural, perishable, i.9.192b1; and nature, ii.1.-193a31, 35, b1, 4, 2.194a13, 21, 22, 24; and privation, ii.1.193b19; and art, ii.2.-194b3, 5; and science, ii.2.194b9, 10, 12; as factor, ii.3.194b26, 195a21, 7.198a24; of factors, ii.3.195a27; mover, moved, ii.7.198a26; and movement, change, iii.1.-201a9, 3.202b24; v.1.224b5, 11, 25, 2.225b22, 24, 2.226a16, v.4; vi.10.240b21; vii.1.-242a69-b41, 4.429a12, 16, 17, 20, b10, 11, 13; viii.5.256b31, 257a22, 8.262a2, 4, 264a25; convey, change to, iii.2.202a9, 3.202a36; geometrical, iii.4.203a15; of place, iii.5.205b32; iv.1.208b13; and matter, iii.6.207a26, 7.207b1; iv.2.209b8, 3.210a21, 5.213a2; not place, iv.1.209a21, iv.2, 3.210b29, 4.211b13, 212a3; and genus, iv.3.-210a18, 19; v.4.227b12; vii.4.249a15; and time, iv.14.223b4; and succession, v.3.226b35; of motion, vii.3.243a16, b6; occurrence of, vii.3.246b15; have, viii.5.-257b10; contrary, viii.6.260a9. *See also* 90

21. ὅλον, *totum,* whole: situation, generality, name, i.1.184a24, 25b11; and part, i.2.185b11-16, 34, 4.187b13-21 (size); ii.3.195a18, 20; iii.4.203a23, b1, 205b21, 6.206b8, 207a27; iv.3.210a16, 17, 25-b21 (in), 4.211a31, b1, 22, 26, 29, 5.212a35, b34-213a1, 8, 8.214b25-27, 10.218a4, 7; v.1.224a25, 26; vi.1.231b2-6, 232a14, 4.234-b21-235b5, 7.237b28-30, 238a22-24, 9.240a24, 25, b1, 10.240b12; vii.5.250a21-25; viii.2.252b15, 3.253a34, b22, 5.257b26-258a5, 8.264a26; definition, i.3.186b25; mixture, i.4.187b5; earth, iii.5.205a11, b21; place, iii.5.205a17; and finite, infinite, iii.5.205a31, 6.207a8-32; viii.2.252b28, 8.263a9; and movement, rest, iii.5.205b6, 19, 20, 23, 24; vi.1.232a6, 14, 3.234b1, 4.234b21-235b5, 7.237b28-30, 8.239a30, 10.240b12, 17; vii.1.241b40, 242a49, b49, 51, 5.250a15, 16; viii.2.252b15, 3.253a34, b22, 5.257b26-258a5, 24-26, 6.259a20, 10.266a21, 267a12; place in, iii.5.205b34; defined, iii.6.207a8-10; cosmos, iv.5.212b9, 16, 9.216b25 (bulge), 217a17, 10.218a33 (revolution), b5, 7 (sphere); viii.1.252a30 (love, strife); measure, iv.12.221a2, 4, 14.224a2; vi.2.233b5; unity, v.3.227a16, 4.228b14; vii.2.245a.3; now, vi.3.234a29; time, vi.5.236a2, 6.236a23, 25, 7.238a23, 24, 8.238b33, 34; vii.4.248b4; viii.8.262a31, 264a5; and change, vi.9.240a22, 26, 29; organism, animal, viii.2.253a18, 4.254b17; circle, viii.9.265b7. *See also* 44

22. μέρος, μόριον, μοῖρα, *pars,* part, aspect: particular, i.1.184a26, 5.189a8; vii.3.247b6, 7; and whole (*see* 21, 150); unitary, i.3.186a17; different, i.3.186b14; defined, i.4.187b15, 16; and contrariety, i.9.192a14; of animal, ii.1.192b9, 4.196-a23, 8.198b23, 28; viii.4.254b23; of a science, ii.2.193b26, 194a20; in definition, ii.3.194b29, 9.200b7; of primary being, iii.5.204a26; and place, iii.5.205a11; iv.1.-208b12, 2.209b27, 4.211a32, b25, 28, 5.212b5, 10-12, 8.216b9; vi.9.240a32, 33; of species, iv.3.210a19; and movement, change, iv.4.211a20, 5.212a33; v.1.224a32, b17, 20, 21, 2.226a20; vi.2.233b13, 4.234b13, 21-235b5, 8.239a27, 29; vii.1.242a48, 2.244-b10, 5.250a21; viii.1.250b27, 252a8, 20, 28, 2.252b14, 3.253b19, 4.254b10, 12, 5.257-b26-258a5, b17, 8.264a26, 10.266a17; and time, iv.10.218a5, 6, 12, b2, 3, 11.220a17, 19, 21, 13.222b8, 13; vi.3.234b6, 6.237a8, 7.237b33, 238a22, 34, 8.238b33, 239a6,

18, 32, 33; vii.4.248b3; of number, iv.12.221a12; of magnitude, v.4.228b25; vi.7.- 237b33; and indivisible, vi.1.231a27, b24; and infinite, vi.7.238a12; sensitive, mental, vii.3.247a6, b1, 248a8; μεριστός, *partibilis, secari posse, in partes dividi posse, divisibilis,* part, divisible: of infinite, iii.5.204a22, 34; whole, iv.10.218a3, 6; vi.4.- 235a10; time, vi.8.239a10, 17; continuous, vi.8.239a22; and divisible, vi.10.240b26; and alteration, viii.3.253b24; κατὰ μέρος, *secundum partem, particularis,* particular: movement, iii.3.202b24

22a. ὁμοιομερής, *similes partes, similares partes,* similar parts: infinite, i.4.187a25; iii.4.203a21; and place, iv.5.212b5

22b. ἀμερές, ἀμέριστον, *impartibilis, partibus vacare,* without parts, partless, iii.5.204a27; vi.1.231a28, b3, 12, 232a17, 2.223b30, 32, 3.234a7, 5.236a21, b12, 6.237a32, b7, 8.239a3, 4, 11, 17, 10.240b8, 11, 18, 31, 241a5, 23; viii.6.258b18, 25, 10.266a10, 267b26

23. εἶναι, *esse, essentia,* be: eliminate, i.2.185b27-32; ways of being, definition, i.3.186a31, b13, 4.188a17, 9.192a25; ii.1.192b11, 2.193b33; iii.3.202b0, 16.6.206a21; iv.3.210b16, 8.216b5, 9.217a24, 11.219b21, 22, 27, 12, 222a8, 13.222a20, b22; v.5.229a18, 29; vi.8.239b2, 9.240a25, 26, 10.240b30; viii.2.253a4, 6.260a2, 7.260b18, 8.263b9; subject in process, i.7.190a11; and becoming, i.8.191a30, b2, 12; ii.3.195b32, 8.199a8, 23; iii.5.205a3, 6.206a31; viii.1.252a33; and not be i.8.191b26, 27; vi.5.235b16, 236a11, 14 (primary time); viii.5.256b10, 6.258b17-27, 259a4, 5; principles, i.9.192b2; physician, patient, ii.1.192b26; nature, ii.1.103a1; together, ii.3.195b18; chance, automatism, ii.4.195b36, 196a12, 30; factors, ii.7.198a14; by necessity, ii.8.198b14; if, then, ii.9.200a15-30, b5-7; iii.5.204b5, 6, 6.206a9; viii.3.254a27-30; possible, iii.2.202a3, 4.203b38, 6.206b18, 207a8; infinite, iii.4.202b35, 36, 203a7, 8, 10, b15, 31-204a2, 5.204b3, 5, 205a2, 3, iii.6, 7.207b11, 16-21, 28.208a2; potential, actual, iii.6.206a14; and thinking, iii.8.208a19; place, iv.1.208a28-30, b1, 26, 209a3, 7, 30, 2.209b16, 210a12, 5.213a10, 7.214a18, 8.214b17, 20, 215a8, 216a25; void, iv. 6.213a13, 20, 23, b22, 28, 7.214a17, 26, b8, 10, 8.214b12, 24, 28, 29, 215a11, 13, 216a21-26, b20, 9.216b23, 28, 217a21, b27; movement, iv.8.214b13, 29, 33, 215a2, 5, 6, 13, 18. 216a23, 9.216b25, 217a12, 12.221a6, 7; viii.1.250b15-23, 262b6, 3.254a28, 7.261a22; time, iv.8.216a4, 10, 10.217b32, 34, 218a6, 17, 11.218b29, 33, 219a2, 3, 9, 10, 29, 12.220a28, 221a5, 10.b28, 13.222b28; vi.5.236a10-15; viii.1.251b19, 252b5; eternal, iv.12.221b5; "in", vi.6.237b21; love, stirfe, viii.1.252a28, 30; *et passim*

23a. ἀπεῖναι, ἀπουσία, *absentia,* absence, ἀπόν, *absens,* absent: and presence, i.7.191a7; ii.3.195a12-14; and whole, iii.6.207a10, 12

23b. ἐνεῖναι, *inesse, inhaerere,* be in, be constituent: in definition, i.3.186b24, 30, 31; in the "one", i.4.187a20; and becoming, i.4.187b23; not everything, i.4.187b31, 34; end, ii.8.199b10, 30; in agent, ii.9.199b28; in vessel, place, iv.1.208b3, 3.210b15, 30, 7.214a33, b1, 8.214b24; void, iv.6.213b18, 9.216b33; in flame, iv.9.217a7; mental, iv.11.219a6; movement, viii.2.252b18, 253a8, 19, 6.259b8; in divisible, continuous, viii.5.258b4, 8.263a28

23c. παρεῖναι, *adesse, inesse,* be present, παρουσία, *praesentia,* presence, παρόν. *praesens,* present; and absence (*see* 23a); context, iv.10.218b19; change, iv.12.220b7; time, iv.13.222b7, 12; movement, vi.1.231b26, 232a2, 16; not incompatibles, viii.7.26-b9

23d. ὑπεῖναι, *subesse,* be underneath, v.2.226a10

24. ἑνότης, *unio, copula,* unity: and time, iv.13.222a19; ἕνωσις, *unio, copulatio,* unity: and time, iv.13.222a20; ἕν, *unum,* one, unit: principle, i.2.184b15, 16, 185a4, 6.189a12, 20, b12, 9.192a35(form); iii.4.203a31; viii.1.252a10; and many, i.2, 3, 4.187a12-26; ii.1.193a24; iii.7.207b7; iv.11.220a12, 12.220a29, 32, b10, 13, 20, 21, 13.222a17; v.5.229a15; vi.4.234b33; vii.1.242b56; viii.1.250b28-31; viii.5.257b27, 258a11, 12, 6.258b11, 259a8; being, i.2.(vs. principles, categories, meanings, aspects), 3(vs. changes, kinds, attributes and possessors, nonbeing, parts, definition, analysis); viii.5.257b3; meanings of, i.2.185b5-25, 3.186a28; v.4.227b2; part, i.3.186a17; the all, universe, i.5.188a20, 6.189b2; iii.5.205a3, 20; viii.1.250b22; and contrariety, i.6.189a12-14, b8, 11-16, 23-27; iii.5.205a5; iv.9.217a22, 25, b11; viii.8.262a5; element, body, i.6.189b19; iii.5.204b14, 15, 22, 28, 205a3, 26; iv.11.220a7; subject, i.7.190a16, b24, 191a12, 8.192a2; iii.201b3, 3.202a35; v.4.227b31, 228a14, b2, 14; vi.4.235a2, 6, 7.238a19; vii.4.249a3; viii. 6.262a1, 10.267a23; form, i.7.190b28, 191a13; iii.3.202a36; v.4.227b29, 228b10; viii.5.257b3; primary being, i.7.191a12 nature, i.9.192a10; iii. 7.207b22; vii.4.248b21; individual, ii.5.196b29; viii.7.260b30, 261a5; factor, ii.7.198a25; actualization of mover and moved, iii.3.202a18-20, 36, b1, 8, 12; distance, iii.3.202b18, 19; and geometry iii.4.203a14, 15; clod, iii.5.205a12; number, iii.7.207b5-8; receptive, place, iv.2.209b13; actually, iv. 5.213a10; present time, iv.10.218a10, 26, 11.218b26, 28, 219a30, 13.222a32, 33, 14.223b4, 11; v.4.227b31, 228b2, 8; vi.8, 239a33; viii.8.262a2, 263b13; state, iv.11.218b31; v.4.228a7, 15; vi.8.239a16; not limit, iv.10.218a23; v.4.228a25; dimension, iv. 10.218a23; unit, iv.14,223b13; division, iv. 14.224a9; place, v.3.226b22; continuity, v.3.227a11, 15, 16, 22, 23, 6.228a21, 22, b14, 229a1, 5; vi.1.231a22, 26, b17; viii.4.255a13; movement, rest, v.4, 5.229a18, 6.231a3; vi.4.234b33, 235a2, 5, 10.240b17, 241b12-20; viii.1.242a65, 67, b38, 41, 56, 63, 4.249b10; viii.2.252b29-34, 4.255a10, 6.259b6, 260a4, 18, 7.260a26, 8.261b27, 36, 264a20, 21, b9, 9.265a28, 10.267a15, 20, 22, b15; and measure, vi.7.238a15; whole, vii.2.245a13; ratio to, vii.4.248b15; equivocal, vii.4.248b19; all things, vii.4.248b25; genus, vii.4.249a22; science, viii.1.251a30; certitude, viii.3.254a35; prime mover, viii.6.258b11, 259a7, 8, 12, 14, 18, 10.267a23, 25

24a. πρὸς ἕν, *ad unum,* (ratio) to one: two, ii.3.194b28 (in an octave); iii.3.202a19 (interval)

24b. ὅπερ ἕν, *quod vere unum, quod est proprie atque essentialtier, unum,* identifiable unity, i.3.186a34; iii.7.207b16

24c. μονάς, *unitas,* unit iii.6.206b31; iv.11.220a4, 12.221a15, 14.223b14; v.3.227a2, 27-32

24d. μόνως, μοναχῶς, *singulariter, uno modo, hoc tantum modo, solummodo,* only, in one way: meaning, i.2.185b31; solution, i.8.191a23; infinite, iii.4.203b18; motion, vi.10.240b31, 241a23; viii.1.251a28

II. CATAGORIES

25. κατηγορία, κατηγόρημα, *praedicatio, appellatio, praedicamantum, categoria,* category, classification: art, ii.1.192b17; of being, iii.1.200b28, 201a1, 2.201b27; and movement, v.1.225b5, 4.227b5; vii.1.242b35

25a. κατηγορῥειν, *praedicare, attribuere,* predicate: "being", i.3.186a33, attribute, i.6.189a32; "form", i.7.190a29

26. οὐσία, *substantia, essentia,* (*ens per se*), "what is", primary being, essential being, full being: and other categories of being, i.2.185a20-b5, 6.189a33, 34, 7.190-a31; ii.1.193a25; v.1.225b5, 2.225b10, 226a23, 28; vii.1.242b35; viii.7.260b19; not "the infinite", i.2.185a32-b5; iii.4.203a5, b33, 5.204a10, 21, 23, 27, 33, 6.206a32, b24; viii.8.263b8; one genus, i.6.189a14, b23; not contraries, i.6.189a29, 33, 34; and becoming, generation, destruction, i.7.190a33, b1, 19, 191a11, 19; v.1.225a16, 18; viii.7.260b12; not matter, privation, i.9.192a6; ii.1.193a10, 16, 20, 25; natural, ii.1.-

192b33; relating to, ii.7.198b9; and change, iii.1.200b34; of place, iv.2.210b13; and the void, iv.7.214a12; and time, iv.10.218a3, 12.221b31; one, v.4.227b21, 228a8, b13; and number, vii.4.249b23; movement, viii.7.261a20

26a. ὅπερ ὄν, (existens) quod vere est, (id) quod est proprie atque essentialiter ens, "what primary is", i.3.186a33, b2, 4-8, 10, 12-14, 33

26b. ὅπερ ὄν τι, quod vere est aliquid, prope atque essentiliter ens quiddam, "whatever is something", i.3.186b14-17, 32, 187a8. See also 1, 4c-d

27. ποσόν, quot, quantitas, quantum, tantum, how many, quantity, so much, extent: beings, i.2.284b23; as a category of beings, i.2.185a20-b5, 16; iii.1.200b35, 201a6; v.1.225b7; vii.4.247b26; infinite, i.2.185a32-b5, 4.187b9; iii.5.204a28, 206a3-5, 6.207a7; vi.2.233a26; elements, i.4.187b12; determinate, i.4.187b36; not independent, i.4.188a10; become, i.7.190a34; principles, i.7.191a3, 4, 21, 9.192b3; know form, ii.2.194b10; factors, ii.3.194b17; change, movement, rest, iii.1.200b34; iv.12.221b19, 20; v.1.225b82.226a25, 30, b16; vi.5.236b16; vii.2.243a37, 5.250a17, 30, 31; time, iv.10.218b17, 12.220b26, 13.222a27, 14.223b17; vi.7.238a29; vii.5.249b30, 250a30, b1; divisible, indivisible, vi.4.235a18, 10.240b13; distance, vi.7.237b30, 238a17, 23; vii.5.249b29, 30; species, vii.4.249b11; potentially, viii.4.255a25, 27, b13, 22

28. ποιόν, qualitas, quale, quality, such, ποιότης, qualitas, quale, quality: as a category of being, i.2.185a20-b5, 16; iii.1.201a1, 5; v.1.225b6; vii.4.249b25; not "infinity", i.2.185a32-b5; infinite, i.4.187b9; not independent, i.4.188a11, 12; become, i.7.190a34; art product, ii.2.194b5; material, ii.2.194b6; process, ii.2.194b6; iv.14.223a30; factors, ii.3.194b17; change, iii.1.200b34; time, iv.10.218b18; and movement, rest, v.7.225b8, 2.226a24, 26-28, b17, 5.229a7, 6.229b26; vi.5.236b18; vii.1.242b35, 2.243a36, 244b5, 6; viii.3.253b12, 4.254b20, 7.261a22; possessors, v.3.-226b20, 227b2; vi.5.236b7; divisible, vi.4.235a18; equality, vii.4.249b3; potentially, viii.4.255a25, 27, b13, 22

29. πρός τι, ad aliquid, related to something, relative; become, i.7.190a35; material, ii.2.194a9; distinctive being, ii.7.198b9; more-less, active-passive, iii.1.-200b29; limitation, iii.4.203b20; place, iii.5.205b33; touch, iii.8.208a12, 13; ratio, iv.8.215b7, 216a16; vii.4.248b15, 5.250b9; time, iv.13.222a25, 27, 14.223a5; intermediate, extremes, v.1.224b32-35; as category, v.1.225b6, 2.225b11, 226a23; vii.3.-246b4, 8, 11; and states, vii.3.247a2, b3; change, viii.1.251b7, 6.259b16

30. ποῦ, ubi, where, πού, alicubi, somewhere: come to be, be, i.7.190a35; vi.5.235b20, 21; not ideas, iii.4.203a9; body, iii.5.205a10, 206a2, 3, 5; iv.7.213b33; move, rest, iii.5.205a15, 17; iv.8.214b18; v.2.226a24; naturally, by compulsion, iii.5.-205b5; and place, iv.1.209a23, 2.209b32, 5.212b9, 14, 27; and void, iv.8.215a20, 9.217a16; not nonbeing, v.1.225a32; as category, v.1.225b6; potentially viii.4.255a25, b11

31. ποτέ, forte, quanod, tandem, tunc, aliquando, ever, perhaps, related to some time, at one time: not, i.2.185b30; iii.6.207a5; incidentally, i.5.188b4; come to be, 1.7.190a35; iii.5.205a4; v.2.226a1, 3; together, iii.4.203a25, 32; movement, iii.4.203a33; v.4.227b26, 228b26; viii.3; and time, iv.13.222a24-29, b8, 10, 13, 28; as category, v.1.225b6

32. συγκεῖσθαι, componi, constare, be put together: materials, i.5.188b18; terms, i.7.189b34, 190a3, 20, 29, b20; magnitude, time, movement, iv.8.215b19, 10.218a2, 7, 8; v.4.229a6; vi.1.231b19, 21, 232a19, 9.239b8, 31; viii.8.261b31; rotation, vii.2.244a2; κεῖσθαι, be proposed, put down, v.1.225b4, 4.228a29; vi.5.236a18

32a. ἀποκεῖσθαι, reponi, rejici, leave, i.9.192b1

33. ἔχειν, habere, have, contain: contention, i.2.185a8; difficulty, i.2; iii.4; iv.1; v.4, 6; viii. 2, 5, 8, 10; pattern, form, order, i.5.188b10; ii.1.193a35, 6.207a26; viii.252a13, 16, 22, 5.257b10; beginning, movement, ii.1.192b14, 28, 33, 193a29, b3, 7.198a28, b1; iii.3.202a30, 4.204a5, 6, 5.205b9, 27; vii.1.241b35; viii.2.252b13, 4.255a25, b30, 5.256b22; tendency, weight, ii.1.192b18, 20; iii.5.205b15, 26; iv.1.208b21, 7.214a1, 8, 8.216a13, 19; science, art, ii.1.192b25, 193a34; nature, ii.1.192b33, 193a1, 9.200a9; iv.1.209a15, 3.210b7; viii.5.258b2; relation, ratio, ii.1.193b16; iv.8.215b, 216a16; aspects, ii.2.193b24 (mathematical); method, ii.2.194a9; consummation, ii.2.194a31; plan, ii.6.197b8, 33; angles, ii.9.200a18; magnitude, size, iii.5.204b1, 8.208a19; iv.1.-209a16, 8, 216b3, 13; viii.6.258b25; number, iii.5.204b8, 18; iv.11.219b3; extension, iii.5.204b20; iv.9.216b32; contrariety, iii.5.204b26; end, iii.6.207a14; all things, iii.6.-207a20; similarity, iii.6.207a20; objection iii.8.208a7; function, iv.1.208b10; place, iv.1.208b24, 25; dimensions, iv.1.209a5; difference, iv.8.215a1; vii.4.249a4, 5; void, iv.9.216b31; and lose, v.6.230b30; vii.3.246b1; kinds, vii.4.249a12; terms, vii.4.249b22; solution, viii.3.253a31; knowledge, viii.4.255a34, b2, 5.257a13, 14; *et passim;* ἔχειν πῶς, *quo modo se habere, aliquomodo esse affectum,* how related, be the case: analogy, i.7.191a10, 11; ii.8.199a18, b3; iv.5.213a1, 8; vii.5.250a8; principles, i.7.191a18; viii.1.252a33; form, ii.2.194b14; chance, ii.4.196a36; necessity, ii.8.198b11; mover, iii.1.201a27; viii.6.258b30, 7.260b3, 10.267b16, 17; remain, rest, unmoved, iv.4.212a24; vi.8.239a15; viii.3.254a24; void, iv.7.213b30; equal, iv.8.216-a14; time, iv.13.222b2, 14.223a16; vi.3.234b6; change, vi.4.234b14, 5.235b14, 236b2, 10.240b25; viii.7.261b25; infinite, finite, vi.4.235b1; excellences, states, vii.3.246b4, 8, 247a2, b3; movement, viii.1.251b2, 5; nature, viii.1.252a17, 20; potential, actual, viii.4.255b4, 5, 8, 12; variously, viii.6.260a7; individual, viii.7.261a5; solution, viii.-8.263a16; καλῶς ἔχειν, *bene sese habere, e re esse,* do well: discuss, i.2.185a19; ii.4.196a36; iv.10.217b30

33a. ἕξις, *habitus,* positive state, disposition: not separate, i.4.188a7; and primary being, ii.1.193a25; not place, iv.2.209b27, 3.210b26; and time, iv.14.223a19; one, v.4.228a8, 15; and qualitation alteration, vii.3.245b7, 246a10, 12, b13, 247a1, b1, 17; εὐεξία, *firma (corporis) constitutis,* good health, wellbeing: as end, ii.3.195a9; relational, vii.3.246b5

33b. ἀπέχειν, *distare,* be absent, distant, far apart, ii.4.197a30; v.3.226b33; vii.4.249a23; viii.8.264b16

33c. κατέχειν, *definere, occupare, continere, obtinere,* occupy: place, iii.5.205a16; iv.1.208b4, 8.216b3, 8; viii.9.265b2

33d. παρέχειν, *adhibere, praebere, ingenerare, facere, afferre, exhibere, efficere,* give, give occasion to: infinity, iii.4.203a12; difficulties, vi.9.239b10; viii.4.255a1; opinion, viii.6.259b3; movement, viii.6.260a9

33e. περιέχειν, *continere,* include, contain, surround, embrace, environ, encompass, comprehend, bracket: contraries, i.5.189a2; factor, ii.3.195a32, 35; all things, iii.4.203b11, 5.205b3, 6.207a19, 25, 29, 31; matter, form, infinite, iii.7.207a35, b1, 208a3, 4; iv.2.209b8; place, iv.2.209b1, 2, 31, 4.210b34, 211a26, 29, 32, 36, b11-15, 212a2, 6, 9, 18, 20, 26, 29, 5.212a31, 33, b9, 13, 16; time, iv.10.218a12, 13, 12.221a17, 28, b4, 33, 222a1, 3; relation to, vii.3.246b6; and movement, viii.2.-253a16, 6.259a3, b11

33f. προέχειν, *habere, praecedere, progredi,* be ahead, vi.9.239b17, 26

33g. συνέχειν, *continuare, continere,* hold together, iv.11.220a10, 13.222a10; v.3.227a12

34. ποιεῖν, *facere, efficere, statuere, agere (in),* present, posit, treat, act upon, produce, construct, make, behave, institute, regard: one, many, i.2.185b31, 4.187-a13, 15, 17, 24; iii.5.205a5; vii.4.428b25; viii.1.250b28; atoms, i.3.187a3; impossible, i.4.188a10; assume, i.4.188a18; contraries, i.5.188a19, 21, 26; and being acted upon, i.5.188a32, 33, 6.189a14, 15, 19, 23, b14, 15, 8.191a35, 36, b1, 3, 7, 8; ii.6.197b12 (chance), 7.198a35; iii.1.201a22, 202a6-9, 3.202a26, 27, 31, b4, 5, 11;

v.2.226a29; viii.1.251b1, 4.255a14; principles, elements, beginning, i.6.189a21, b5, 6; iii.4.203a19, 20, 5.205a26; viii.3.254a17, 5.256b26, 7.260a20; change, i.7.191a7; nature, i.9.192a11, 12; ii.8.198b17, 29; art, ii.1.192b28, 29, 2.194a33, b7; abstractions, ii.2.193b35; in inquiry, ii.3.194b20; factor, ii.3.194b31, 195a22, 4.196a16; iii.4.203b12; with result, ii.5.196b35, 6.197b6, 7; viii.5.257a8; animals, ii.8.199a21, 26; man, iii.2.-202a11; view, iii.4.203a2; number, iii.4.203a7, 6.206b32; problem, iii.4.203b23; infinite, iii.5.204a33, b25, 6.206b28, 30, 7.208a4; vi.7.238b21; investigation, iii.5.204b2; place, iv.1.208b30, 4.211a6, 7, 8.216b14; parts, iv.4.24b22; evidence, iv.6.213b21; void, iv.8.214b23; bulge, iv.9.217a16; present, iv.11.218b26; and time, iv.13.222-b26; viii.8.263b9; as category, v.1.225b7, 2.225b14, 226a24; movement, vi.10.241a4; vii.2.243b1; viii.4.255b30, 10.267a1, 3, 10, 19; conditions, vii.3.246b19; kinds, vii.4.-249b10; light, heavy, viii.4.256a1; actual, viii.7.260b2; treat as two, viii.8.262b6, 263a24, 25, 29; intermediates, viii.8.264b33; attention, viii.9.265b18; soul, viii.9.-265b33; and power, viii.10.266a27. See also 189b

35. πάσχειν, *sustinere, affici, pati,* be related, be acted upon, happen, interact: names, definitions, i.1.184a26; and act upon, (See 34); i.7.190b33; failing, ii.1.193a6; change, ii.1.193a17, 18; motion, iii.5.205b29; body, iv.11.219a5; movement, iv.12.-220b27; viii.4.255b31; by time, iv.12.221a30, b6; as category, v.1.225b7, 2.225b14, 226a24; alteration, vii.2.244b5, 12, 245a1, 3.245b5, 13, 16, 247a4, 12; and power, viii.10.266a29

35a. πάθος, *passio, affectio,* attribute, modification, something undergone, quality, state, subjection, property, condition, aspect, phase: not infinite, i.2.185a34; iii.4.202b33, 5.204a19; not independent, i.4.188a6, 13; and primary being, iii.1.-193a25: and movement, iii.3.202a24; v.1.224b11, 14, 4.227b26; viii.1.251b28, 7.260-a27, 261a35, 8.262a4; and material, iv.2.209b10; vii.3.246a2; of body, iv.3.210b27; vii.3.246b9; to change, iv.9.217b26; of number, iv.12.221a12; and time, iv.14.223a18; viii.1.251b28, 8.263b15; one, v.4.228a8; of quality, vii.2.244b6; event, vii.4.248a15; compare, vii.4.249b5, 7, 15, 18, 26

35b. πάθημα, *passio, affectio,* attribute, modification, πάθησις, *passio,* undergoing, iii.3.202a23, 27, 32, b3, 20; iv.8.216b5; viii.7.260b8

35c. τὸ παθητικόν, *passiuum, patibile,* passive, capable of being acted upon, susceptible: relative, iii.1.200b30, 201a23, 3.202a23, b27; viii.4.255a35; things in contact, iv.5.212b32; quality, v.2.226b29; vii.2.244b5; body, vii.3.246b20

35d. ἀπαθής, *impassibilis, impatibilis,* not acted upon, without attributes, insusceptible, impassive, ἀπάθεια, *non passio, impatibilitas,* imperviousness, iv.5.-212b32, 8.216b15, 9.217b26; v.2.226a29; vii.3.246b19, 20; viii.4.255a13, 5.256b25

36. λέγειν, *dicere, asserere, inquam, pronuntiare, loqui, tradere, intelligi, verbi gratia, explicare, vocare, referre,* say, mean, predicate, give an account, define, describe, formulate, socalled, maintain, explain: in dialectic, i.2; problems, i.2, 6; ii.8; various meanings (see 6, 6a, 141a); "all things one", i.2-6; of primary being, i.2, 3, 6, 7; iii.5; single meaning (see 24d, 105); part, whole, i.4; vi.9; contraries, i.5; v.1, 2, 5, 6; viii.7; becoming, i.7-9; v.1; viii.7; form, 1.7; principles, i.7, 9; iii.5; viii.1;potential, actual, i.8; iii.1, 6; matter, i.9; "not incidentally", ii.1; nature, ii.1; viii.1; art, ii.1; ideas, ii.2; definitions, ii.2; iii.3.202b12; explanatory, factor, ii.3-6, 9; vii.1, 9; necessity, ii.9; change, movement, iii.2, 3, 5; iv.9, 10, 14; v.1, 2, 4, 5; vi.4-6, 10; vii.1, 5; viii.1-10; element, iii.4; reference, iii.4; iv.4, 5, 8; vi.2, 6, 8; viii.3; infinite, iii.5-7; and place, iv.1-4; void, iv.6-9; principle, iv.6.213b9; time, iv.11-14; vi.3, 6, 8; viii.8; substantive, v.1; carried, v.2; continuous, v.3; vi.2; primary, vi.5.235b33, 236a11; material, vii.3.245b10, 14, 16, 246a1, 3; faster, vii.4; much, vii.4; equal, vii.4; mind, viii.5; solution, truth, viii.8.262b28, 29, 263a22, b3; *et passim;* λέξις, *dictio, oratio,* expression, i.2.185b28; λογικός, *rationabilis,* *logicus*, dialectical, iii.3.202a22; λογικῶς, *rationabiliter, logice*, logically, abstractly, dialectically, iii.5.204b4; viii.8.264a8

36a. εἰπεῖν, *dicere, declarare*, declare, speak of, suggest: being, i.3; harmony, i.4; reference, i.5-7; ii.8; iii.1, 2; iv.5, 7, 14; vi.10; vii.2, vii.5, 6, 8; poet, ii.2; factor, ii.3, 4, 8; good, ii.3; movement, change, iii.3, 5; v.1; vi.5; viii.1, 3; place, iv.2; void, iv.6, 8; time, iv.10, 11; viii.8; rest, vi.8; alteration, vii.3; necessary, viii.5; ὡς εἰπεῖν, *ut est dicere, ut ita dicam*, that is to say, virtually: principles, i.7.190b36; void, iv.8.216a8; sciences, viii.3.253b1

36b. εἴρειν, *dicere, praedicere, commemorare*, say: reference, i.3, 5-9; ii.1-3, 5, 6; iii.1-3, 5, 6; iv.5, 7-14; v.1-6; vi.1-3, 5, 6, 9; vii.1, 3; viii.1, 3-6, 8-10; infinite, iii.6.207a15, 8.208a23; place, iv.1.209a27, 2.210a13; movement, rest, viii.1.252a20, 8.264a23; συγείρειν, *continuum, continuare*, refer together, viii.8.262a16

36c. διαρθροῦν, *dearticulare, distinguere*, articulate, organize, ii.8.199b10 σμνείρειν

37. καλεῖν, *vocare*, call: "elements," i.4.187a26, 5.188b28; iii.5.204b33; "activity", iii.3.202a28; "mind", iii.4.203a31; "local motion", iv.1.208a32; v.2.226a33; viii.7.260a28; "void", iv.9.217b22; "rest", v.2.226b14; "Achilles", vi.9.239b14; "passive", vii.2.244b5; προσαγορεύειν *appellare*, call, name: indiscriminately, i.1.-184b13; differently, i.4.187b3; principles, i.5.188a21; whole, part, iv.3.210b2; material, iii.3.245b13, 16

37a. ὄνομα, *nomen*, name, word and definition, i.1.184b10; talk about, ii.1.193a8; "automatism", ii.6.197b29; common, iii.1.201a13; number, iii.7.207b9; "void", iv.7.213b30; "transformation", v.1.225a1; "alteration", v.2.226a27; "continuous", v.3.227a12; ὀνομάζειν, *nominare, denominare*, name: opposite, i.5.188b11; change, v.1.224b7; παρωνυμιάζειν, *denominare*, name, vii.3.245b12

37b. ὁμώνυμος *aequivocus, homonymus*, in another sense, same term, equivocal: "activity", iii.3.202a28; "end',, v.4.228a25; material, property, vii.3.245b16; and comparability, vii.4.248b9-249a25; συνώνυμος, *univocus, synonymus, aequivocus, homonymus*, in the same sense, designated by a single term, vi.3.234a9; vii.4.248b7; viii.5.257b12 (quality)

37c. ἀνώνυμος, *innominatus, vacare nomine, nomine carere*, nameless, v.2.226-a30, 33; vii.4.249b24

38. σημαίνειν, *significare, indicare*, designate, mean, indicate: "circle", i.1.184b11; "white", i.3.186a26; "being", i.3.186a33, b3, 4, 5, 11, 12, 187a2, 4; "come from," i.8.191b9, 26; "quantity", iii.5.206a4; "void", iv.7.213b30; time, iv.12.221a12, 13.222b22; "continuous", v.3.227a12; and comparability, vii.4.248b10, 13, 249a21

38a. σημεῖον, *signum, argumentum, argumento, punctum*, clue, evidence, indication, point, moment: nature, ii.1.193a12; chance, ii.6.197b3, 22; infinite, iii.4.202b36, 203a12, 6.207a2; time, iv.11.219b3, 12.221b5; viii.8.262b2, 263b10, 12, 264a3; on orbit, vi.9.240b3; motion, viii.8.262a6, b12; middle, viii.8.262a23, 29, b4, 7, 25, 263a24, 31

39. καθ' ὅ, καθό, *secundum quod, quatenus*, with respect to which, ii.1.192b25; vi.5.236b1

39a. κατά, *secundum, de, consentaneum, ratione, in*, according to, in, in the order of, as regards: subject, attribute, i.2.185a32, 3.186a33, 34, b34, 4.188a8, 6.189a30; iii.5.204a24; vii.4.249a6-8; nature, ii.1.192b35, 193a1, 2, 32, 8.199a18, 19, 35, 9.200a16; v.6.230a18, 6-231a17; vii.2.243b1, 3.246a15; viii.1.252a12, 4.254b27; art, ii.1.193a32, 34, 2.194b7, 3.195b24, 8.199a17, 19, 33, b1, 9.200a35; definition, ii.1.193b15; matter, ii.2.194a14; meaning, ii.3.195a6; potentiality, actuality, ii.3.-195b16, 20; iii.6.206a16; iv.5.2.12b4, 13.222a18, 14.223a20, 21; vii.2.244b11, 3.247b4; categories, iii.1.200b33, 34, 201a5-7, 6.207a7; iv.1.208a32, 10.218a25; v.1.225b8,

2.225b10, 6.229b27; vi.2.223a26, 5.236b18; viii.7.260a27, 28; respect, iii.1.201a21, 6.207a24; iv.13.222b6; vi.8.238b35; viii.5.257b10; part, iii.4.203b1; iv.3.210a29, b1; vii.3.247b6; addition, subtraction, iii.4.204a6, 7, 6.206b3-33; analogy, ratio, iii.6.-207a4; iv.8.216a15; magnitude, size, iii.7.207b23; vii.4.248b24; viii.7.260a27; place, path, position, iv.1.208b16, 23, 2.209a31; v.3.226b22; vi.7.238a6, 8.239a25, 30, 34, 35, b3, 9.239b6; viii.7.260a28, 8.262a29; being, iv.3.210b16; viii.7.260b19, 261a19; contrariety, iv.9.217b24, 25; time, before, after, iv.11.219b2, 220a25, 14.223a9; vi.6.236b21; genesis, v.3.227a24; derived sense, vi.3.233b33; movement, change, vi.4.235a30, 31, 5.236b2; viii.3.253b31; contradiction, vi.5.235b13, 30; viii.7.261b8; end, beginning, vi.5.236a10, 13; viii.6.260a11, senses, vii.3.247b19; viii.3.254a26; theory, viii.1.250b22; truth, viii.3.254a25; other, viii.5.257a31; *et passim*

40. καθ' ἕκαστον, *singulare, particulare, secundum unumquodque*, καθ' ἕκαστα, *singulare*, particular, individual, severally: in science, i.1.184a24, b12; and sense perception, i.5.189a6; factor, ii.3.195a32, b13, 17, 26; movement, change, iii.3.202b27; vi.5.233b19; whole, iii.6.207a11; part, vi.1.232a12, 7.238a25, 8.239a27; viii.8.263a8; infinite, vi.2.233a23; moment, vi.6.237a15; generation, destruction, viii.7.261b5; ἕκαστον, *unumquodque, quaeque, res*, each: subject matter, i.1-184a12; ii.2.194b12, 3.194b18, 19, 23; iii.1.200b23; instance, i.3.186b3; whole, i.4.187b6; ii.1.193a10, 29; not everything in everything, i.4.187b27; part, i.4.188a4; vi.7.237b31, 238a26; viii.5.257b28; change, process, i.7.189b32; ii.8.199a10; iii.3.-202b24; v.5.229a25, 35; being, i.7.190b19; ii.1.192b13, 17, 193b7; anything, i.9.190a31; ii.8.198b13; art, ii.1.192b27, 30; material, ii.1.193a17; as limit, ii.2.193b32; explain, ii.3.195b22, 4.196a13, 7.198b9; kind, category, ii.4.196a32, 8.199b17; iii.1.201a3, 10; v.2.226a25; power, iii.1.201b8; mover, moved, iii.3.202a34; vii.1.242a61, 62, 64, 65, b51, 57; viii.1.251a18, 6.258b12, 30, 259a4, b16, 260a10; movement, iii.3.202b29; iv.14.223a33, b13; viii.1.251a11, 13, 5.257b12; element, iii.5.204b19; body, iii.5.205a11; iv.1.208b11, 2.209b2, 8.214b13, 215a5, 216b15; vi.9.240a13; place, iii.5.206a7; iv.1.208b19, 25, 209a13, 4.211a27, 28, 5.212b30; number, iii.7.207b10; infinite, iii.7.207b26; together, etc. v.3.226b20, 227b2; subject, vii.4.249a1; point, viii.9.265a34; ἑκάτερον, *unumquodque, utrumque*, each: individual, i.1.184b14; part, i.2.185b15, 3.186b13, 32, 34; iv.3.210b10; vi.4.234b26, 27, 31, 235a1, 28, 33, 5.235b37, 236a1; contrariety, i.6.189b20; iv.9.217b16; v.1.224b31, 5.229b20; viii.5.-257b14, 16; great, small, i.9.192a8; v.2.226a30; mathematics, ideas, ii.2.194a2; matter, form, ii.2.194a17, 18; chance, automatism, ii.4.196b7, 6.198a3; element, iii.5.205a27; motion, movement, iii.5.205b29; iv.14.223b2; vii.5.250a25; kinds, iv.14.-223b5, 224a3; limit of, v.3.227a11; time, now, vi.3.234a8, 20; change, vi.5.235b12, 13; terms, vii.4.249b25; nature of, viii.6.259a28; points, viii.8.262a20

41. ἄτομον, *individuum, indivisible, (specialissima sc.species), atomum*, atom, indivisible: magnitude, i.3.187a3; lines, iii.6.206a17; present, time, iv.13.222b8; vi.5.235b33, 236a6, 8.230a9, 11, 10.241a25; viii.8.263b26-264a4; species, v.4.227b7, 30; vii.4.249b21; viii.5.257b4; and continuous, vi.2.232a24, 233b17, 19, 24, 26, 30; and processes, viii.5.257a1, 9.265b29

42. ἴδιον, *proprium*, pertinent, peculiar, special, own: objection, i.3.186a23; to kinds of change, i.7.189b32; after common, iii.1.200b24; place, iv.2.209a33; name, v.2.226a33; animate, viii.4.255a7

43. καθόλου, *universale, universaliter*, generality, universal: in science, i.1.184-a23-26; iii.1.200b22; excess, deficiency, i.4.187a16; and reason, i.5.189a5, 7; movement, change, iii.3.202b23; vi.5.235b29; viii.8.264a21, 265a8; infinite, iii.5.204a34; and particular, vii.3.247b6; and nature, viii.5.257a34

44. ὅλως, *omnino*, (not) at all, generally: nonbeing, being, i.3.186b9; iv.6.213-a30; viii.2.252b9, 8.263b25; excess, deficiency, i.6.189b10; material, i.7.190b25;

ANALYTICAL INDEX OF TECHNICAL TERMS 197

ii.3.195b9, 9.200a25, 28; change, i.8.191b33; vi.4.235a35; number, ii.3.194b28; iv.-12.221a12; maker, ii.3.194b31, 195a22; animal, ii.3.195b; image, ii.3.195b8; chance, automatism, ii.4.195b35; conduct, ii.6.197b2; moved, movers, ii.7.198a27; art, ii.8.-199a15; nature, ii.8.199b14; relative, iii.1.200b30; movement, change, iii.3.202b19; iv.7.214a27, 8.216b25; v.2.225b15, 226a19, b10; (unmoved) 6.230b10, 27, 32; vi.9.-240a30, 10.240b31, 241b6; vii.3.246b12, 247b12, 5.250a15; viii.5.258a15, 6.259b4, 7.261a15, 10.255a28; infinite, iii.5.205a1, 8, b24, 6.206a27; viii.10.266a24; "in", iv.3.-210a16-24; growth, iv.7.214b3; void, iv.9.217a7; time, iv.10.217b32, 12.221b25, 29; vi.6.237a4, 8; states, v.4.228a8; continuous, vi.2.233a25, b17, 8.239a22; beginning, principle, vi.5.236a14; viii.1.252a32; opinion, viii.2.254a27; generation, viii.5.257b10, 7.261a13. See also 21

45. ἐπὶ μικρόν, aliquantum, paululum, briefly, i.2.185a19; ii.2.194a20

46. φάναι, dicere, inquirere, aio, affirmare, assertio, firmare, declare, describe, speak of, allege: being, i.2, 3, 5, 6; iv.6; viii.3; attribute, i.3; great, small, i.4; one, many, i.4, 6, 8; becoming, i.7, 8; vii.3; viii.1; matter, privation, i.9; iv.4; nature, natural, art, ii.1, 8; factor, chance, ii.3-7; reference, iii.1.200b35; iv.9.-216b26; infinite, iii.4-6; place, iv.2, 4; void, iv.6-9; viii.9; time, now, iv.11-14; vi.3, 9; rest, vi.8, 9; altered, vii.2; viii.3; excellences, vii.3; movement, vii.4; viii.1-5, 7-9; φάσις, affirmatio, affirmative, vi.10.241a28

46a. φάσκειν, dicere, statuere, affirmare, declare: principles, elements, i.2.184b17, 6.189b16; movement, iii.2.201b21; vii.1.241b42; viii.8.265a4; infinite, iii.5.204a13; void, iv.1.208b26, 6.213a20, 21, b3, 7.214a23, 8.216a22; other, vii.4.249a1; mind, viii.5.256b25

46b. ἀποφαίνεσθαι, enunciare, pronuntiare, pronounce, iii.5.204a32; iv.-2.209b16

46c. ἀξιοῦν, dignum esse, censere, velle, existimare, putare, postulare, require, assume, deign: place, iv.4.201b34; solution, iv.7.204b7; overtaking, motion, rest, vi.9.239b26, 240a3; vii.3.254a15, 8.263a5, 7; alternation, viii.1.252a24; to search, viii.1.252b1. See also 60d, 67

46d. διδόναι, dare, concedere, give, admit, grant: absurdity, i.2.185a11, 3.186a9; explanation ii.3.194b34; finite, vi.9.239b29; time, vi.9.239b, 32; viii.8.263b20, 10.266b14; ἀποδιδόναι, assignare, explicare, reddere, demonstrare, derive, deal with; from principles, i.6.189a16; explanation, ii.7.198a23, 32, b5; definition, iv.4.211a8; motion, iv.8.217a5; viii.9.265b19; ἐνδιδόναι, acquiescere, assentire, yield: to argument, i.3.187a1; παραδιδόναι, tradere, hand down: concerning time, iv.10.218a32

47. κατάφασις, affirmatio, positive assertion, v.1.225a6, b4

48. ἀπόφασις, negatio, negative, v.1.225a19; vi.10.241a29

49. ἀντίφασις, contradictio, opposite, contradiction: not possible, i.3.187a5; and change, v.1.224b29, 225a12, b1, 3.227a9; vi.5.235b13, 16, 6.237b2, 9.240a19, 27, 10.240b22, 241a27, 28; viii.7.261b8; ignorance, viii.4.255b5; rest in, viii.7.261b11. See also 13, 50

49a. ἀπαντᾶν, contradicere, occurrere, object, reply, iii.8.208; iv.6.213b3; viii.3.2.253b13, 8.263a4

50. ἐναντίον, contrarium, contrary: principles, i.2.184b22, i.4 (excess and deficiency, separation from the "one" or the "mixture"), 5 (arguments: primary contraries, pattern of processes, familiar or intelligible), 6 (number of principles), 7 (terms and subject of change); differ, i.3.186a21, 7.191a18; viii.8.262a5, 264b8; exemplified, i.4.187a16, i.5, 6.189a18, 19, 22, 24; iii.5, iv.5, 6; differentiae, i.4.-187a19; infinite, i.4.187a25; coming into being, change, movement, rest, i.4.187a32, 5.188b22, 25, 6.189a18; ii.1.193b21; iii.5.205a6; v.1.224b29, 31, 33, 225a12, 32, b3, 4,

2.226a7, b1-8, 15, v.5, 6; vi.6.237b1; 10.241a27, 30, 32; 10.241a27, 30, 32; viii.1.-251a29, 2.252b11, 7.261b16, 8.261b33, 262a6, 11, 264a18, 28, 29, b12, 14, 15; genera of, i.5.188a24; primary, prior, i.5.188a28, 6.189a18; pass into, i.5.188b23; vii.3.247b14; and intermediates, i.5.188b24; v.3.226b23, 227a910; not one, i.5.189-a12, 6.189a12, b9; contrariwise, i.6.189b15; iv.14.223a8; vii.3.246b20; viii.6.260a9; not primary being, i.6.189a29, 33; v.2.225b11; not subject, i.7.190b33, 35; to the divine, i.9.192a17, 19; mutually destructive, i.9.192a21, 22; iii.5.205a25; viii.8.262-a11; explain, ii.3.195a12, 13; events, theory, ii.4.196b5; and potentiality, iii.1.201-a35; elements, iii.5.204b14; to given views, iii.6.206b33; iv.8.214b29, 21ba21; viii.2.-252b7; smaller, greater, iii.7.207b4, 16; and material, iv.9.217a22, b18; iv.12.222a7, 13.222b6, and eternal time, generation, destruction, v.1.225a33, 2.22ba7; and place, v.3.226b32; vi.10.241b3; viii.4.255b12, 6.260a8, 8.261b34, 264b15; mixture with, v.4.229a3; directions, extremes, vi.9.239b34, 240a15; viii.7.261b1, 17; and science, viii.1.251a30; and alteration, viii.3.253b30, 7.260a31-33, 261a35; rest in, viii.7.261b3; changes, viii.7.261a36b8, 11, 8.264b26. See also 13, 49

50a. ἐναντιότης, *contrarietas*, contrariety: in the "one", i.4.187a20; and elements, i.6.189b4; and movement, v.5.229a24

50b. ἐναντίωσις, *contrarietas*, contrariety: one in one category, i.6.189a13, b18-27; and action, i.6.189a24; incidental, i.7.190b27; aspects of, i.9.192a14; elements, iii.5.204b27; and material, iv.9.217a23, b24; and movement, v.2.226a26, 5.229a23; viii.7.261b14; of place, v.6.230b11; viii.8.261b36

50c. ἀντεστραμμένως, *e contrario, modo converso, modo contrario, vice versa*, inversely: addition, iii.6.206b5, 27, 207a23; time, iv.8.215b31; vi.z.233a9; ἀντιστρόφως, *conversim, reciproce*, conversely, viii.9.265b8

51. ἀμφισβήτησις, *ambiguitas, dubitatio*, playing fast and loose with, viii.253a34; ἀμφισβητεῖν, *ambigere, in dubium revocare, dubitare, controversari*, fly in the face of, clash, viii.3.253b29, 254a8

52. ὁμολογεῖν, *confiteri, fateri*, acknowledge, agree, admit: one, many, i.2.186-a1; becoming, i.4.187a35, 8.191b36; movement, vii.1.242a46; viii.1.251a12; ὁμολογουμένως, *confiteri, omnium concessione*, admittedly, viii.8.263a10

53. σεμνότης, *dignitas, excellentia*, dignity, iii.6.207a19

54. ἀναλογία, ἀνάλογον, *analogia, proportio, similitudo*, analogy, proportion: contraries, i.5.189a1; persistent nature, i.7.191a8; in movement, iv.8.215b3, 29, 216a7, 11.219a17; vii.5; viii.10.266b12, 19; differences, vii.4.249a25

55. οἰκεῖον, *proprium, accomodatum, proprietas*, appropriate, resident, proper, intimately related: question, i.2.185a21; factor, ii.3.195b3; to natural science, iii.4.203a1; place, iv.4.211a.5, 5.212b34, 8.215a17; viii.3.253b34, 35, 4.255a3; to time, iv.14.224a16; nature, vi.10.241b1; conditions, vii.3.246b9, 10, 247a4; arguments, viii.8.264a7, b2

55a. κυρίως, *proprie*, chiefly, strictly, above all: becoming, i.8.191b7; same, iii.3.202b20; infinite, iii.4.203b22, 6.207a4; whole parts, iii.6.207a11; vi.9.240a24; "in", iv.3.210a24; place, iv.4.212a23; v.2.226a34; movement, viii.6.259b7, 7.261a24, 9.266a1

55b. τρόπος, *modus*, manner, sense, type, way, interpretation, point of view: in investigation, i.1.184a19; ii.7.198a34; iii.1.200b16; relation, i.1.184b10; "all things one", i.2.185b7, 3.186a4; of argument, proof. i.3.186a22; ii.8.198b34; vi.8.239a19; viii.8.213a4; definition, i.3.186b27; iii.2.202a1, 3.202b28; physical accounts, i.4.187-a12; of existence and origination, i.4.188a15; differ, i.5.188b35, 6.189b13, 9.192a8; analysis, i.7.190b22; principles, i.7.190b35, 191a5, 21; becoming, i.8.191a36, b19, 27; "nature", ii.1.193a28, 30, b3; in science, ii.2.194a9; "explanatory factor", ii.3.194b23, 195a7, 10, 15, 27, b29, 4.195b33; (chance), 6.198a2; "infinite", iii.4.204a3; "what participates", iv.2.209b13; "in", iv.3.210a15, 17; potentiality, iv.5.213a4; void, iv.6.-213a12, 25, b15, 7.214a7, 12.217b28; movement, rest, v.5.229b8, 16, 6.230b11; vi.4.234b21, 235a25, 9.240b6; vii.2.243b18, 245a7; viii.4.254b24, 5.257a33, 6.260a4, 9.265b30; completion, vii.3.246a20; of throwing, viii.5.257a3; solve, viii.10.267a18; τρόπον τινά, quodammodo, quomodo, in a way, ii.9.200a16; iii.6.206b17

56. ἁρμονία, harmonia, harmony: pattern, i.5.188b15

56a. ἁρμόττειν, consonare, in tune, be consonant, i.5.188b12, 13, 7.190b32; iv.1.209a9, 8.214b23

56b. ἐφαρμόττειν, convenire, congruere, apply, fit, exhibit a harmony, iii.1.201-b14; v.4.228b25, 229a6

56c. ἀναρμοστία, inconsonantia, harmoniae privatio, disharmony, i.5.188b14

57. ὁμοιότης, similitudo, analogy, similarity: good fortune, ii.6.197b9; infinite, iii.6.207a4, 20; differences vii.4.249a24; alterations, vii.4.249b4

57a. ὁμοειδής, similis species, quae formam similem habent, ejusdem speciei, unius formae, similar in form, of the same kind, homogeneous: Anaxagoras, i.4.188a13; factors, ii.3.195a29; and infinite, iii.5.205a13; place, iii.5.205b21; member, vii.4.249b24

57b. ὁμοιότροπον, simili modo, simile, in a similar way, viii.1.251a31

57c. ὁμοίως, similiter, non aliter, peraeque, similarity: search, i.2.184b22, 185a5; change, i.5.188b8, 17; iv.9.217a31; vi.5.235b11, 18, 236b2, 16, 9.240a26, 10.241a28; viii.7.261b3, 25; contraries, i.6.189a23; great, small, i.9.192a7; art, nature, ii.1.192b27, 8.199a18, b2, 29, 9.200a7; material, ii.1.193a19; sciences, ii.2.194a24; viii.3.253b4; factors, ii.3.195b6; chance, ii.4.196a5, 8.198b21, 27; categories, iii.1.-201a7; movement, rest, iii.3.202a18, 21; iv.8.214b31, 14.223a3; v.1.225a29, 6.230a16, 23, 231a16; vi.1.232a18, 3.234b6, 4.235a16, 7.238a25, 8.239a15; vii.2.246b7, 13, 245a9, 4.248a20; viii.1.252a5, 2.252b21, 34, 3.253a27, b14, 254a33, b2, 4.255a23, 5.257b10, 8.264a14, 9.265b23, 10.266b32, 267b6; part, whole, iii.4.203a23; elements, iii.5.204b31; infinite, finite, iii.5.205b16; vi.4.235b1; viii.8.263a14; number, iii.7.207-b9; place, iv.1.208a27, b20, 2.209a35, b26, 8.215a23; viii.8.262b3; void, iv.8.215b19; equal, iv.8.216a14; time, iv.10.218a31, b4, 13, 11.219b16, 12.220a31, b22; vi.4.235-a22, 6.237a6; perish, vi.6.237b13, 16; alteration, vii.2.244b8; viii.3.253b23, 7.260a30, 32; relations, vii.3.246b7; states, vii.3.246b20, 247b9, 248a1, 4.249b7; ratio, vii.5.250a8; actualization, viii.7.260b3, 10.267b16, 17; point, viii.9.265a33

57d. παραπλήσιος, similis, affinis, analogous, related, ii.9.200a16; iv.6.213a14; v.4.228a7; **πλησιάζειν**, proximum esse, propinquum esse, proximare prope accedere, come into close relations, viii.1.251b3

57e. εἰκός, merito, decere, except, ii.4.196b4; **εἰκότως**, merito, really, iv.2.209-b17, 8.214b23

57f. ἀνόμοιος, dissimilis, unlike, iii.5.205a19, 20; viii.7.260a31; **ἀνομοιότης**, dissimilitudo, unlikeness, vii.4.249b23

57g. ἴσος, aequalis, tantumdem, equal parts, i.4.187b33; iii.6.207a17; angles, ii.9.200a17; elements, iii.5.204b13, 16, 17; place, iv.4.211a27, 28, 33; bodies, iv.6.-213b11, 22; vi.9.239b6, 240a3; movement and change, vi.8.215b2, 23, 216a, b9, 16, 9.216b26, 27, 217a16, 17; vi.1.231b22, vi.2, 4.235a1, 3, 8, 10.241a9, 11, 12; vii.1.242b47, 66, vii.4, 5, 8.262b10; time, numbers, iv.14.223b1-12, 224a3; vi.2.7.237-b29, 238a24, 9.239b35, 240a1-17; vii.4, 5; viii.1.252a31, 10.266a17, 20, b4; and infinite, finite, vi.7.238a13, 15, 25; ambiguous, vii.4.249b19, 249a1; force, vii.5; angles, viii.1.252b3; and measure, viii.7.261b19; effect, viii.10.266a26; **ἄνισος**, inaequalis, unequal: movement, iii.2.201b20, 22; bodies, iv.6.213b12; vi.9.239b34; and infinite, finite, vi.7.238a14, 15, 25; alteration, vii.4.249b19. See also 70

58. συμβλητός, *comparabilis*, comparable, vii.4

58a. ἀσύμβλητος, *incomparabilis, inter se inferri nequire*, incommensurable, incomparable, iv.9.217a10; vii.4.248b7

58b. συμβάλλειν, *profici, conferre, comparare*, contribute, compare, iv.4.212-a12, 8.216b15; vii.4.249b5

58c. βάλλειν, *projicere*, throw, viii.4.225b28

58d. ἀποβολή, *remotio, abjectio, amissio*, loss, v.5.229b13; vii.3.245b8, 246b13, 247a5; ἀποβάλλειν, *abjicere, rejicere*, lose, v.6.230b29,30; vii.3.246a18

58e. καταβάλλειν, *projicere, dejicere*, knock down, vii.5.256a24

58f. ὑπερβάλλειν, *excellere, superare, exsuperare*, exceed, be greater: not, i.4.187b30; iii.6.206b18, 19, 21, 29; element, iii.5.204b18; magnitude, and infinite, finite, iii.7.207a34-b21; viii.10.266b3, 15, 20; ratio, iv.8.215b22; multiple, vi.2.223b3; movement, vi.4.235a7

59. δηλοῦν, *demonstrare, declarare, ostendere, monstrare, significare*, clarify, show: mathematical, physical, ii.2.194a7; place,iv.1.208b10, 22; transformation, v.1.225a1, 2; substantive, v.1.225a7; privation, v.1.225a4

59a. δῆλον, *manifestum, perspicuum, ostensum, constat, palam*, plain, clear: in science, i.1.184a14, 21; from induction, i.2.185a13; fallacy, i.3.186a11; impossibility, i.3.186b11, 4.187b20; principles, i.5.188a27, 6.189a20, 7.191a4, 21; ii.3.194-b20; becoming, i.7.190b10, 23, 191a20, 8.191b3, 9; abstraction, ii.2.194a2; chance, ii.5.196b19, 197a5, 6.198a8; factors, ii.7.198a15; end, ii.8.199a17, b31; mistakes, ii.8.199a35; movement, change, iii.1, 2; v.1.2, 4-6; vi.2, 5, 7; vii.1.2; viii.1-9; infinite, iii.4.203b3, 5.205a9, 27, b18, 6.206a10, 13, 25; vi.4.235b5, 7.238a27, b1; viii.10.266-a26, b27; place, iv.1.208b1, 6, 209a8, 25, 3.210b9, 17, 22; void, iv.7.214a10, 31, 8.214b15, 216b17, 21, 9.217a6, 14, 27; time, iv.12.221a7, b9, 27, 13.222a31, b19, 23; vi.3.234a23, 6.236b23; viii.8.263b9; union, v.3.227a26; rest, v.6.230a15, 231a8; vi.3.234a34; viii.3.254a14, 6.259a27, 260a14; continuous, vi.1.231b13, 20, 2.233a4, b15; incomparables, vii.4.248b25

59b. ἄδηλον, *immanifestum, incertum, obscurum*, obscure, not evident: not possibility of, inability to discriminate, ii.1.193a6; chance, ii.4.196b6, 5.197a10; not movements, iii.3.202b24; viii.4.254b28; displacement, iv.8.216a29; time, iv.10.-218a32

59c. σαφές, *certum, clarum*, evident, clear, apparent, ἀσαφές, *incertum, obscurum*, obscure: in scientific procedure, i.1.184a17, 19, 20, 21; iv.5.213a5, 6

59d. φανερόν, *manifestum, patet, constare*, evident, exhibit, evident, obvious, visible: definition, i.3; being, i.3; part, whole, i.4; patterns, i.5; principles, elements, i.5, 6; becoming, i.7; nature, ii.1; factors, ii.3, 7; chance, ii.5 ends, ii.8; necessity, ii.9; movement, change, rest, iii.1, 3; vi.2, 5-7, 10; vii.1, 2; viii.1-8, 10; infinite, viii.10; iii.5-7; place, iv.3-5; void, iv.7-9; time, now, iv.10-14; vi.2, 3, 5, 6; viii8; between, v.3; continuous, v.3; successive, v.3; becoming, vi.6; vii.3; stop, vi.8; alteration, vii.3; viii.3; disappear, viii.3.

59e. συγκεχυμένα, *confusa*, confused situation to be analyzed, i.1.184a22

60. ὀρθός, *rectus*, right, justifiable: attributes, not independent, i.4.188a6; not things similar, i.4.188a13; becoming, i.9.192b1; chance, ii.5.197a11, 18; procedure, ii.8.199a34, b1; viii.1.252a33, b1, 2; place, iv.1.208b29; time, number, iv.13.222-b19, 14.224a2; angle, viii.1.252b2; movement, viii.2.252b31; "mind", viii.5.256b24

60a. εὔλογος, *rationabilis, rationi, rationi consentaneus, optima ratione*, justifiable, with good reason: principles, i.5.188a27; iii.4.203b4; chance, ii.5.197a12, 31; infinite, iii.7.207b1; place, iv.5.212b30; time, iv.12.220b24; movement moves, viii.5.256b13, 23, 9.265a27

60b. ἄλογον, *irrationabile, a ratione alienum, absurdum, sine ratione,* unreasonable, incongruous, absurd: infinite in infinite, i.4.188a5; not persistent being, i.6.189b5; function, iii.3.202b1; not place, iv.5.212b34; contrary, v.6.230a4; before movement, viii.1.251a21; axiom, viii.1.252a24; movement, viii.4.255a10, 5.257a14

60c. παράλογον, *extra rationem, a ratione alienum,* incalculable: chance, ii.5.197a18

60d. ἄξιον, *dignum,* well: wonder, ii.4.196a28, 36; ask, iv.14. 223a16; ἀξιολόγως, *rationabiliter, cum dignitate,* well, iii.4.203a1. See also 46c, 67

60e. ἱκανός, *sufficiens, satis, sufficient:* one passive element, i.6.189b19; one contrary, i.7.191a6; nature, i.9.191b36; factors, ii.3.195b30; not, state without reason iii.5.205b8; evidence, iv.13.222b22; principle, viii.1.252a33; see, viii.3.254a35; one mover, viii.6.259a12; solution, viii.8.262a16, 18, 21

61. ἄτοπον, *inconveniens, absurdum, ineptum,* absurd, foolish, unreasonable: Melissus, i.2.185a11, 3.186a9, 13; "all things one", i.2.185a30, 3.186a8; undertake an impossibility, i.4.188a9; not to know essential attributes, ii.2.193b28; chance, ii.4.196a7, 19, b1, 2; movement, rest, ii.8.199b12, 26; iii.3.202a34, b6; v.6.229b32, 231a7; vii.3.246a4, 4.248a19, 249a11; viii.2.253a3, 5.257a24; and infinite, iii.5.204a32, 205b2, 6.207a30, 7.208a3, 8.208a15; viii.8.263..a12; void, iv.7.214a4, 8.216b11; completion, vii.3.246a18; and time, viii.7.261b23

61a. γελοῖον, *ridiculum, derisoriom,* ridiculous, absurd: to prove, ii.1.193a3; poet ii.2.194a31; in vain, ii.6.197b28;

61b. εὐηθικός, *stultus, fatuus,* trivial, iv.10.218b8

61c. πλασματῶδες, *figmentum, commentitium,* fanciful, viii.5.256b23

62. συλλογισμός, *syllogismus, conclusion,* συλλογίζεσθαι, *syllogisare, concludere, ratiocinare,* reason: wrongly, i.2.185a10, 3.186a6, 8; about colors, ii.1.-193a7; void, iv.7.214a2, 4; movement, rest, vi.9.239b32

62a. πρότασις, *propositio, premise,* ii.7.198b7

62b. συμπέρασμα, *conclusio,* conclusion: does not follow, i.3.186a 24, 25; from assumptions, ii.3.195a18, 7.198b8, 9.200a21

62c. παρασυλλογίζεσθαι, *paralogisare, vitiose argumentari, male ratiocinari, captiose ratiocinari,* commit a fallacy, i.3.186a10

62d. παραλογισμός, *deceptio, rationis fallacia,* fallacy, ii.9.239b5, 240a2

63. δεικνύναι, *demonstrare, probare, ostendere, determinare, fidem facere,* expound, prove, demonstrate, determine, ἐπιδεικνύναι, *demonstrare, demonstrate,* show: erroneously, i.2.185a15, 3.186a5; forms, i.9.192b2; nature, ii.1.193a3, 4; and infinite, iii.7.207b33; viii.10.266b21, 267b21, 24; place, iv.4.211a11; void, iv.6.213-a22, 25, 26, 31, b12, 15, 7.214b7, 10; continuous, vi.2.232a23; distance, vi. 2.232b26; now, vi.3.234a3; time, vi.3.234a11, 10.241a2, 19; divisible, indivisible, vi.4.235a31, 34, b4, 5.236a13, 29, 6.236b32; viii.5.257a34; change, vi.6.237a24; stop, rest, vi.8.-238b28, 239a5, 7, 19; mover, vii.1.242b54, 55; generation, viii.3.254b6, 6.259a15, 27, 28, 7.260b24, 261a28

63a. ἀπόδειξις, *demonstratio, proof,* demonstration, ἀποδεικνῦναι, *demonstrare,* demonstrate, vi.2,233a7, b14, 6.237a35, 7.238a32, b16, 8.238b26, 10, 240b8; viii.1.252a24

64. θέσις, *positio, thesis, situs,* position, convention: arbitrary view, i.2.185a5; as a genus of contraries, i.5.188a23, 24; place, iii.5.205b34; iv.1.208b16, 21, 23, 25, 11.219a16; viii.4.254b24; and succession, v.3.226b35; τιθέναι, *ponere, referre, concedere, collocare, proponere,* suppose, treat, assume, propose, insist, classify: principles, i.5.188b29; iii.4.203a3, b4; factors, i.5.188b35; movement, iii.2.201b19, 24, 29, 34; v.5.229b15; viii.1.252a24, 4.254b35, 8.262b17; infinite, iii.4.203b31;

place, iv.1.209a14, 19; void, iv.6.213a16, 8.214b25, 27, 216a28; possibility, vii.1.-243a31; viii.5.256a11; well-being, vii.3.246b6; mover, viii.5.257a30, 6.259a30; stop, viii.8.262a31; stage, viii.8.265a1

64a. ἐκτιθέναι, *ponere, exponere,* abstract, vi.4.235a28

64b. ἀντιτιθέναι, *contra, ponere, opponere,* oppose, v.5.229a16; viii.4.254b35; ἀντίθεσις, *oppositio,* opposition, v.1.225a11, 13

64c. διάθεσις, *dispositio, affectio,* stage, arrangement, disposition, διατιθέναι, *disponere,* dispose: opposite, i.5.188b11; and primary being, ii.1.193a15; body, vii.3.246b9, 247a3

64d. ἐντιθέναι, *imponere,* place in, insert, iv.8.214b19, 21, 216a33

64e. περιτιθέναι, *circumponere,* place around, iii.4.203a13

64f. πρόσθεσις, *oppositio, adjectio, appositio, additio,* addition, position: and growth, i.7.190b6; infinite, iii.4.204a7, 6.206a15, b3-33, 207a23, 7.207a33; and movement, viii.8.264b11; προστιθέναι, *apponere, adjicere, addere,* add: infinite, iii.6.206b6; parts, viii.10.266a19; power, viii.10.266b2

64g. συντιθέναι, *componere, conjungere,* put together: knowledge of composite, i.4.187b12; beings, i.5.188b10; terms, i.7.190a12; in becoming, i.7.190b11; and infinite, iii.5.204b11; forces, weights, vii.5.250a26; and movement, viii.9.264a21; σύνθεσις, *compositio,* composition, predication: pattern, i.5.188b16, 21; materials, i.7.190b8; factor, ii.3.195a21; nonbeing in, v.1.225a21

64h. ὑπόθεσις, *suppositio, hypothesis,* assumption, ὑποτιθέναι, *subjicere, supponere, concedere, apponere, ponere,* take as basic, assume, suppose: changeful, i.2.185a12; "being", i.3.186b4; "nature", i.6.189a 28; iii.4.203a17; principle, i.6.189b1; and conclusions, ii.3.195a18; vii.1.242b73, 243a1; and necessity, ii.9.199b34, 200a13; time, iv.11.219a30; change, movement, vi.10.240b28; vii.1.242a44, 57, 2.244b5; viii.1.250b22, 3.253b5, 7.260b24, 261a29; alternation, viii.1.252a25, 30

65. λαμβάνειν, *recipere, accipere, sumere, concipere, opinari,* assume, grasp, get, take, derive, select: false, i.2.185a9, 3.186a7, 11, 25, 26, 32, 4.188a13; vi.2.-233a22; principles, elements, i.4.188a18, 5.188b31, 189a1, 6.189a26; not chance, i.5.188a31, 34; ii.4.196a6; change, i.7.190a13; vi.5.235b19; form, i.7.191a11; ii.1.193b1 opinion, i.8.191a33; power, ii.1.193a13; viii.10.266b1, 21; why, ii.3.194b19, 20; good, evil, ii.5.197a28; categories, iii.1.200b35; movement, iii.2.201b33; vii.1.241b37, 242a37, 38, 62, 65, 2.243b14; viii.1.251a24, 4.255a20, 5.257b22, 34, 6.259a32, b6, 7.261b18; initial terminus, beginning, iii.4.203b9; viii.5.257a31; part, iii.5.204a22, 205b20, 6.206b1, 7, 18, 207a3, 5, 8, 7.207b12, 8.208a21; iv.9.217b6, 13, 14, 10.218b3; vi.2.233a35, 4.235a33, 6.237b6 (first), 7.237b28, 238a6, 10, 11, 22; 10.266b11-14; potential, iii.6.206a19; one after another, iii.6.206a28; infinite, iii.6.206a30; dignity, iii.6.207a18; place, iv. 3.210a14, 4.210b32, 212a8; void, iv.6.213a20, 7.213b30, 8.216a6; time, iv.10.218a1, 25, 11.219a2, 25, 220a12, 12.221a27, 13.222a25, 14.223b27; vi.2.-233a12, 4.235a18, 6.237a4, b2, 7.238b16, 9.239b31; vii.1.242b42; viii.1.251b23, 24; different, iv.11.219b20; health, v.4.228a11; vii.4.249b6; body, vi.2.232b7; orbit, vi.9.240b2; possible, vii.1.242b49, 67, 243a30; excellence, state, vii.3.246a13, b1, 247a18, b16; species, vii.4.249b11; determinate, viii.6.259a10

65a. λῆψις, *acceptio, assumptio, susceptio, acquisitio,* gain, acquisition, v.5.229-a25, b13; viii.3.245b8, 246b13, 247a5, b10

65b. ἀπολαμβάνειν, *accipere,* recover, attain; loan, ii.5.196b33; end, viii.7.261a18

65c. διαλαμβάνειν, *distinguere, permeare per, occupare, interrumpere, interjicere, intercipere,* break, distinguish, iv.6.213a33; v.4.228b6; viii.4.254b29, 8.264a20

65d. ἐναπολαμβάνειν, *comprehendere, accipere,* intercipere, divide, let in, iii.4.203a11; iv.6.213a27

65e. καταλαμβάνειν, *reperire, j ungere,* comprehendere, conjungere, meet, catch, overtake, ii.4.196a4; vi.9.239b15, 20, 27, 28

65f. μεταληπτικόν, *receptivum, vim habere recipiendi,* what participates, iv.2.209b12, 14; μεταλαμβάνειν, *accipere,* sumere, substitute, vi.2.233a5

65g. περιλαμβάνειν, *comprehendere, accipere,* contain, keep: whole, parts, i.1.184a26; question only, ii.7.198b15; parts, iii.6.206b9, 10

65h. προσλαμβάνειν,*accipere, rursus sumere,* assumere, take in addition, prolong: part, iii6.206b9; not transformation, expansion, iv.9.217a28, b9; path vi.9.239b19

65i. ὑπολαμβάνειν, *accipere, existimare, arbitrari, opinari, concipere,* accept, recognize: doctrine, i.4.187a27; knowledge, i.4.187b12; element, ii.1.193a23; chance, ii.4.196a19; place, iv.1.208a29; b29; void, iv.6.213a12, 15; movement, vii.1.241b39; alteration, vii.3.245b6; principle, viii.1.252a34; better, viii.7.260b22

66. δέχεσθαι, *recipere, accipere, suscipere,* admit, hold, iv.6.213b6, 7, 21, 7.214a11. See also 12

67. ἀξίωμα, *dignitas, axioma,* axiom, viii.1.252a24. See also 46c, 60d

68. ἀβέβαιος, *incertum, inconstans,* unstable, ii.4.197a30, 31

69. σχεδόν, *fere,* perhaps, i.5.188b26; ii.3.195a3; iv.4.211b6, 6.213b28; viii.3.-253b6, 254a11

70. ἴσως, *fortassis, fortasse, forsitan, forte,* perhaps: discuss, i.2.185a19; part, whole, i.2.185b11; necessity, ii.9.200b4; actualization, iii.3.202a22; infinite, finite, iii.4.202b33, 5.204a34; vi.7.238b9; cosmos, iv.5.212b18; void, iv.9.217a18; movement, v.1.224b15, 5.229a29, 6.231a16; vi.10.241b15; viii.2.252b31, 253a13, 15; one, v.4.228-a15; becoming, vii.3.246a6, b14; consequence, viii.5.256b12; nonbeing, viii.7.261b12. See also 57g

71. διόρισις, *determinatio, distinctio,* differentiation, iv.6.213b26. See also 72e

72. ὁρίζειν, *determinare, definire, finire, terminare,* determine, define, bound specify, divide: quantity, i.4.187b36; what is, ii.1.193b2; explanation, ii.4.196a2, 5.196b28; sawing, ii.9.200b5; continuous, iii.1.200b18; movements, iii.3.202b24; body, iii.5.204b6; iv.1.209a5, 2.209b4; part, whole, iii.6.206b6, 7, 9, 12, 20; vs indeterminate, iii.6.207a32; iv.2.209b8; viii.10.266b20; member, magnitude, size, limit, iii.7.-207b13, 20; vi.7.238a16; viii.10.266b3; fast, slow, iv.10.218b15; vi.2.232a27; time, movement, change, iv.10.218b17, 11.218b30, 32, 219a22, 23, 25, 29, b12, 220a9, 11, 12.220b16, 17, 221a1, 3, 13.222a25, 14.223b15, 16; vi.6.237a5, 9, 10.241b12; viii.10.266b18; points, viii.9.265a29

72a. ὁρισμός, *definitio,* definition: analytical, i.1.184b12; and explanation, ii.7.198a17; beginning, ii.9.200a35; primary time, vi.6.236b24; motions movement, vii.2.244a7; viii.1.251a12; ὁριστικός, *definitivus,* definitory, i.3.186b23

72b. ὅρος, *terminus, definitio,* definition, limit, boundary, terminus: mathematical entities and "ideas", ii.2.194a2; not attain to, ii.8.199b6; time, iv.14.223a6; of species, vii.7.249a27; of generation, destruction, viii.7.261a34

72c. ἀόριστος, *infinitus, indefinitus, indeterminatus,* indeterminate, indefinite: factor, ii.5.196b28, 197a8, 9, 20, 21, 6.198a5; movement, iii.2.201b24, 28; privations, iii.2.201b26; vs. determine, iii.6.207a31; iv.2.209b9, 210a8; circle, viii.9.256a32

72d. ἀφορίζειν, *determinare, definire,* separate, determine: not infinite, iii.8.208a6; succession, v.3.226b35

72e. διορίζειν, *determinare, definire, discernere, explicare, distinguere, dis-terminare,* determine, discriminate, distinguish: principles, i.1.184a15, 9.192b3; individual, i.1.184b14; in change, i.7.190a13; potential, actual, i.8.191b29; iv.5.213-

a4; one, many, i.9.192a36; nature, natural, ii.2.193b22, 3.194b16; form, ii.2.194b15; factors, ii.3.195b29, 4.196a10, b9; movement, rest, iii.1.200b15, 2.201b17; v.4.227b19, 5.229a7, 6.229b24; viii.1.251a8, 3.253b2, 254a17, 8.262a1; infinite, iii.4.204a2, 6.206-a12; place, iii.5.205a28, b34; iv.1.208b18, 7.214a16; viii.4.255b17; "in", iv.3.210b9; void, iv.6.213b24, 26, 27, 9.217b27; present, iv.10.218a9; time, iv. 14.222b30; vi.3.-234a4; continuous, etc., v.3.227a14; vi.1.231a22; rate, v.4.228b27; moved, viii.5.257b6; mover, viii.10.266a11, b17. *See also* 71

72f. προσορίζειν, *determinare, adjungere,* secure, definiteness, vii.1.252m27

72g. ἀδιοριστῶς, *indistincte, indefinite,* without further determination: name, i.1.184b11

73. χωρίζειν, *separare, abstrahere,* separate, abstract, distinguish, detach, disjoin, divide: not primary and dependent being, i.2.185a31, 3.186a30 (but distinct); ii.1.193b4; iv.2.209b23 (matter form), 8.216b6, 7; accident, i.3.186b22, 28; parts, i.4.188a3; viii.5.257b31; physician, patient, ii.1.192b26; mathematical considerations, ii.2.193b33-194a1; iii.5.204b7; v.3.227a28; form, matter, ii.2.194b12, 14; iv.7.214a15, 9.217a24; automatism, chance, ii.6.197b33; number, iii.4.203a6; not the infinite, iii.5.204a8, 7.207b13 (number); place, iv.2.209b27, 30, 4.211a3, b31, 212a1, 5.212b6, 8.214b20, 26, 216a25, vi.1.231b6; not interval, iv.6.213a32; and void, iv.6.213b15, 16, 19, 7.214a30, 8.214b12, 20, 216b20, 9.216b30, 31, 33; motion, vii.2.244a10; active, passive, viii.4.255a14

73a. χωρίς, *seorsum, separatum,* distinct, separate, apart from: nature, i.6.189-b20; great, small, i.9.192a8; in geometry, iii.4.203a14; question, iii.5.205a2; place, iv.1.208b18, 8.214b27; v.3.226b18, 22; matter, form, iv.2.209b20; not parts, iv.3.-210b3; viii.3.253b21; not extension, iv.7.214a30; movements, iv.14.223b10; now, vi.3.234a7; definition, viii.1.251a12

73b. ἀχώριστος, *inseparabilis,* not independent: attributes, i.4.188a6, 12; void, iv.7.214a19

74. παρά, *extra, praeter, apud, secus, juxta, alienum,* independent, besides, contrary to by means of, other than, destructive of, distinct from: not dependent being, i.2.185a31, 3.186a30; being, one, i.3.187a7; iii.5.205a5; necessary, usual, ii.5.196b14, 20, 197a20; nature, ii.6.197b34; iv.8.215a3, 4; v.6.230a18-231a17; viii.3.-254a10, viii.4; not movement, iii.1.200b32, 201a2, 3; factors, iii.4.203b12; not infinite, iii.5.204b23, 32, 34; place, iv.1.208b28, 209a13, 4.211b9, 17, 212a4, 5.212b15, 16, 7.214a23, 8.216b15; not whole, parts, iv.3.210a17; measure, iv.14.233b33; argument, vi.9.239b22, 31, 240a18; differences, genus, vii.4.249a22; present, viii.1.251b25; science, viii.3.253b7; movers, viii.6.259a4

75. διαιρεῖν, *dividere, sejungere, percipere, distinguere, scindere, explicare,* analyze, divide, tear, detach, split: into elements, i.1.184a23; definition, meaning, i.1.184b12, 3.186b14; iii.7.207b34; into parts, i.4.187b15; iii.5.205b30; iv.5.212b35, 10.218a22; viii.5.257a1, 8.262a24; in various ways, i.4.188a14; materials, i.5.188b18; food, ii.8.198b25; infinitely, iii.1.200b20, 5.204a25, 6.206b5, 207a22, 7.207b16, 27; iv.220a30; vi.1.231b15, 16, 6.237a33, b21; viii.3.253b22, 5.257b33, 34; actual, potential, iii.1.210a9; magnitude, plurality, iii.5.204a11, 6.206a11; vi.1.231b20, 2.232a23, 233a4, 7, 12, b6, 23, 9.239b19, 23; and continuous, iv.4.211a29, 31, 35, b1, 3; vi.1, 2.232b25, 6.237b11; viii.5.257a33, 34, 8.263a23, 26; medium, iv.8.215a31, b11, 216a18, 19; viii.10.267b14; greater, iv.8.215b16; time, iv.11.220a5, 12.220b26, 13.222a14, 16; vi.1.231b20, 232a21, 2.233a3, 7, 12, b26, 27; vi.3.234a10, 11, 13, 18, 4.235a12, 6.236b26, 27, 237a10, 26, 8.238b32, 239a6, 31, 10.241a16, 19; viii.8.-263b10, 28; categories, v.1.225b5; movement, change, v.4.228a21, b18, 5.229b8; vi.1.4, 5.236a29; vii.1.242a39, 40, 48; viii.3.253b22, 10.267a24; vs. indivisible, vi.2.-

233b18, 29, 3.234a21, 30, 5.235b34, 35, 236a4, 21, 22, b7-15, 10.241a22; stones, viii.3.253b16, 19; mover, moved, viii.4.254b31, 255a16, 20, 5.257a33, 34, 258b1-3; διαίρεσις divisio, division, dividing, subtraction, section, disjunction: into parts, i.2.185b34; sawing, ii.9.200b5; magnitude, iii.4.203b17; 6.206a17, 26; infinite, iii.4.-204a7, 6.206a15, b4, 17, 19, 27, 7.207a35; vi.2.233a20, 25, 27, 5.236a27, b15, 6.237a8; viii.8.263a21; and time, magnitude, iv.11.220a19, 13.222a18, 14.224a9; vi.2.233a10, 12, 16, 20, 3.234a16, 4.235a15, 6.237a7; viii.8.262a30; nonbeing in, v.1.225a21; movement, vi.4.235a7, 9, 15; viii.4.254b34, 8.264b29; color, vii.4.249a5. See also 118d

75a. ἀδιαίρετον, indivisibile, individuum, indivisible, undivided: unity, 2.185b8, 16-19; part, i.2.185b15; viii.5.258b2; limit, i.2.185b18; and the "All", i.3.186b35; and infinite, iii.5.204a11, 24, 27, 7.207b6, 8; vi.2.233a25, 27; state, iv.11.218b31; and continuity, vi.1.6.237a31; now, vi.3.233b34, 234a4, 23, 31, 9.239b9; end, beginning, vi.5.236a13, 16, b11; and movement, vi.10.240b13, 241a7-26; first mover, viii.10.-267b25

75b. καθαίρεσις, divisio, detractio, annihilatio, division: infinite, iii.6.206b13, 29, 31, 207a23, 8.208a21

76. διαφέρειν, differre, interesse, differ: in shape, size, i.2.184b21; iii.4.203b2; in species, i.2.184b22; vii.4.249a16, b12, 14; viii.8.262a5; alternatives, i.2.185a26; views, i.4.187a23; appearances, i.4.187b2; contraries, i.5.188b30, 35, 6.189b24; aspects, i.9.192a3; natural, ii.1.192b12; sciences, ii.2.193b23; arts, ii.2.194b3; instruments, actions, ii.3.195a2; potential, actual, ii.3.195b16; chance, automatism, ii.6.197a36, 198a2; place, iv.1.208b22, 8.216b9; and movement, iv.8.215a26, 27, b28, 216a11; v.4.228b18, 5.229a31; vii.4.249a16, b12, 14; viii.4.254b19; vs. same, iv.14.224a6-8, 13, 15; argument, v.4.228a13; vi.9.239b19; infinite, vi.7.238b20; in sensible traits, vii.2.244b5; οὐθὲν διαφέρει, differt nihil, nihil interest, makes no difference: harmony, i.5.188b15; becoming, i.8.191a36; abstraction, ii.2.193b34; good, ii.3.195a25; in the eternal, iii.4.203b30; infinite, iii.7.207b33; vi.7.238a30, b3; vii.1.242b64; place, iv.8,216b14; movement, change, iv. 10.218b19; v.2.226b6; vi.7.237b34 (rate), 238a26; vii.2.244a13, 4.248a22; measure, vi.2.233b4; contraries, viii.7.261b8; intermediates, viii.8.264b33

76a. διαφορά, differentia, differentia, difference: contraries, i.4.187a19, 7.191-a18; sensible, i.6.189b7; chance, automatism, ii.5.196b31; of place, iii.5.205b32; iv.6.213b8; viii.8.262a6; point, iv.1.209a11; not in void, iv.8.215a1, 6-12; of kind, iv.14.224a6, 7, 13; v.4.227b8, 228b30; as quality, v.2.226a28; movement, v.4.228b22, 29; vi.10.240b15; vii.4.249a21; and comparability, vii.4.249a4; and time, viii.1.252a15

76b. ἀδιάφορον, indifferens, non dividitur (in species), quod non habet differentiam, do not differ, undifferentiated, vii.1.242b37, 4.249a20; viii.8.262a2

77. ἀκολουθεῖν, sequi, consequi, imitari, correspond to, accompany, follow, iv.11.210a11, b15, 23, 220a6, 9, 12. 220b24, 14.223a14; v.4.228b30; vi.4.235b1, 5.235b10; ἐπακολουθεῖν, sequi, follow, vii.2.243a19, 20; παρακολουθεῖν, consequi, follow, iv.12.221a24; συνακολουθεῖν, sequi, proficere, consequi, go along, i.5.188b26; viii.10.267a26

78. διαλέγεσθαι, disputare, disceptare, disserere, engage in, dialectic, i.2.185a6, b19; διάλεκτες, locutio, lingua, conversation, iii.5.204a16

79. ἐρωτᾶν, interrogare, ask, viii.8.263a4, 7, 15, 19

80. ἐριστικός, litigiosus, sophisticus, contentious: argument, i.2.185a8, 3.186a6

80a. ἐλέγχειν, argumentari, redarquere, refute, iv.6.213a25; ἐξελέγχειν, arguere, refellere, refute, iv.6,213a23

81. σοφιστής, sophista, sophist, iv.11.219b20

III. PRINCIPLES OF EXPLANATION

82. ἀρχή, *principium, initium*, principle, beginning, starting-point, source: in science, i.1.184a11, 13, 16, 23, 2.185a2, 15, 9.192b2; ii.9.200a30; viii.1.251a7, 3.-253b2; number of, i.2.184b15-25, 1.6, 7 (subject and contrary terms of change), 9(matter, form, privation); viii.1.252a10; and fact, i.2.185a4; of an inquiry, arguments, i.2.185a20, 9.192b4; viii.3.253a22, 254a16, 5.257a32, 6.260a11, 7.260a20, 8.262a19; of what has originated, i.3.186a12, 13; infinite, i.4.187b10; contraries, i.5.188a19, 21, 22, b28, 189a10; i.7; iii.2.201b25; and derivation, i.5.188a27, 6.-189a19; iii.4.203b6; viii.3.254a33; not of principle, i.6.189a30, 31; of beings, i.6.189-b13 (one, excess, deficiency), 7.190b17; iii.4.203a3 (infinite); viii.6.259b26, 27; differ, i.6.189b25; formal, i.9.192a34; iii.2.202a10; of movement, rest, change, ii.1.192b14, 21, 193a29, b4, 7.198a29, b1, 8, .99b16; iii.1.200b12; vi.5.236a14; vii.1.241b35; viii.2.252b20, 3.253b6, 8, 4.254b16, 255a25, b30, 5.256b22, 26, 257a29, b20, 6.259a13, 7.261a26, 9.265b19, 33, 266a7; of production, ii.1.192b28; knowledge of, ii.3.194b22; physical, nonphysical, ii.7.198a36; seed, ii.8.199b7; end, prime, ii.9.200a21, 22, 35; viii.7.261a13; infinite as, iii.4.-203a3, b4-10, 5.204a21, 27; of genesis, things, iii.4.203a30, b1; mind, iii.4.203a31; and time, iii.5.206a10; iv.11.220a11, 13, 15, 13.222a12, 33-b7, 14.223b28; viii.-1.251b21, 25; after, v.3.226b34; and nature, viii.1.252a7; and invariance, viii.1.-252a32, 35, b4; and unmoved movers, viii.6.258b20, 32, 259a22, 33; external, viii.6.259b14; of celestial bodies, viii.6.259b30, 10.267b7; dense, rare, viii.7.260b8; and middle, end, viii.8.262a20, 25, 26, b6, 8, 24, 263a1, 24, b2, 14, 264b27, 9.265a27-b16. *See also* 95b-e

82a. ἐξ ἀρχῆς, *ex principio*, (*in*) *primis* (*constitutionibus*), original, first, initial, ii.8.199b5; vi.7.238a25; vii.1.242b54, 3.247b10. 16

82b. ἀρχιτεκτονικός, *architectonicus*, architectonic, ii.2.194b2-4

82c. ἄρχειν, *dominari*, rule, ii.7.198a20

82d. ἄρχεσθαι, *incipere, principare, praeesse, initium, considerationis fieri, coepisse, incoepisse, exoriri*, begin, preside: inquiry, i.9.192b4; iv.6.213a19; viii.-1.251a8, 3.254a18; arts, ii.2.194b1; movement, change, iii.4.203a32; vi.5.236a9, 6.-236b36, 9.240a5; viii.8.262a24, 264a13, 9.265a31; time, iv.11.219a3; slowly, v.2.-226b12

82e. τὸ ὑπάρχον, *quod est*, actual: attribute, i.3.186b1; possible, vii.1.242b67; conduct, ii.6.197b2, 8, 12; contingently, necessarily, ii.9.199b34, 200b1, 3; categories, iii.1.201a3; and movement, functioning, iii.2.202a5, 3.202a34, b9, 21; v.2.-225b32; viii.1.250b14, 251a10, b4, 4.254b9, 11, 5.265b5, 7.261a15, 17, 18, b15, 21, 9.265b27; validity, iii.4.203b5; infinite, iii.6.206b30; condition, iii.6.207a5; place, iv.1.208a34, b32, 4.210b33, 212a4; explanatory import, iv.1.209a20; part, iv.3.210b10; not in void, iv.8.215a18; other things equal, iv.8.215a28; time, iv. 10.218a30, 12.221b3; vi.3.234a19, 5.236a11(primary); together, etc., v.3.226b21, 227a29, b2;

82f. ὑπάρχειν, *inesse, esse, existere, adesse*, belong to, happen, apply, fact, prevail, occur: attribute, and possessor, i.3.186a31, b19-22; ii.1.192b35; iii.3.-202b15, 5.204a29; iv.4.211a23; viii.4.255a28, 6.259b30; everything, i.4.187b27; contraries, i.5.188a29, 6.189b20; v.2.226b7, 6.230b31; "being", i.7.191a1; and becoming, destruction, i.8.191b22; v.1.225a28; viii.1.250b18; and nature, ii.1.192b22; viii.6.259a12, 7.260b23; arrangement, ii.1.193a15; materials, ii.2.194a35, b8, 9.200a25; uses, ii.5.196b21, 8.198b28; think, ii.5.197a29; limit, vi.2.233b14; qualitative alteration, vii.3.245b4, 8; object of knowledge, vii.3.247b5; health, vii.4.-249b7; love, strife, viii.1.252a8; mover, viii.7.260b5; infinite, viii.8.263a14

82g. προϋπάρχειν, *praeesse, antecedere,* exist before, viii.1.251a20, 7.260a30

82h. ἐνυπάρχειν, *inesse,* inhere, exist, be in, be present: part, i.3.186b24, 4.187b15; vi.10.240b10; contraries, i.4.187a32; v.2.226b8; source, i.4.187a37; everything, i.4.187b22, 32; infinite, finite, i.4.188a2; vi.4.235b3, 6.237b14, 7.238b15; not privation in product, i.8.191b16; constituent, i.9.192a30, 32; ii.1.193a10, 3.194b24; now, vi.3.233b35; mover, vii.2.243b14

83. αἰτία, *causa,* basic (explanatory) factor, condition, ground: in science, i.1.184a11, 13; ii.9.200a33; viii.1.252b4; of origination, generation, i.5.188b34; viii.-7.261a2; of beings, i.7.190b17; ii.1.192b8; nature, ii.1.192b21, 6.198a12, 8.198b11; viii.1.252a12, 34; essential, incidental, ii.1.192b31, 5.196b25, 26, 28, 197a25, 6.-198a7, 9, 8.199b23; four varieties of, ii.3 (and six lesser varieties), 7 (and their relations in comprehensive answers to the question "why"); iii.7.207b34; iv.1.-209a20; chance, automatism, ii.4.195b31, 33, 196a9, b6, 5.196b12, 197a5, 8, 13, 21, 22, 33, 6.198a2, 6, 11.8.199b23; determinate, ii.4.196a2, 4, 6, 13, 31, 34, b9; external, ii.5.197a2, 6.197b20, 36; viii.6.259b8, 11; efficient, ii.6. 198a4; necessity, ii.8.198b12, 15; final, ii.8.199a29, 32, 9.200a33; of definition, iii.2.201b24, 28, 32; formal, iii.2.202a10; infinite, finite, iii.4.203b12, 7.207b5, 35; vi.7.238a34; rest, 5.-205b14, 17; viii.1.251a26; place, iv.1.209a19; viii.4.255b15; of perplexity, iv.4.-211a10; void, movement, iv.7.213b32, 214a24, 8.214b15, 16, 215a26, 29, 216a8, 17, 9.216b35, 217a1, b22; contrariety, v.5.229a23; no absolutely first part, vi.6.-237b7, 8.239a20; incomparability, viii.4.249a11, 14; alternation, viii.1.252a23, 25; movement, walking, viii.2.253a13, 4.255a8, 20, 31, 5.257a28; prior, viii.5.257a30; and prime mover, viii.6.258b23, 26, 30, 259a4, 6, b17, 260a8, 14, 9.265b24; and circle, viii.9.265b2; soul, viii.9.265b33; συναιτία, (*materia*) *cum* (*forma*) *causa* (*est*), contributing factor, i.9.192a13

83a. αἴτιος, *causa,* responsible: physician, ii.1.192b24; decision, ii.3.194b30; man, ii.3.195a36, b2; builder, ii. 5.197a14; mind, nature, ii.6.198a6; time, iv.12.221b1, 13.222b20; αἰτιώτερον, *magis causa,* accounts, better for, viii.5.257b16

83b. αἰτιᾶσθαι, *dicere causam esse, causam esse ponere, causam adscribere,* blame, attribute, ii.3.195a12, 4.19ba25

84. ὕλη, *materia,* (*causa materialis*), material: great-and-small, i.187a18; one, i.4.187a19; and qualitative alteration, i.7.190b9; viii.3.246a1, 7; and becoming, i.7.190b25, 191a10; ii.7.198a21; v.2.22ba10; and privation, i.9.192a3-6; and form, i.9.192a22; ii.2.194a12-27, b9, 13, 8.199a31; iii.6.207a26, 28, 7.207a35; iv.2.209b9, 3.210a21, 5.213a1, 2, 6; defined, i.9.192a31; and nature, ii.1.193a29, b7, 2.194a12-27, 8.199a31; and art, ii.2.194a33, b1, 4, 8; factor, ii.3.195a8, 17, b9, 7.198a21, 24, 32, 9.200a33; necessary, ii.9.200a6, 10, 14, 25, 27, 31; in definition, ii.9.200b8; infinite, iii.6.206b15, 207a22, 26, 28, 7.207a34, b34-208a4; not place, iv.1.209a21, iv.2, 3.210b28, 31, 4.211b7, 29, 33, 36, 212a3, 9; not the void, iv.7.214a13, 15; and contraries, iv.9.217a22, 24, 26-b16, 23

85. ὑποκείμενον, *subjectum, suppositum,* subject matter, referent, persisting in change, subject in process, fact, substantive: of predication, i.2.185a32, 3.186a34, 7.191a35, 36; and accident, attribute, i.3.186b18, 4.188a8, 6.189a30; element, material, i.4.187a13, 19, 6.189b6, 7.190b20, 9.192a31; principle, i.6.189a31; and contrary terms, i.7; iii.1.201b2; source, i.7.191b2,10; aspect, i.7.190b13,16; as one, i.7.190b24; not a contrary, i.7.190b34; and essential being, i.7.191a20; and primary being, ii.1.192b34; iii.5.204a24; and nature, ii.1.192b34, 193a29; factor, ii.3.195a20; of infinite, iii.7.208a1; and place, iv.4.211a6, 9; and void, iv.9.217a21; and time, iv.14.223a1; and change, movement, v.1.225a1-20, 2.225b17, 20, 21, 226a17, 5.229a31, 32, 6.229b30, 230a11; continuous, vi.2.232b25; quality, viii.2.-244b6

85a. ὑποκεῖσθαι, *subjici,* be, exist, persist in change, i.7.190a15, 34, b3, 14, 191a4, 8, 17, 8.191a31, 9.192a10, 29

86. ἐξ οὗ, *ex quo, id ex quo,* source, ἐξ ὧν εἶναι, *ex quibus esse, ex quibus constare,* consist: beings, i.2.184b23, 6.189a26, 7.190b23; demonstration, i.3.186a5; the All, i.3.186a19; definition, i.3.186b24; elements, principles, ii.5.188b24, 26; source of product, i-7.190b4; matter, form, ii.1.193b5; explanatory factor, ii.3.- 194b24, 19; altars, ii.6.197b10; continuum, vi.1; 2.232a24; shaped, altered, vii.3.- 245b10

87. τί ἐστιν, *quod quid est, quid est, quid sit, quidditas, (causa formalis),* what is: principles, i.7.191a22, 192a35, b3; nature, natural, ii.1.193a2, b2; and essential attributes, ii.2.193b27; and matter, ii.2.194a13, b10; in first philosophy, ii.2.194b14; chance, automatism, ii.4.195b35, 6.198a1; as explanation, ii.7.198a16, 25, 32, b3; movement, iii.1.200b14, 2.201b33, 3.202b23; v.1.224b10; infinite, iii.4.- 202b36, 7.207b26, 8.208a23; place, iv.1.208a29, 33, 209a3, 13, 30, 2.209b17, 18, 4.210b32, 211a8, b6, 5.213a10; void, iv.6.213a13; time, iv.10.218a31, 11.219a2; movement, rest, v.2.226b16; together, etc., v.3.226b18, 227a32

88. ὁ λόγος ὁ τοῦ τί ἦν εἶναι, *ratio quae aliquid erat esse, definitio quidditatem explicans, quod quid erat esse, quidditas,* definition stating "what-it-meant-to-be-something": same, i.2.185b8,9; iii.3.202b12; and matter, iii.2.194a21; explanatory factor, ii.3.194a26, 195a20, 7.198b8

88a. εἶναι, with the dative, *huic esse, essentia ejus, esse ipsius, esse ipsi,* be: good, bad, i.2.185b21, 22; such, so much, i.2.185b25; differ in, i.2.185b33; white, i.3.186a29, b7; man, uneducated, i.7.190a17, 191a2; bronze, unshaped, potentiality, i.7.191a2; iii.1.201a31, 32; Polyclitus, ii.3.195a35; infinite, iii.5.204a23; empty, full, place, iv.6.213a19; before, after, iv.11.219a21; now, iv.11.219b11, 14; musician, v.1.224a23; lover, strife, viii.1.252a26; light, heavy, viii.4.255b16

89. ἰδέα, *idea,* idea: covert abstractions, ii.2.193b36; "nowhere," iii.4.203a8

89a. παράδειγμα, *exemplum, exemplar,* pattern: explanatory factor, ii.3.194b26

90. λόγος, *ratio, definitio, disputatio, sermo, argumentatio, verbum, proportio, sententia,* definition, and names, i.1.184b10; vii.4.248b17; of infinite, i.2.185b2; iii.- 1.200b19; of part, whole, i.2.185b13; and being, i.2.185b19, 32, 3.186b15-33, 7.- 190a16, 191a13; iv.3.210b17; and unity, diversity, i.3.186a28; iii.3.202a20; viii.8.- 262a21; and form, ii.1.193a31, b2, 5, 3.194b29, 9.200a35; iv.3.210a20; and necessity, material, ii.9.200b4, 7, 8; and potentiality iii.1.201a33; of activity, ii.3.202b22; of body, iii.5.204b5; of movement, viii.8.263b13; and priority, viii.9.265a23; reasoning, and principles, i.2.185a1; viii.3.253b3; and sense perception, i.5.188b22, 189a4, 6-8; viii.3.253a33, 8.262a19 and place, iv.1.209a9, 24; viii.8.262b4; and continuity, vi.1.231b1; and finite, vi.7.238a20; and time, viii.1.252a32; discussion, and dialectic, i.2.185a6; of time, iv.10.217b31; of movement, vi.9.239b14; argument, and contention, i.2.185a8, 3.186a8; Heraclitean, i.2.185b20; and objection, i.3.186a22, 187a1; and consequences, i.6.189b1; and chance, ii.4.196a14, 8.198b33; and the "one", iii.5.205a4; and infinite, iii.8.208a5; vi.2.233a21, 7.238a31, and place, iv.3.210b9, 8.214b20; and "void", iv.7.214b4, 9, 8.214b23, 9.217a8; and unity, diversity, v.4.228a12; and changes, v.6.230a26; vii.7.261b10; and continuity, vi.1.231a31, b18, 2.233a13 and time, vi.5.236a2; viii.8.264a1; and movement, vi.9.239b10, 18, 21, 30, 240a17, 10.241a25; vii.4.248a23, 5.250a20; viii.1.251b28, 3.253b14, 254a16, 31, 5.256a22, b1, 6.258b13, 8.262a19, 263a5, 7, 23, 264a7, b2, 265a7, 10.267a8; and terms, vii.4.248b16, 249a22; view, of being, i.2.185b24; of infinite, iii.4.203a2; logic, and facts, i.5.188a31, b29; and priority, v.3.227a20; pattern, of process, i.5.188b10, 16; viii.1.252a19; reasonable, i.6.189a21, b18; iii.7.- 207a33 (infinite); term, and analysis, i.7.190b22; talk, about words, ii.1.193a9;

calculation ii.5.197a19; form, and matter, ii.9.200a15; and place, iv.1.209a22; account, of movements, iii.1.201b14, 3.202a29; viii.8.263a11; of the infinite, iii.7.-207b27; of moving body, iv.11.219b20, 220a8; of condition to which a thing changes, vi.5.236b6; ratio, proportion, analogy, iii.6.206b8, 10; iv.8.215b6, 12, 16, 20, 22, 25, 216a9, 15, b12, 16, 20, 22, 25; vi.2.233a4, b22; viii.1.252a13, 14; treatment, of infinite, iii.6.207a27; principle, and division, iii.7.207b32; theory, of universe, viii.1.250b23; of movement, rest, viii.3.254a11. *See also* 20

91. μορφή, *forma,* form: not, i.7.190b15; of primary being, i.7.190b19, 191a10, 11; iii.1.201a4; as factor, i.9.192a13; ii.7.198b3; and nature, ii.1.193a30, b4, 11, 18, 19, 8.199a31; not place, iv.2.209b3, 210a6, 3.210b31, 4.211b7, 11, 212a9; and alteration, vii.3.245b7, 246a1, b16

91a. σχῆμα, *figura,* shape, pattern: elements, i.2.184b21; iii.4.203a22, b1; as a genus of contraries, i.5.188a24, 25; v.4.227b4; shaped, unshaped, i.5.188b19, 20, 7.190b5, 15, 191a2; unity, i.6.189b9; and nature, ii.1.193b9; of celestial bodies, ii.2.193b29; in mathematics, ii.2.194a5; iv.14.224a5, 8, 9, 11; similar, same, iv.8.-216a14, i4.224a8, 9, 11; projectile, iv.8.216a19; and alteration, vii.3.245b7, 9, 246a1, 2; of path, vii.4.249a18

92. κοινός, *communis,* universal, common: science, i.2.185a3; doctrine, opinion, i.4.187a27; iv.6.213a21; becoming, i.7.189b31; in natural science, iii.1.200b22, 25; and categories, iii.1.200b34; name, iii.1.201a13; v.2.226a27, 30, 32, 33; vii.2.249b25; body, iii.4.203a34; problem, iii.4.203b23; iv.7.214b7; argument, iii.5.204b31; movement, iv.1.208a31; vii.2.244a1; place, iv.2.209a32; subject, v.2.228a1; mover, moved, vii.2.243a34; moment, viii.8.263b12

93. μεθεκτικόν, *participativum, habere vim participandi,* what participates, **μετέχειν,** *participare, obtinere,* participate, iv.2.209b35, 210a1, 10.218a3

94. μιμεῖσθαι, *imitari,* imitate: art, ii.2.194a21, 8.199a16

95. ὅθεν, *causa movens, motiva,* (*causa efficens quadruplex: perficiens, praeparans, adjuvans, consilians*), efficient, factor, ii.3.195a23

95a. ὅθεν ἡ κίνησις, *unde motus, id unde motus perfectus est,* agent: factor, ii.3.195a8, 7.198a26

95b. ὅθεν ἡ ἀρχὴ τῆς κινήσεως, *unde principium motus,* by which movement is started, ii.6.198a3; vii. 3. 243a32

95c. ἀρχὴ κινήσεως, *principium motus,* means: factor, ii.3.195a11

95d. ὅθεν ἡ ἀρχὴ τῆς μεταβολῆς ἡ πρώτη ἡ τῆς ἡρεμήσεως, *unde principium mutationis primum aut quietis,* the agent whereby a change or a state of rest is first produced, ii.3.194b29

95e. ἡ ἀρχὴ τῆς μεταβολῆς ἡ στάσεως ἡ κινήσεως, *unde principium mutationis aut status aut motus,* the factor whereby a change or state of being is initiated, iii.3.195a23. *See also* 82

96. τὸ οὗ ἕνεκα, (*quod*) *cujus causa* (*fit*), *id cujus causa, id cujus gratia,* (*causa causarum*), (*causa finalis*), end: science, ii.2.194a28, 30, 31; nature, ii.2.-194a28, 30, 31, 8.199a32, b19; what, whom, ii.2.194a36; factor, ii.3.194b33, 195a24, 7.198a24, 25, b4; and means, ii.9.200a11, 22; form, ii.9.200a15, 34; perfection, iv.3.210a23; and movement, vii.2.243a32

96a. ἕνεκα, *propter, causa,* for the sake of, with the result that: means, end, ii.2.194a28, 34, b8, 12, 3.195a2, 8.199a25, 28, 32, b19, 9.200a6, 11; action, events, ii.5.196b33, 35, 36, 197a4, 16, 6.197b16-18, 23, 24, 26-28, 30, 32, 8.198b22, 27, 199a8, 15, b2, b21, 22; and explanation, ii.7.198a20, 9.200a33

96b. ἕνεκά του, *propter aliquid, alicujus causa,* to a purpose, to some end, have uses: and chance, ii.5.196b17-196a6, 35, 6.197b19, 8.198b29, 199a4, 5; and nature, ii.7.198b4, 8.198b10, 17, 28, 199a7, 11, 12, 17, 26, b4, 10, 27, 32; and art, ii.8.199b1, 30; and necessity, ii.9.200a8, 10, 19, 26

97. χάριν, *gratia,* for the sake of: knowledge, ii.3.194b17

98. ἀγαθόν, *bonum,* good: vs. bad, vs. not good, i.2.185b22, 23; vii.1.242b37; better, i.4.188a17, 5.189a3, 6.189a15; ii.7.198b8, 8.198b17; iii.6.207a15; viii.1.252-a19, 3.254a32, 6.259a11, 7.260b21.22; devine, i.9.192a17; ending, ii.2.194a33; factor, ii.3.195a24, 26; luck, ii.5.197a25, 28; "in", iv.3.210a23

98a. εὖ, *bene,* well, vii.3.246b9, 247a3

98b. ἀρετή, *virtus,* virtue, vii.3.246a11-248a19

98c. κακία, *malitia, vitium,* vice, vii.3.246a11-248a19

99. καλός, *bonus, pulcher,* beautiful, excellent, i.9.192a23; iv.4.211a11, 12.221b1; κάλλος, *pulchritudo,* beauty, vii.3.246b7

99a. καλῶς, *pulchre,* well, rightly: defined, iii.2.201b16; not void, iv.7.214a14; and motion, viii.8.265a3; problem, viii.10.266b27

100. τέλος, *finis,* end, completion, result, perfection: of science, ii.2.194a27; viii.3.253a32; nature, ii.2.194a28, 8.199a25, 31, b17; passage, movement, change, ii.2.194a29, 32, 8.199a8, b6; iii.3.202b24; vi.5.236a10, 12, 9.239b13; viii.8.262b30, 10.267a9; we, ii.2.194a35; factor, ii.3.194b32, 36, 195a2, 10, 24, 25, 5.197a1, 7.198b3, 8.199a32; and necessity, ii.9.200a14, 20, 22, 27, 34; initial, iii.4.203b9; limit, iii.6.207a14; not place, iv.1.209a22; "in", iv.3.210a23; of race-course, vi.9.-239b34; and beginning, middle, viii.8.262a20, 25, 26, b5, 24, 32, 263a25, b1, 9.265a27-b16

100a. τελευτή, *finis,* consummation, ending, stop, end: for the sake of, ii.2.194a31; conclusion, iii.4.203b9; and time, iii.5.206a10; iv.11.220a11, 13, 16, 13.222a12, 33-b7, 14.223b27; viii.1.251b21, 22, 25, 8.263b14. *See also* 14

100b. τελευτᾶν, *perficere,* finish, stop, viii.8.262b8, 9.265a31

100c. ἐπιτελεῖν, *perficere,* complete: art, ii.8.199a16; change, vi.5.236a8, 11; shape, vii.3.245b9

100d. τελειοῦσθαι, *perfici,* be completed, vii.3.246a20; viii.7.260b33, 261a17

100e. τελείωμα, *perfectio,* completion, vii.3.246a17; τελείωσις, *perfectio,* completion, vii.3.246a13, b2, 247a2, b2

100f. ἀτελής, *imperfectus,* incomplete, uncompleted, undeveloped, iii.1.201a6, 2.201b32; v.4.228b14; viii.5.257b8, 7.261a13, 36, 9.265a22, 23

100g. ἀτελεύτητος, *inconsumabilis,* endless, iii.4.204a5

IV. NATURE AND CHANGE

101. φύσις, *natura,* nature: science of (see 179); not one unchanging being, i.2.184b26, 185a18; writers on, i.4.187a35; ii.2.193b29; iii.4.202b35, 203a16; of beings, i.4.187b7, 6.189a27, 29, 8.191a25; ii.1.193a22; viii.7.261a19; elements, i.6.-189b2, 21, 9.192a30; iii.4.203a17; procedure, according to, i.7.189b31; persistent, i.7.191a8, 8.191b34, 9.192a10; v.2.226a17; and tendency, i.9.192a19; defined, ii.1. (in relation to art, natural beings and changes, matter, form), 2.193b22, ii.8. (teleology in relation to necessity and chance, deliberate and organic activity, imperfect and incidental results, visible artists); iii.1.200b12; have (*see* 33); in a subject (*see* 85); according to (*see* 39a); obvious (*see* 59d); and privation, ii.1.193b19; matter, form, ii.2.194a12, 16, 27, 28 (means, end), 8.199a30; and art, ii.2.194a22, 8.199a16, 18, 19, b1, 30, 32; as factor, ii.4.196a30, 6.198a6, 10, 12, 8.198b10, 199b32; and end, ii.5.196b22, 7.198b4, 8.198b17, 24, 199b15, 30, 32, 9.200a9; viii.7.261a19; contrary to, ii.6.197b34; viii.3.254a10, viii.4; and necessity, ii.9.200a16; and change, movement, rest, iii.1.200b12-15; iv.8.215a; v.-226b25; viii.1.250b16, 251a6, 3.253b5, 8, 6.259a28, 9.265b25; and infinite, iii.4.-203b33, 7.207b22; whole, complete, iii.6.207a14; and place, iv.1.2008b18, 209a4, 15, 3.210b20; and void, iv.6.213b25, 27; of time, iv.10.217b32, 218a31; vs. violent, v.6.230a18-231a17; vii.2.243b1; and size, vi.10.241b1; and excellence, vii.3.246a15; allay agitation, vii.3.248a3; single, vii.4.248b21; and order, viii.1.252a12, 16; and explanation, viii.1.252a34; treatise on, viii.5.257b1; and better, viii.7.260b23; and genesis, viii.7.261a14; prior, viii.9.265a22; φύσει *naturae, natura, a natura, naturaliter, quod est naturae, quod natura constat,* naturally, by its own nature: intelligible, i.184a17, 20 (vs. to us); changeful, i.2.185a13; produced i.5.188b25; ii.1.193a1, 8.199a12; beings, i.7.190b17; ii.1.192b8, 9, 11, 13, 193a2, 10, b3, 6; events, ii.6.197b33, 8.198b34, 199a6; factors, ii.6.198a4; ends, ii.8.199a7, 26, 30, b15 (defined), 29; motion, movement, iii.5.205b27; iv.4.211a4, 212a25, 8.215a6, 12, 13; viii.1.252a18, viii.4; place, iv.1.208b25, 5.212b33; bodies, iv.8.214b14; change, iv.13.222b16; vs. violent, v.6.230a25, b14; performances, viii.1.250b15; order, viii.1.252a12; determinate, viii.6.259a11

101a. φυσικός, *naturalis, physicus,* physicist, physical, natural philosopher, natural, concrete: principle, i.2.184b17, 7.191a3; viii.3.253a35, b5; problems, i.2.-185a18; concerning unity, i.3.186a20; two types, i.4.187a12; concerning becoming, i.4.187a28; forms, i.9.192b1; ii.2.194b10; and artistic, ii.1.193a33, 2.194b8, 8.199b3; scope of ii.2. (in relation to other sciences, art, materials, forms, means, ends), 7.198a28, 9.200a32; iii.4.203b3, 204a1; iv.6.213a12; bodies, ii.2.193b24, 32; iv.1.-208b8, 8.215a5; viii.4.255b6; change, movement, ii.3.194b21; viii.3.253b9, 6.259b8; explanation, ii.7.198a22, 23, 36; necessity, ii.8.198b11, 9.200a31; always ii.8.199-b25; agent, iii.1.201a24; argument, iii.5.204b10; one, elements, iii.5.205a5; and plan, iv.1.208a27; opposites, iv.14.223b26; vs. violent, v.6.230b4; disturbance, vii.3.247b17; treatise, viii.1.251a9, 3.253b8, 10.267b21; pattern, viii.7.261b25

101b. φύεσθαι, *nascere, innascere,* grow, arise, ii.1.193b17; plant, i.4.187b16, 19, 7.190b4; ii.1.192b10, 4.196a29, 35, 8.199a24, 27, b10, 11; viii.7.261a16; unity, viii.1.241a30; φυόμενον, *quod nascitur,* growing thing, ii.1.193b17

101c. πέφυκε, πεφυκός, *innatum est, naturaliter constitutum est, aptum natum est, natura comparatum,* is natural, functions naturally, course: path, order, i.1.184a16; vs. chance, i.5.188a32; produce, i.6.189a23; tend, i.9.192a18; as means, ii.6.197b26, 27; hot, cold, ii.8.198b13; in stages, ii.8.199a10, 11, 14; be carried, ii.9.200a2; span, iii.4.204a4, 5; place, iii.5.205a10, b1, 4, 6; viii.4.255b15; change, movement, rest, iii.5.205b8, 10, 13; iv.12.221b13; v.2.226b13, 14, 3.226b23; vi.3.-234a32-b3, 8.238b23, 239a14; viii.1.252a6, 2.252b10, 3.253a27; fit, iv.4.211b19, 20; displacement, iv.8.216a30; count, iv.14.223a25; together, etc., v.3.226b21; unity, v.3.227a15; preserve, perish, vii.3.246b10; altered, vii.3.246b18; active, passive, viii.4.255a14; circular motion, viii.9.265b15; mover, moved, viii.10.267a4

101d. διαφύεσθαι, *geminari, genitus,* be diffused, viii.1.250b31

101e. ἐκφύεσθαι, *enascere, pullulare,* grow, viii.3.253b15

101f. ἔμφυτος, *innatus, ansitus,* natural: tendency, ii.1.192b19

101g. πρόσφυσις, *adnascentia, naturalis copulatio,* growing together, v.3.227a17

101h. φυσιολόγος, *philosophus, naturalis, physicus, physiologus,* physicist, iii.4.203b15, 5.205a26, 6.206b23; iv.6.213b1; viii.8.265a3

101i. συμπεφυκότα, *simul apta nata, copulata in unam naturam,* forming an organic unity, iv.5.212b31; v.3.227a25; σύμφυσις, *copulatio, insertus, naturalis copulatio, consertus,* organic unity, natural union, iv.5.213a9; v.3.227a23; συμφύεσθαι, *adnatum esse, naturaliter copulari,* become naturally united, v.3.227a25; σύμφυτος, *naturalis, innatus,* part, viii.1.253a12; συμφυής, *consitus, copulatus,* unitary, viii.4.255a12, 15

102. σῶμα, *corpus,* body: persisting, i.4.187a13; limited, i.4.187b25, 26, 35; not infinite, i.4.188a2; iii.4.203a8, b26, 29, 5.204b, 205, 206a1, 2, 7, b23, 25, 8.208a9; simple, ii.1.192b10; iv.1.208b8; natural, ii.2.193b24, 32; elements, ii.3.-195a17; iii.4.203a29, b1; common, iii.4.203b1; defined, iii.5.204b5, 20; and place, iii.5.206a1; iv.1.208b27, 28, 209a, 2.209a32, b26, 4.211a5, b9, i4, 17, 18, 212a31, b7, 10, 19, 25-29, 6.213b8; viii.6.259b19; character, state of, iv.3.210a34, b1, 5, 27; vii.3.246a10, b4, 248a4; and movement, change, iv.4.211a20, b2, 4; v.1.224-a25, 2.226a12, 4.228a8; vi.10.240b10; vii.2.243b13, 244b12, 245a6; viii.2.253a16, 4.254b18; and the void, iv.6.213a29, 32, 33, b2, 18, 20, iv.7, 8; viii.9.265b29; size, iv.9.217a26; and time, iv.11.219a5; differ, vii.2.244b5; alteration, vii.3.247b4; σωματικός, *corporeus, corporals,* corporeal: not place, iv.1.209a15, 5.212b26; not air, iv.4.212a12; and the void, iv.7.214a12; motion, vii.1.242b60; pleasures, pains, vii.3.247a8

102a. ὄγκος, *moles, corpus,* bulk, mass: minute, i.4.187a37; and place, void, iii.4.203b28; iv.1.209a3, 6.213a17, 8.216b6, 15, 9.217a17, 32, b9; and motion, vi.9.239b34, 240a5

103. στοιχεῖον, *elementum,* element: in science, i.1.184a11, 14, 23, 2.184b25; number of, i.2.184b25, 4.187a13, 26, 6.189b16, 27; earth, i.5.188a22; ii.1.193a19, 2.193b30; iii.5.205a12, b11, 14, 16; iv.1.208b21; 31, 2.209a35, 14.223a18; v.4.-228b31, 6.230b14, 17, 23, 231a12; water, i.2.184b18, 3.186a17, 4.187b24-34 (and flesh), 188a16, 6.189b3, 8; ii.1.193a19, 8.198b20; iii.4.203a18, 5.204b25, 28, 31, 205a24; iv.1.208b2, 6, 209a10, 2.209b25, 26, 210a10, 4.211b3, 5, 14-36, 5.212a32, b35, 213a1-4, 7, 6.213b12, 7.214b1, 3, 4, 9.216b26-28, 217a13-20, 26-33; vii.4.248-b14, 24; viii.10.267a4, 16, b14; air, i.2.184b17, 4.188a16, 6.189b6, 7; ii.4.196a21; iii.4.203a18, 5.204a26, 31, b16, 17, 25, 27, 31, 6.206b24; iv.1.208b3, 6, 209a10, 2.209a34 b24, 26, 210a9, 4.211a24-26, b14-36, 212a12, 5.213a1-4, 6.213a26, 27, 30, 31, 7.214b1, 3, 9.216b26-28, 217a13-20, 26-33; vii.2.245a5, 6, 8, 4.248b14, 5.250a21; viii.10.266b31, 267a3, 15, b14; fire, i.5.188a21, 6.189b3; ii.1.192b36, 4.196a18; iii.5.204b16, 28, 205a4, 12, 24; iv.1.208b20, 9.217a1, b14; v.4.228b31, 6.230b14, 15, 231a11, 15; vii.2.244a12; viii.1.251a29, 252a18; earth, water, air, fire, i.6.189b4, 19; ii.1.192b10, 11, 193a21, 22, 3.195a17; iii.5.204b34, 35, 205a26, 27; iv.1.208b9, 5.212b20-22, iv-8; viii.3.253b33, viii.4; other, i.4.187a14, 6.189b3; iii.4.203a18, 5.205a27; contraries, i.5.188b28; letters, ii.3.195a16; and infinite, iii.-4.203a18-20, 5.204a15, b13, 24, 33, 205a1, 30; and unity, iii.5.205a5; and place, iv.1.209a14, 15, 17; and animal, viii.4.254b10

104. συστοιχία, *coordinatio, classis,* list: of contraries, i.5.189a1; iii.2.201b25

105. ἁπλοῦς, *simplex,* absolute, simple, abstract, invariant, ἁπλῶς, *simpliciter,* inherently, strictly, generally intelligible, i.1.184a18 (vs. to us); generation, becoming, (*see* 116, 116a); meaning, i.3.186a24; not nonbeing, i.3.187a6; beings, i.5.188b9, 7.190b2; v.1.225a22, 25; terms, i.7.189b33, 190a1, 3, 9, 12; bodies, ii.1.192b10; iv.-1.208b9, 8.214b13; art products, ii.2.194a34; factors, ii.3.195b15; not chance, ii.5.-197a14, 34; consequence, ii.7.198b6; not better, ii.7.198b9; not necessity, ii.9.199b35; definition, iii.1.201a32; actuality, movement, change, iii.2.201b35, 3.202b27; v.1.-224a24, 4.227b21, 29, 228a20, b2, 10, 11(one); viii.6.258b15, 260a18, 9.265a10, 266a4; and infinite, iii.5.204b11, 22, 24, 205a30, b35, 6.206a9; viii.8.263b7; and void, iv.9.217b21; number, measure, iv.12.220a27, 221n19, 14.223a33; destruction, v.1.225a18; viii.7.261b4; contraries, v.2.226b3, 6.229b24, 28, 230b8; species, v.4.-227b13; distance, time, vi.2.233a15; and nature, viii.1.252a17, 19; prime mover, viii.6.260a17

106. στέρησις, *privatio, destitutio, (ipsa carentia formae, vel contrarium formae, quod subjecto accidit),* privation: as contrary, i.7.191a14; ii.1.193b20; iii.2.201b26; v.1.225b3, 6.229b26; come from, i.8.191b15; and matter, i.9.192a3-6, 27; and form, ii.1.193b19; iii.1.201a5; and movement, iii.2.201b34; iv.12.221b13; and infinite, iii.7.208a1; and place, iv.1.208b27; and the void, iv.6.213a18, 7.214a17, 8.214b18, 215a11; rest, v.2.226b15, 6.229b25; viii.1.251a27, 8.264a27

107. μένειν, *manere,* remain, endure, rest, continue, persist, be stationary, διαμένειν, *permanere,* remain, abide, endure, ὑπομένειν, *permanere, subjicere,* persist: principles, i.6.189a20; subject in process, i.7.190a9-13; material, i.9.192a13; ii.1.193a16; where, iii.5.205a15, 18; earth, infinite, iii.5.205b12, 15, 17, 19, 20, 24; segment, iii.6.206a1; not infinite numbering, iii.7.207b14; part, iii.8.-208a21; natural, contra-natural, v.6.230b23, 26, 231a7; and movement, v.6.230b32; vii.2.244a13; viii.6.259b26, 27; in place, iv.4.211a4, 212a23, 24, 51212b33; viii.3.-253b32, 34; container, iv.4.211b15, 20; and the void, iv.8.214b22, 26, 215a30; and time, iv.10.218a22, 11.218b31, 13.222a13; viii.7.261b24, 8.264b4; rate, vi.7.-238a6; prime mover, viii.6.260a18; circle, viii.9.265b7

107a. μονή, *mansio, status,* rest, iii.5.205a17, b2; v.5.229a8, 6.229b28, 31, 230a10, 20, b15-18, 28, 231a16

108. σώζειν, *salvare, tueri, servare,* save, protect: do justice to arguments, i.6.189b1; ship, i.3.195a14; horse, ii.6.197b15, 16; organism, ii.8.198b30; in order to, ii.9.200a7

109. κίνησις, *motus, motio,* movement, process, passage, course: beginning of, ii.1.192b14, 193a30, b4, 7.198a29, b1; iii.1.200b12; v.1.224b1; vii.1.241b35; viii.-1.252b20, 2.253a16, 19, 3.253b6, 8, 4.254b16, 255b30, 5.256b21, 26, 257a28, 6.259a13, 8.263a1, 9.265b7; mathematical abstraction from, ii.2.193b34, 194a5; and end, ii.2.194a29; v.1.224b1; viii.7.260b33, 8.263a1; art, ii.2.194b6; and vortex, ii.4.196a27; have, ii.7.198a28; iv.14.223b26; and necessity, ii.9.200a31; and nature, ii.1.200b12-25 (and infinite, place, void, time), 4.202b31; viii.3.253b6, 8, 9, 9.265b25; kinds of, and categories, iii.1.200b32, 201a2, 8; iv.1.208a31; v.1.224a29, b4, 25, 225a13-16, 26, 32, b7, 2.225b10-16, 226a23-b17; vi.5.236b17; vii.2.243-a36, 40, 4.249a11, 21, b12, 13; viii.1.251a14, viii.3.4.254b19, 24, 5.256b31, 257a6, 7, 20, b26, 6.259b6, 7.260a27, b32, 261a21, 30-32, b1, 8.264a22, 25, b29, 265a5, 8, 9.265b19, 266a8, 10.267a16; defined, iii.1. (in relation to being, actual functioning, the potential as such, process), 2 (as to genus, "indefiniteness," contact, conveying of form), 3 (as to what is acted upon, identity and diversity of actualization of mover and moved, kinds of "movement"); v.1.224b11; viii.1.251a9, 5.257b8; and rest, iii.5.205a18, 19; v.1.225a33, 2.226b16, v.6; viii.1.251a27, 2.252-b18, 24, 253a9, viii.3.7.261b17, 19, 8.264a23, 25, 27, 31, 9.265a27, 10.267a12, 13; and infinite, iii.7.207b22, 8.208a20; vi.7.238b19, 20, 10.241b10, 17; vii.1.242a65, b44, 46, 56, 65; viii.8.263a11, 10.266a23; and magnitude, iii.7.207b23; viii.7.260-a27; and time, iii.7.207b25; iv.10.218a33, b5, 9, 12, 18, iv.11, 12.220b14-222a9, 13.222a29-33, 14.223a2-b17; vi.2.232b20, 6.236b32, 8.239a4, 12; vii.1.242b43, 44, 51, 69, 4.249a8; viii.1.251a19, b10-13, 27, 28, 252b5, 6.259a7, 7.261a30, 8.263a11, 9.266a6, 8, 267b25; and place, iv.1.208a31, 2.210a4, 4.211a13, 14, 6.213b4, 7.214-22; vii.1.242a51, b60, 2.243a18, b2, 10, 244a9; viii.2.253a14, 7.260a27, 9.265b20, 25, 26a1, 10.266a16; and void, iv.6.213b4, 6, 7.214a22, 24, 26, 27, 29, 8.214b17, 29, 215a16, b21, 216a9, 9.216b25, 34, 217a1, 12; viii.9.265b26; natural, violent, iv.8.215a1, 5, 6; v.6.230a18-b6, 10, 17, 231a6, 10; viii.3.254a9, 4.254b19; differentiated from change, iv.10.218b19; v.1.225a34-b3, 2.226b17, 5.229a31, 32, b10, 14, 6.230a9, 15, 16; classification of, v.1 (essential vs. accidental, differentiation from change, relation to categories), 2 (not literally a subject, not essentially a terminus, not to infinity, not generated or moved, three kinds, the "immovable"), 3 (related distinctions), 4 (single or diverse generically or specifically or numerically, continuous, completed, uniform), 5 (movements mutually opposed), 6 (rest vs. movement, contrary rests, rest vs. changelessness, natural vs. violent movement and rest); not of movement, v.2.225b15-226a23; unity, diversity of, v.4; vi.10.241b17; vii.1.242a64-68, b41, 4.249a20, 21, b12, 13; viii.2.252b30, 33, 4.255a11; one, continuous vs. successive, v.4.228a19-b15; viii.2.252b28-253a2, 6.-259a20, b22, 23, 260a4, 19, 7.260a22, 25, b19, 261a9, 30, 8.262a1, 13, 263a3, 27, 264a20, b1, 7, 29, 265a8, 10.267a13, 21; uniform, v.4.288b15-299a6; vi.7.-238a6; viii.10.267b4; contrariety of, v.5.229a7, 9, 31-b1, 21, 6.229b23, 25, 32; viii.1.251a28, 29, 7.261b15-22, 8.261b34, 262a6, 17, 264a17, 28, 33; continuity of, vi.1 (infinite divisibility of line, movement, time), 2 (faster vs. slower body in relation to time and distance, vs. Zeno), 3 (moments as indivisible, movement and rest in time), 4 (divisibility of moving thing and its movement, of movement as identified and as process), 5 (no end, or primary time of completion or of beginning, or other "absolutely first *part*", of change, with qualified exception of qualitative change), 6 (no partless part of commensurate time, of changing, of having changed of a stage in the continuous), 7 (finite movement, time, extension, moving body, vs. infinite), 8 (coming to rest without partless part of end, of beginning, of path), 9 (division not prior to motion, overtaking, time of flight, times of passing, opposite states, rotation), 10 (the indivisible as without motion essentially its own, single movements as having definite limits); continuous, divisible, vi.1.231b18-232a22, 4.234b21-235b5, 10.240b13-17; vii.1.-242b60; viii.8.263a27; arguments about, vi.9.239b10; and the indivisible, vi.10.-240b32, 241a3, 5; series of, vii.1. (moved movers, the unmoved mover, unity of series by contact), 2 (togetherness of agents and things acted upon in local motion, in qualitative alteration, in increase or decrease), 3 (alteration of objects is by sensible things, is not change of shape or of state), 4 (the question of movements incomparable in speed, with application to paths, to alterations, to generations and destructions), 5 (proportions between agents, objects, distances, and times of movements); of mover, moved, vii.1.242a59, 62, b58; viii.4.254b19; unrest, vii.3.248a2; comparable, vii.4.248a10, b6, and proportions, vii.5.250a3, 6; eternity of viii.1 (movement as ungenerated and indestructible, opposed theories, no absolutely first movement, time as eternal, no absolutely last movement, no ratio between infinite periods of rest and of movement,) 2 (objections from definite termini, starts, self-movers), 3 (elimination of alternatives to classes of beings inert, in movement, and admitting of movement and rest), 4 (classification of movements as due to agents), 5 (arguments concerning the first mover as self-moved or umoved), 6 (the first mover as necessary, eternal, one, immovable, uniformly operative), 7 (local motion as the primary kind of movement and as alone continuous), 8 (circular motion as alone continuous and infinite, turning back as requiring a pause, a line or a time as infinitely divisible only incidentally, dialectical and general arguments), circular motion as the primary kind of local motion, circular vs. rectilinear motion, local motion vs. other processes), 10 (the first mover not a finite or infinite magnitude with finite power, not like a mover of projectiles, but without magnitude or parts and at the circumference of the universe); being vs. generation, destruction of, viii.-1.250b11-29, 251a21, b11, 27-30, 252a4, b5, 2.252b8, 253a9, 5.256b12, 13.257b25, 6.258b10, 259a15, b4, 26, 260a24, b5, 9.265a27; self-movement, viii.2.252b14, 20, 24, 4.255a11; and explanation, viii.2.253a13, 5.257a28, 6.259a6, b8, 10, 260a8, 9.265b33; and science, opinion, viii.3.253b1, 254a28, 29, 9.265b17; prior, primary, viii.7.260a23, b6, 17, 261a4, 19, 27.9.265b9, 17, 266a7; circular, viii.8.264b18, viii.9.10.267b9

109a. κινεῖν, *movere*, move, induce movement, act upon, κινῶν, *movens, quod movet*, mover, agent, κινεῖσθαι, *moveri*, move, be moved, undergo movement, be acted upon, κινούμενον, *mobile, motus, quod movetur*, moved, subject to movement, acted upon: principle, i.2.184b16; ii.7.198a36; viii.9.265b21; natural beings, i.2.185a13; the All, i.3.186a17 (within itself), 5.188a20 (not); iii.5.205b6, 8; iv.5.212a34, b1, 14, 6.213b13; viii.2.253a28; nature, ii.1.192b21; viii.9.265b25; as means, ii.3.194b35; first mover, ii.7.198a19, 24, 33; v.1.224a34; vii.1.242a53, b59, 72, 2.243a32, 14, 245a8, b1; viii.1.251a24, viii.5.6, 7.260a25, b29, 261a26, 9.265b23, 266a9, viii.10; moved movers, ii.7.198a27, 29; iii.1.201a24, 25, 2.202a3, 6, 3.202b26, 30, 31; vii.1; viii.2.253a17, 18, viii.5, 6.259b15, 16, 260a6, 16, 7.261a1, 6, 25, 10.266b30-267a20, 25, b10, 14; moved, ii.7.198a30; viii.1.251a32, 7.260a8, 8.262a3, 16, 10.267a10, b5; unmoved mover, ii.7.198b1, 2; iii.1.201a27; viii.6, 9.266a9, 10.267a25, b16; from, to, ii.8.199b16, 27; v.1.224a35, b2, 7, 12, 2.225b17, 226a13, 18, 6.230b29, 32; vi.5.236b11; and realization, iii.1.201b6, 2.-201b30; v.1.224b26, 225a21; vii.3.247b5; viii.1.251a13, b1-6, 2.252b12-16, viii.4, 5.257b7, 9, 7.261a21; and necessity, iii.2.201b21; viii.5.257a25, b20, 34, 8.265a4; and form, kind, iii.2.202a9, 11; v.1.224b5; vii.4.249b12; mover, moved, iii.3.202-a17, 21, 26, 29-31, 35; iv.4.203a33; v.1.224a30-b8, 24, 25, 2.225b14; viii.1, 2, 5; viii.1.251a20, 24, b1-6, 2.252b19, 20, 22, 23, 253a2, 3, 10, 12, 14, viii.4 (classified) 5, 6.259a18, b15, 33, 260a3, 5, 7.260b3, 261a26, 9.265b34, 10.266a13, 16, b27, 267a11, 18, 19, b5, 8, 17; where, iii.5.205a15, b10; vi.9.239b6; and magnitude, iii.7.207b24; iv.4.212a11; vi.7; viii.6.258b25, 10.267a23, b18; and place, iv.1.209a22, 2.210a7, 4.211a17, 19, 35, 36, b4, 26, 212a16, 5.212a33, b10; v.1.224b5, 225a31; vii.1.242a51, b59, 2.244a6, b1; viii.3.254a1, viii.4, 5.256b19, 7.261a24, 8.264b11, 12, 9.266a1-5; and void, iv.7.214a25, 31, 8.214b30, 33, 215a14, b24, 216b17, 9.217a8, 18; viii.9.265b24; and time, iv.10.218b12, 16, 11.219a10, b30, 220a14, 12.221a1, b13, 17, 19, 21, 28, 13.222b23, 24, 14.222b31, 33, 223a2, 3, b1; v.1.224a35; vi.2.232a26, b15, 18-31, 3.234a24, 31-b9, 4.234b22, 235a13-b5, vi.6, 7, 9.239b6, 33, 240a2, 10 (Stadium), 10.241a15; vii.4.248a12, 16, 24, 249a9, 19; viii.2.252b15, 16, 253a7, viii.3, 6, 7.260b30, 8.264b11, 12, 10.266a10-b24, 267b23, 24; essentially, accidentally, v.1.224a26-b8, 22-26, 225a26, 2.226a19; vi.-10.240b9, 11, 18; vii.1.242a43, 44; viii.4, 5.256b8, 28, 257b1, 33, 6.259b21, 28, 8.263b6; and nonbeing, v.1.225a21, 26, 30; viii.2.252b16; and immovable, unmoved, v.2.226b10-16; viii.2.252b12-16, 253a2, 5, viii.3, 6.259a28-b1, 5, 260a3; continuously, v.3.226b27, 31; vi.1.231b21-232a17; viii.3.254b3, 6.259b14, 22, 8.-263b6, 264a9, 16, b17, 22, 25, 265a9, 10.266b29, 267a13, 20, b10, 17; and unity, v.4.227b24-26, 28, 31, 228a9, b22; viii.2.252b34; naturally, violently, v.6.231a9; viii.3.254a9, viii.4; divisible, indivisible, vi.4.234b22-235b5, 9.239b11, 10.240b9, 30-241a26; vii.1.242a38-49; viii.5.257a33, 258b1, 8.263a8; no partless part, vi.6.236-b25-237a9, vi.8; and stop, rest, vi.8, 9.240a32; vii.1.241b44-242a37, 41-46; viii.1.-250b27, 252a1, 21, 2.252b15, 28, viii.3, 6.259a23-26, b8-10, 260a10-13, 7.261b1, 8.264a23, 25, 31, 9.265b1, 10.266b33-267a20, b1; rotate, vi.9.240a30, b7; infinite, finite, vii.1.242b45; viii.2.252b12, and viii.5, 8.263a6, 10.266a10-b27, 267b23; by something sensible, vii.3.247a11; and path, vii.4.248a20, 249a17, 19; viii.8; nothing, viii.1.250b12, 23, 27, 5.256b9, 9.265a19; self, viii.5.257a27-258b9, 6.258b23, 259a2, b1, 13, 16, 17, 20, 7.261a24, 26, 9.265b33, 34, 10.266b29, 267a20; begin, end, viii.8.264a13, 9.265a31, b19; sensible things, viii.8.265a4

109b. κινητός, *mobilis*, movable, moved: relative, iii.1.200b31, 32; viii.1.251a23, b5; mover, iii.1.201a24, 2.202a4; viii.5.257a15, 16, 19, 20; realized as, iii.1.201a29,

2.202a8; viii.1.251a10, 18, 23, b5, 32, 5.257b6, 8; movement in, iii.3.202a14, 17; v.1.224b26; body, iv.4.212a7, 5.212b8, 19, 29; and time, iv.14.223a19; essentially v.1.224a29; αὐτοκίνητος, *ipsum mobile, sui movendi vim habere*, self-moved, viii.5.258a2

109c. κινητικός, *motivus, movendi vim habens, motor*, a possible mover, what can impart movement, agent: relative, iii.1.200b31, 32, 3.202a14-16; viii.1.251a24, b4, 4-255a21-23, 5.257a15, 20; by contact, iii.2.202a8; first, iv.3.210a22; being of, viii.1.252a1, 253a5, 6.258b15, 21

109d. ἀντικινεῖν *contra movere, invicem movere*, move in return, viii.5.257b21, 23

109e. ἀκίνητος, *immobilis*, independent of movement, unchanging, immovable, unmoved: principle, i.2.184b16; being, beings, i.2.184b26, 3.186a16; ii.7.198b2; viii.3, 6.258b27, 259a2, b5, 7.261a16; and explanation, ii.7.198a17, 29, 30; mover, iii.1.201a27; viii.5.256b20, 24, 27, 257b23, 24, 258a1, 5, 7, 9, 19, 29, b5, 9, 6.258b12-14, 21, 259a1, 27-b1, 21, 24, 32, 260a6, 15, 17, 9.266a9, 10.267-a25, b2, 16, 18; and infinite, iii.5.205a13, b7; place, iv.4.212a18-20; the All, iv.6.-213b13; and void, iv.9.217a9, 10; and movement, rest, iv.12.221b12; v.1.224b12; vi.9.239b7; viii.2.252b24; defined, v.2.226b10-16; and cycle, viii.1.251a3; first, viii.-6.259a13, 21; ἀκινησία, *immobilitas, motus vacuitas*, unmoved state, pause, iii.2.-202a4, 5; v.4.228b3, ἀμετακίνητος, *immobilis*, not capable of being transported, iv.4.212a15; δυσκίνητος, *graviter mobilis, dificulter mobilis*, hard to move, v.2.-226b12

109f. κίνημα, *momentum, terminus motus*, discrete atomic movement, vi.1.232a9, 10; 241a4

109g. δίνη, *volutatio, conversio, revolutio*, vortex, ii.4.196a26; iv.7.214a32

109h. ἀνακάμπτειν, *reflecti, recurrere*, revert, turn back, ἀνάκαμψις, *reflexio, recursus*, turning back, viii.5.257a7, viii.8, 9.265a21

110. ἠρεμία, *quies*, rest, ἠρέμησις, *quietatio, procreatio quietis*, state of rest, ἠρεμεῖν, *quiescere*, rest, ἠρεμίζεσθαι, *quiescere, tendere ad quietem*, coming to rest: nature, ii.1.192b22; agent of, ii.3.194b30; and movement, iii.2.202a4, 5; v.2.225b29, 226a7, b14, 15, v.6; vi.1.232a12-17, 6.236b29, 7.238a18, 21, 8.238-b23-26, 9.239b5, 240a3, 31, 32, b6; viii.1.250b27, 251a25, 26, 252a9, 21, 2.252-b15, 23, 28, 253a9, viii.3, 5.257a26, 6.259a23-26, b10, 260a10-13, 7.261b2, 10-12, 19, 8.264a22-33, 9.265a27-b16, 266a3, 10.267a13; and place, iv.4.211b30, 212a9, 5.212b20; viii.8.262b3; and void, iv.8.214b32, 215a20; and time, iv.11.220a18, 12, 221b8-28; v.1.228b4, 5; vi.3.234a31-b9, 236a17-20, 8.239a10-b4, 10.240b29, 30; vii.1.-242a35-49; viii.1.250b25, 29, 252a14; and nonbeing, v.1.225a29; thinking, mind, vii.3.247b10, 11, 248a2

110a. παύειν, *pausare, desinere, cessare, repausare*, stop, pause, v.4.228a16; vi.6.237a12; vii.1.242a35, 36; viii.1.251b32, 8.262a8, 10, 226b33, 267a1-12, 19; ἄπαυστος, *incessabilis, indeficiens*, pauseless, unceasing, viii.1.250b24, 6.259b25

111. ἵστασθαι, *stare, consistere, progressus ad quietem, ad quietem progredi, ad quietem tendere, progressus ad statum, sisti*, stop, coming to a standstill: in taking a point as two, iv.11.220a13; and movement, v.6.230a4, b22, 24-26, 231a5; vi.8.238b23-239a10; viii.5.258b5, 8.262a18, b17, 264a20, 10.267b1; series, vii.1.-242b17; ἱστάναι, *facere stare, sisti*, come to an end, continue at rest, check, stop: separation, i.4.187b31, 32; arrow, vi.9.239b30; animal, viii.4.255a7; series, viii.5.257a26; movement, vii.5.257b7, 8.262a8, 10, 263a30; στῆναι, *stare, consistere, residere*, stand, stop, pause: tripod, ii.6.197b17; and infinite, iii.5.205a19, 7.207b8; viii.5.256a29, 257a7; and void, iv.8.215a19; and motion, v.2.226a35; vi.9.240a4; viii.8.262a14, 27, 32, b6, 8, 25, 263a2; thinking, vii.3.247b11; στάσις, *status*, *quies*, rest, standstill, pause: beginning of, ii.1.192b14; vs. movement, v.4.-228b6; viii.7.261b17, 8.264a21, 9.265a26, 27

111a. ἀντιμετάστασις, *transmutatio*, *mutatio*, displacement, iv.1.208b3; ἀντιμεθίστασθαι, *commutari*, *transferri*, take place, iv.2.209b25, 4.211b27; ἀντιπερίστασις, *antiperistasis*, mutual replacement, iv.8.215a5; viii.10.267a16, 18

111b. ἀφίστασθαι, *distare*, be distant: fail to distinguish, i.8.191b10; rectilinear motion, viii.9.265b14; ἀπόστασις, *distantia*, remoteness, iv.14.223a5, 8

111c. διάστημα, *spatium*, *interrallum*, interval, dimension, extent, iii.3.202a18; iv.1.209a4, 2.209b6, 4.211a33, b7, 9, 17, 19, 212a3, 11, 5.212b25, 27, 6.213a28, 32, 7.214a5, 11, 20, 30, 8.216a35, b16, 9.216b32, 14.223a1; vi.7.237b35, 238a7, 16; διάστασις, *dimensio*, *extensio*, distance, extension, direction, iii.3.202b17, 5.204-b20, 206a6; iv.1.208b14; διίστασθαι, *distare*, *extendi*, be separated by distance, span, extend, iii.3.202b18, 5.204b21, 22; iv.8.215b7

111d. ἐξιστάναι, *distare facere*, *de statu suo dimovere*, become displaced, iv.8.216a28, 12.221b3; ἐκστατικός, *remotivus*, *vim habere de statu dimovendi*, *destructivus*, change from a former condition, iv.13.222b16; ἐξίστασθαι, *cedere*, *exire*, *distare*, *mutare*, *recedere*, give up, deviate from, vi.5.235b9; viii.7.261a20; ἔκστασις, *remotio*, *recessus*, *amotio*, loss, disruption, vi.10.241b2; vii.3.246a17, b2, 247a3

111e. ἔνστασις, *importunitas*, *objectio*, assault, viii.3.253b2

111f. ἐπίστασις, *consideratio*, give pause, ii.4.196a36, stop, viii.8.262a24

111g. καθίστασθαι, *statuere*, *esse*, *redegere*, *sisti*, arrange, compose, i.4.187a30; ii.4.196a27; vii.3.247b17 (mind), 248a2; καθῆσθαι, *sedere*, sit, ii.6.197b17, 18

111h. μεθίστασθαι, *transmutari*, *transferri*, *distare*, *cedere*, change, be displaced, pass to, iv.4.211a17, b17, 18, 21, 26, 212a4, 8.216a31, 216b1. 13, 9.217a18; μετάστασις, *transmutatio*, *transferri*, displacement, iv.4.212a10, 8.216a30

111i. συνίστασθαι, *componi*, *existere*, *constare*, *subsistere*, be constituted, ii.1.192b13; viii.4.254b20, 31; συνιστάναι, *subsistere*, *constare*, constitute, be organized, ii.1.193a36, 8.198b30, 199b5; iv.1.209a21; viii.1.250b15, 9.265a16

111j. ὑφίστασθαι, *sustinere*, *subesse*, obstruct, vii.4.255b24

112. ἀποβαίνειν, *evenire*, result: good, bad, ii.5.197a26

112a. βαδίζειν, *ambulare*, *vadere*, *ire*, *proficisci*, *transire*, *progredi*, walk, progress, be on the road, i.2.185b29, 30; ii.6.197b24, 25; iv.13.222b10; v.1.224a23, 4.227b32, 228a16, 17, 26; vi.1.21b30-232a11; viii.2.253a10, 4.255a8, 5.257b8; to infinite, v.2.225b34; vii.1.242a34

112b. περιπατεῖν, *ambulare*, *deambulare*, walk, ii.3.194b33

113. ἄγειν, *ducere*, push, pull, vii.2.244a4

113a. ἀνάγειν, *reducere*, *referre*, *inducere*, *revocare*, *adducere*, reduce, refer, carry upward: contraries, i.6.189b27; to principles, ii.3.194b22; viii.1.252a34; explanation, ii.7.198a16, 18, 23, 32, 8.198b12; vapor, ii.8.198b19; earth, iii.5.205b12; motion, vii.2.243a18, b15, 4.249a10

113b. ἐνάγειν, *ducere*, *inferre ferre*, move, vii.4.248a15

113c. ἐπαγωγή, *inductivo*, induction, i.2, 185a14; v.1.224b30, 5.29b3; vii.2.-244b3; vii.1.252a24; ἐπακτικῶς, *inductivo*, *per inductionem*, by iv.3.210b8

113d. προάγειν, *procedere*, *progredi*, *apponere*, *adducere*, *ingredi*, advance, proceed, transport, i.1.184a19; ii.2.194a31; viii.1.251a22

113e. συνάγειν, *inducere*, *copulare*, *conducere*, *contrahere*, *conjungare*, attract, contract, combine: not contraries, i.6.189a24; material, iv.9.217a12-16; love, viii.-1.252a27, 29

114. ἔρχεσθαι, *venire, coire, pervenire,* proceed, go, come: out of, into, ii.1.1934, 35, 197a3, 15, 16, 6.197b15, 16, 8.199b6, 20-22; factors, ii.7.198a24; and time, iv.13.222a22, b13; vi.9.239b16; viii.8.262b15, 264a15

114a. ἀπέρχεσθαι, *abscedere, removeri, discedere, abire,* depart, ii.8.199b21; viii.1.251a32, 8.262b15, 26

114b. διέρχεσθαι, διεξέρχεσθαι, *transire, pertransire, meare, venire, disserere, dividere, devenire, conficere,* divide, span, go over, pass through: distinction, i.8.191b10; infinite, iii.4.204a3; vi.7.238a33, b9, 10, 241b11; viii.8.263a6, 9, 13, 17, 20, b4, 7, 9.265a19; void, iv.8; time, iv.10.218a33; and continuity, vi.1.-232a3, 4, 10, vi.2; and movement, vi.9.239b13; vii.4.248b1; and rest, viii.9.265b6

114c. ἐξέρχεσθαι, *exire,* go out, iv.1.208b3

114d. ἐπέρχεσθαι, *aggredi, pertractare, venire, agere,* proceed, i.7.189b31; iii.1.200b16, 8.208a5; iv.10,217b29; viii.5.256a22

114e. τὸ παρελθόν, *praeteritum,* past, iv.10.218a9, 13.222a28, viii.1.251b22; παρέρχεσθαι, *transagere, pertransire,* transpire, pass, iv.13.222a11, 13.222b9, 13, 14, 223a9, 10; vi.9.240a14

114f. προέρχεσθαι, *pervenire, progredi, procedere, devenire,* progress, proceed, i.9.192a9; vii.6.259a

114g. ἰέναι, *ire, abire,* proceed: infinitely, iii.6.206b29; iv.3.210b27; viii.5.-256a17, 29; to less, v.2.226b4; to prime, viii.7.261a13; διεξιέναι, *transire, pertransire,* traverse, exhaust, iii.5.204a14, b9, 6.206b9, 11, 7.207b29; iv.8.216a6; vi.1.232a10, 7.238a23, 9.239b29; viii.10.266b13; διιέναι, *transire, pertransire,* span, iii.4.204a4; vi.2, 7; vii.1.242b70, 4.248b4; viii.8.263a5; εἰσιέναι, ἐπεισιέναι, *intrare, ingredi,* enter, vi.6.213b23, 7.214b2; viii.6.259b12; ἐπιέναι, *facere, aggredi,* approach, i.3.186a4; i.3.186a4; προιέναι, *procedere, progredi,* proceed, i.1.184a24; ii.8.199a24; vi.6.259a29; συνιέναι, *coire,* contract, iv.6.213b16, 18, 8.216b24

114h. ὁδός, *via,* path, course, road, way: order, i.1.184a16; of reasoning, i.8.191a26, b32; nature, ii.1.193b13, 14; Thebes, Athens, iii.3.202b13; and time, iv.12.220b30, 31

115. μεταβολή, *mutatio, permutatio,* change, transformation: sudden, i.3.-186a16; and contraries, contradictories, i.7.191a7; v.2.225b24, 3.226b27, 227a8; vi.9.240a19, 10.241a26, 31, b12; viii.2.252b11, 7.260a33, 261a33; denied, i.8.191-b33; tendency to, ii.1.192b18; and elements, ii.1.193a26; beginning, end of, ii.1.-193a30, 3.194b29; iii.1.200b12; v.1.224b8, 225a1; vi.4.234b11, 18, 5.236a7-15, 6.237a19, 20; viii.2.252b10; exploration of, ii.3.194b22; kinds of, and categories, iii.-1.201a2, 8; v.2.226b2, 17, 3.226b32; vi.5.235b29, 236b16; vii.3.246b12; and movement, rest, iii.2.202a18, 23; iv.10.218b20; v.1.224b15, 225a34-b3 (differentiated), 2.225b24, 5.229a31, 32, b10, 13, 6.230a8, 19; viii.3.254a7, 7.261a12, b11; and time, iv.10.218b9, 11, 14, 20, 11.218b21, 31, 33, 12.220b6, 13.222b16, 21, 27, 14.222b31, 32, 223a14; vi.5.236b2; viii.2.252b10; essential, accidental v.1.224b18, 27, 225a8, 11, 13, 2.226a21; not of change, v.2.225b16, 21, 24, 28, 34; and divisibility, vi.4.-235a35; and continuous, vi.5.235b24; viii.7.261b6, 14, 8.265a1, 11; and pause, vi.-6.237a13; no generation of, vii.3.247b13; viii.1.251a19; prior, viii.1.251a27, 28, b10, 30; and prime mover, viii.6.258b14, 259a5, 260a2, 10.267b4, 6; opposite kinds of, viii.7.261b4, 8, 21, 8.264b27

115a. μεταβάλλειν, *mutare, commutare, immutare,* change: factor, ii.3.-194b31; subject of, iii.1.200b33; v.2.225b18, 21, 27, 32, 33, 226a6, 11, 21; vi.5; from, to contrary, iii.5.205a6; v.2.226b4, 6, 5.229a19, 23, 25, b17, 6.231a1; vi.5.-8.239a23; viii.3.253b28, 254a12, 13, 4.255b22, 8.264a19; place, iv.1.208b5, 8, 2.-210a6, 4.211a16, 22, 23, b23, 212a16, 5.212a35; v.2.226a35, 4.227b16; vi.7.238a5,

9.240b1, 5, 10.241b9; viii.6.259b18, 7.260b13, 14; content, iv.4.211b14; air, water, iv.9.216b26, 217a17; and time, iv.10.218b11, 12, 11.218b22, 14.222b33; vi.2.232-a28, 29, 5.235b30-236b18, 8.239a24, 30; viii.3.253b26; essentially, v.1.224a21, 24, b9, 30, 225a3; and categories, movement, v.1.225b12, 13; viii.1.251b9, 7.261a22; and "between," v.3.226b24, 25; divisible, vi.4.234b10-20, 235a36, b3; no absolutely first part, vi.5.235b6-30 (of end), 30-236a7 (of time of end), 7-27 (of time when anything was first changing), 27-36 (of changing thing), b1-18 (of essentially divisible respects); no partless part, vi.6.236b19-25 (of commensurate time), 237-a9-17 (of changing), 17-b3 (of having changed), 3-9 (of a stage in the continuous), 10.240b20-241a6 (having its own motion) and contradictories, vi.9.240-a20; and infinite, vi.10.241b7-9; previously, viii.1.251a26, b7; move, rest, viii.3.-253a30, 6.259b15, 260a16, 17, 7.261b3; fire, viii.4.255b4; and be, not be, viii.6.-258b19; and prime mover, viii.6.260a5, 10.267a26; in opposite ways, viii.7.261b6, 8.267b26

115b. μεταβεβληκός, *mutatus*, have changed, v.2.225b28

115c. μεταβλητός, *mutabilis*, capable of change, viii.1.251a15

115d. συμμεταβάλλειν, *communicare, una mutari, simul mutari*, change together, viii.5.256b17, 10.267b2

115e. ἀμεταβλησία, *immutatio, mutationis vacuitas*, changelessness, v.6.-230a10, 11, 13, 16

116. γίγνεσθαι, *fieri, contingere, accidere, tendere*, become, originate, come into being, be produced, occur: known, i.1.184a22; one, many, i.2.185b26; iv.-5.213a10; v.3.227a11, 15, 16, 4.228a25, b1; vii.2.245a13; and starting-point, i.3.-186a12, 15; iii.4.203b9, 20; "from," i.4.5, 8 (solutions of problems of "coming from what is and from what is not"); ii.4.196a32; iii.4.203a24, 28, 29, 34; iv.2.210a10, 7.214b3, 9.216b27, 28, 217a14, 24 (potential, actual), 27-b7; viii.4.255b9, 7.261-b25; pattern, from, i.5.188b17, 18; iii.4.203a15; naturally, i.5.188b25; ii.6.197b33, 8.199a12-15, 9.200a16; contraries, i.6.189a18; v.5.229b11; viii.3.253b31; meanings of, i.7.190a31; absolutely, i.7.190a32, b5, 8.191b13, 9.191b36; v.2.226a2; and modes of speech, i.7.189b32-190b29; and being, i.8.191a27-31; ii.3.195b32, 8.199a7, 30; iii.5.205a3; vi.5.235b28; viii.1.252a33, 8.263b11, 26; product, i.9.192a13; ii.3.194b24; iii.6.206a32; matter, i.9.192a25-34; ii.1.193a14, b10, 7.198a20, 9.200a25; factor, ii.-192b24, 32, 6.198a6, 7, 9.200a11; man, vii.1.193b8, 12, 2.2.194a31; not falsity, ii.-2.193b35; clear, evident, ii.2.194a1; viii.4.255a19, 6.259a23, 260a11; by chance, automatism, ii.4, 5, 6.197b21, 35, 8.199b14, 19; always, usually, ii.4.196b11, 8.198-b35, 199b24; to some purpose, ii.3.194b36, 5.196b17, 30, 197a35, 6.197b19, 23, 24, 8.198b27, 29, 8.199a24, b27, 9.200a6, 19; contrary to nature, ii.6.197b34; successive, ii.7.198a34; iii.6.206a22, 207a6; by necessity, ii.8.198b14, 20, 9.200a1, b1, 2; mistake, ii.8.199a33, b7, 11; first, ii.8.199b8; iv.1.208b31; all things, iii.5.205-a4; viii.1.251a17, b16; potentially, actually, iii.6.206a25, b5; viii.4.255a35, b1, 9, 21; infinite, iii.7.207b14; vii.1.242a54; in place, iv.2.209b25, 210b11, 4.211b28, 212-a10; viii.8.262a29, 31, b; void, iv.9.217a4; past, events, time, change, iv.10-14; v.1, 4; vi.3, 5, 9, 10; viii.1-3, 7, 8; not generation itself, v.2.226a1-14, 18; continuous, v.4.228a22; vi.2.233a9; movement, movable, rest, v.6; vi.7, 10; vii.1.2; viii.1-3, 7-10; no partless part, vi.6.237b9-22; not forever, vi.7.238a19; division, vi.7.238a28; viii.8.263a9; alteration, vii.2.245a2, vii.3; viii.3.253b25; worlds, viii.-1.250b19, 251a19; continuously, viii.6.258b26; individual, viii.7.260b32, 231a2, 5-7, 13; and endure, cease, viii.7.261b23, 8.264b35; beginning, end, viii.8.262a25; not impossibility, viii.9.265a19; *et passim*

116a. γένεσις, *generatio, ortus*, generation, origination, becoming, productive process: absolute, i.3.186a14; ii.1.193b21; v.1.225a14, 16, 2.226a1; viii.1.261b4; of things similar, i.4.188a13; accounts of, i.5.188b35; viii.1.250b16; in general, i.7.-189b30, 191a3; denied, i.8.191b13; viii.7.261a14; exploration of, ii.3.194b21, 4.196-a9, 7.198a33; and necessity, ii.9.200a1; defined, iii.1.201a14; of things, iii.4.203-a29, 30, b19; and infinite, iii.6.206a32, 8.208a8, 10; and time, eternal, iv.12.222-a8, 13.222b20, 21, 14.223a21, b26; v.6.230b22; viii.7.261b22, 8.263b31; and change, movement, v.1.224b10, 225a13-20 (defined), 26, 33, 35, 2.225b25, 5.229a10, b12, 6.230a14, 18, 10.241a29, b16; vii.2.243b9; viii.1.250b21, 3.254a10, 7.260b12, 31, 261-a2-15, 34, b4, 8.264b32, 265a7; not of generation, v.2.225b15, 35, 226a1, 7, 12, 14-16; order of, v.3.227a24; natural, violent, v.6.230a26-28, 31, b7, 24; and alteration, vii.3.246a3, 9, b12, 247a18, b2, 4, 8-12; comparability of, vii.4.249b19, 20; and prime mover, viii.6.258b17, 260a2; and dense rare, viii.9.265b13

116b. ἐγγίνεσθαι, inesse, ingredi, come to be in, arise: place, iv.208b5, 4.211b9; states, vii.3.247a15; movement, viii.2.252b19, 253a8, 16, 6.259b4, 10.267a9, 13

116c. ἐπιγίγνεσθαι, fieri, generari, come to replace, viii.6.259a2

116d. γεννᾶν, generare, gignere, generate: by thickening and thinning, i.4.187a15; not contraries, i.6.189b21; man, ii.2.194b13, 7.198a27; not elements, iii.-5.204b24; time, viii.1.251b18; light, heavy, viii.4.256a1; and from, viii.5.257b10; and movement, viii.7.261a3

116e. γεννητός, generabilis, generationi obnoxius, subject to generation, iii.1.201a14, 4.203b8; iv.12.221b29

116f. προσγιγνόμενον, appositum, be added, vii.2.245a13; προσγίγνεσθαι, adjici, accedere, be added, viii.7.260a32

116g. ἀπογίγνεσθαι, abscedere, abesse, recedere, break loose, cease to be, viii.8.262a29, 32, b1, 7, 16, 20, 30

117. φθείρεσθαι, corrumpi, interire, pass, be destroyed, be corrupted, perish, cease: into, i.5.188b3, 13, 22; denied, i.8.191a27; matter, i.9.192a25-34; aspects, ii.1.193a27; not together, ii.3.195b20; seed, ii.8.199b6; element, iii.5.204b19, 26, 28; in generations of man, iii.6.206b2; in place, iv.1.209a2; and time, iv.10.218-a14-17, 20, 13.222b17, 24, 14.223a31; vi.6.237a11; viii.7.261b24, 8.263b22, 24; and change, movement, v.1.224b9, 2.226a9, 10, 4.228b5; vi.5.235b28, 236a5, 6; vii.3.-246b15; viii.1.251b12, 13, 252a1, 2, 3.254a12, 13, 7.260b27, 8.262a11, 264b2, 5, 9.265a27; no partless part, vi.6.237b13-19; not forever, vi.7.238a19; body, vii.3.-246b10; worlds, viii.1.250b19; continuously, viii.6.258b27; movers, viii.6.259a2

117a. φθορά, corruptio, interitus, destruction: denied, i.8.191b33; viii.3.254a11; tend to, i.9.192a20; exploration of, ii.3.194b21, 4.196a9; viii.1.250b17; ending, iii.-4.203b9, 19; contraries, contradictories, iii.5.205a24; viii.7.261a34, b12; and infinite, iii.6.206a32, 8.208a10; and place, iv.2.210a11; and time, iv.12.221b1, 222a8, 13.222b19, 25, 14.223b26; viii.7.261b22; and change, movement, v.1.224b8, 225-a18 (defined), 32, 33, 35, 2.225b25, 5.229a10, b13, 6.230a15, 17; vi.10.241a29; vii.2.243b9; viii.1.250b21, 252a3, 7.260b12, 261a11, 34, b4, 8.264b32, 265a7; natural, violent, v.6.230a26, 27, 30, b6, 9; fault, vii.3.246a16; comparability of, vii.-4.249a19; and prime mover, viii.6.258b18, 260a2; and dense, rare, viii.9.265b31

117b. φθαρτός, φθαρτικός, corruptibilis, interitui obnoxius, perishable, destructive: contraries, i.9.192a22; forms, i.9.192b1; moved, ii.7.198a31; iii.1.201-a15; viii.6.259b31; and time, iv.12.221b28; and change, viii.1.252a1, 2; and generation, viii.7.261a9; movement, viii.9.265a22, 23

117c. ἄφθαρτος, incorruptibilis, non interitui obnoxius, indestructible: matter, i.9.192a28; moved, ii.7.198a31; infinite, iii.4.203b8; movement, viii.1.251b29; priority of, viii.9.264a24

117d. ἀπολλύεσθαι, *perdi, perire*, become spoiled, perish, vanish: crops, ii.8.198b21, 22; organisms, ii.8.198b31; not place, iv.1.209a1, 2.210a10

118. ἀναιρεῖν, *destruere, evertere, resecare, consumere, tollere, auferre, absumere*, subvert, break up, deny, eliminate, abolish, refute, exhaust, use up: principles, i.2.185a2; body, i.4.187b26; becoming, i.8.191b13, 31; viii.3.254a10; be, not-be, i.-8.191b27; chance, ii.4.196a14; nature, ii.8.199b14; indivisible lines, iii.6.206a17; infinite, finite, iii.6.206b11, 7.207b28; vi.7.238a27, 28

118a. ἀφαιρεῖν, *auferre, removere, eximere, detrahere*, eliminate, remove, derive, subtract: "is", i.2.185b27; body, i.4.187b27, 35; product, iii.4.203b20; mathematics, iii.207b27; properties, iv.2.209b10; part, vi.4.234b35; viii.3.253b16, 20, 5.-258a13, 28, 31; hindrance, viii.4.255b20, 26; stage, viii.8.265a1; time, viii.10.266a20; magnitude, viii.10.266b3

118b. ἀφαίρεσις, *remotio, ablatio, abstractio*, subtraction, separation, extraction: not definitely, i.4.187b33; Hermes, i.7.190b7; finite, vi.7.238a28

118c. ἐξαιρεῖν, *removere, eximere*, disregard, iv.11.218b26

118d. καθαιρεῖν, *minuere, detrahere*, divide, vi.6.237b9. *See also* 75

119. αὔξη, αὔξησις, *augmentum, accretio*, growth increase, expansion: and movement, ii.1.192b15; iv.4.211a15, 16, 5.212b8; v.2.226a31, 4.228b21; vi.10.241-b16; viii.6.259b9, 7.260a29, b33, 261a10, 35; defined, iii.1.201a12-14; and infinite, iii.5.204b4, 6.206b28, 32, 7.207b29, 8.208a22; vi.10.241a32, 33; and void, iv.6.213-b5, 19, 7.214b4; and time, iv.14.223a20; natural, violent, v.6.230a24, 25, b1; not continuous, viii.3.254a9, 8.265a10; αὐξάνειν, αὔξειν, *augmentari, augeri*, grow, magnify, increase: by addition, i.7.190b7; vii.2.245a13; consequences, i.8.191a32; crops, i.8.198b18, 20; ratio, iii.6.206b10; and infinite, iii.7.207b24, 8.208a17; vi.6.-237b9; and place, iv.1.209a27; viii.7.260b14; and void, iv.7.214b2, 5, 9; and time, iv.14.223a31; and change, movement, v.2.225b19; vii.2.245a11-17; viii.3.253b13, 5.256b34, 257a4, 5, 24, 7.260b14, 26, 261a23, 9.265b28, 266a3; and proportions, vii.5.250a28-17; by like, unlike, viii.7.260a30; συναύξεσθαι, *simul augmentari, simul augeri*, grow along with, iv.1.209a28, 5.212b24

119a. φθίσις, *decrementum, diminutio, deminutio*, decline: and movement, ii.1.-192b15; iv.4.211a15, 16; v.2.226a31, 32, 4.228b21; viii.3.253b22, 6.259b9, 7.261a10, 35, 9.265b28; defined, iii.1.201a12-14; natural, violent, v.6.230a24, b1; and infinite, vi.10.241a33, b1; φθίνειν *detrimentum pati, minuare*, decrease, decline, decay, v.2.225b20; vii.2.245a14, 15; viii.3.253b13, 7.260b14, 7.261a23, 8.265a6, 9.266a3

120. ἀλλοίωσις, *alteratio, variatio*, qualitative change, alteration: starting-point of, i.3.186a15; being of, i.3.186a18; and becoming, i.4.187a30, 7.190b8; and movement, ii.1.192b15; iii.2.202a35; v.1.224a30, 2.226a11, 18, 4.227b6, 228b3, 20, 229a5; vi.10.241a31, 32; vii.4 (comparability), 5.250a28-b7 (proportions); viii.5.-257a5, 6, 8, 24, b4. 7.260a29-b4, 27, 32, 261a10, 22, 34, 8.265a5, 9.26528, 266a3; defined, iii.1.201a11, 12, 3.20b25; v.2.226a26, b2; and infinite, iii.7.207b24; viii.-3.253b23-26; and material, iv.4.211b31; and void, iv.7.214a28, b2; and time, iv.-14.223a31, b8, 9, 20; natural, violent, v.6.230a21, b3, 6; alter, altered, vii.2.242-b2-245a11; by sensible things, vii.2.245b3, 5, 247a7, 248a6-9; not change of shape or of state, viii.3; and generation, destruction, vii.3.246a1-9, b10-17, 247a6, 18, 19, b1-248a9; viii.8.265a7; potential, actual, viii.1.251a14; and continuous, viii.3.-293b30, 8.264b3, 31, 265a10

120a. ἑτεροίωσις, *alteratio, variatio*, change, iv.9.217b26

121. φορά, *loci mutatio, latio, ferri*, spatial change, local motion: distinctions, iii.1.201a7; defined, iii.1.201a15; natural, violent, iii.5.205b27; v.6.230b12, 13; vii.-2.243b1; and "movement," iv.1.208a32, 4.211a15; v.2.226a18, 33, 4.227b5, 6, 228-a29, 229a2, 6, 5.229b7; vii.2.243a39, vii.4; viii.5.257b3, viii.7, 8.264a11; and place, iv.1.208b8, 4.212a7, 5.212b7; vi.7.238b22; in circle, iv.4.212a22, 9.217a20; viii.8.-265a11, 9.265a13, 14; and void, iv.6.213b5, 8.214b13, 16, 215a12, 13, 17, 9.217-a4, b25, 26; viii.9.265b26; and time, iv.11.219b30, 220a2, 3, 6, 7, 14.223a2, b8, 10, 12, 21, 30; of boat, vi.10.240b11; and limits, vi.10.241b2, 15, 20; kinds of, vii.2.243a16, 4.249a16; orbit, viii.6.259b31, 9.265a14, 26; primary, viii.7, 9.265a13, b17

121a. φέρεσθαι, *ferri*, be carried, move naturally, violently, ii.1.192b36, 9.200a2; v.6.230b12, 24, 25, 27, 231a12; viii.1.252a18; always, iii.5.205a14; and earth, iii.5.205b11; to place, iv.1.209b11, 19, 2.210a2, 4.211a4, 212a10, 25, 5.212-b29, iv.8, 9.217a1-8, b22; v.2.226a18, 34, 4.228b20; vi.9.239b12; viii.3.253b31, 32, 4.255a9, 8.264a12, 30, 31, 9.265b13; and time, iv.11.219b17, 22, 23, 30-22, 220a2, 4, 6, 7, 14, 14.223a32; vi.1.232a22, 9.239b7, 240a3; viii.8.264a15-17; and infinite, finite, vi.7.238a11, 10.241b8; viii.9.265a18; and rest, pause, vi.8.239b4, 9.239b7, 30; viii.8.262b10, 12, 23, 24, 9.265b5; mover, moved, vii.2.243a11, b2; viii.5.256-b33, 257a4, 8, 9, b3; actual, viii.1.251a14; and orbit, path, viii.6.259b31, viii.8, 9.265b6; and change, viii.7.260b27; projectiles, viii.10.266b27

121b. μεταφορητός, *transmutabilies, qui transferri potest*, movable, transportable, iv.2.209, 4.212a15

121c. περιφέρεια, *circulatio, circumferentia*, circumference, orbit, arc, iv.9.-217b2, 12, 14.223a3; vi.9.240b2; viii.8.264b25, 9.265b5

121d. περιφορά, *circulatio, conversio*, revolution, iv.10.218b1-3; v.4.227b20; viii.9.265b9

121e. τὰ συμφέροντα, *expedientia, quae conferunt*, adaptation, ii.8.199a24

121f. ῥοπή, *inclinatio, momentum*, tendency, iv.8.216a13, 19

122. πίπτειν, *incidere, cadere*, fall: classification, ii.3.195a15, 4.196b9; vii.2.-243b16; stone, ii.6.197b31; and place, iv.4.211b18, 7.214a23

122a. καταπίπτειν, *cadere*, fall, ii.6.197b17, 18, 30, 31

122b. συμπίπτειν, *accidere, contingere*, come about, ii.8.198b27; σύμπτωμα, *casus*, coincidence, ii.8.199a1, 3, 4

123. ἀνάγκη, *necesse, necessitas*, necessity: procedure, i.1.184a19; ii.1.193b15; principles, i.2.184b15, 7.191a16; argument, i.3.186a32; being, becoming, i.3.186-b16, 28, 4.187a33, 5.188b12, 7.190a33, 8.191a28; ii.8.198b14, ii.9; v.2.225b35; vi.-5.235b15, 6.237b10, 19; viii.7.261a1, 8.263b25, 26; conclusion, i.4.187a36; iii.1.201-b12; whole, i.4.187b13, 36; iv.10.218a4; matter, i.9.192a28; talk about words, ii.1.193a8; vs. usual, chance, ii.5.196b12, 197a1, 7.198b6; factors, ii.5.197a8, 6.198-a11, 8.198b19, 24; sphere of chance, ii.6.197b3; first, ii.8.199b8; status of, ii.9 (in relation to origination, materials, uses, form, theoretical and practical reasoning, definition); and contingency, ii.9.199b34-200a1, 12, 17, 24, b2, 3; and movement, ii.2, 3, 5; iv.8, 11, 12; v.1, 2, 4-6; vi.2, 4, 6, 7, 10; vii.1, 2, 4, 5; viii.1 ,3-10; all things, iii.4.203a32; infinite, finite, iii.4.203b21, 22, 5.204a19, 205b28, 7.207b8; vi.2.233-b13, 7.238b2, 21; viii.10.266a28, b16; elements, iii.5.204b13; plan, iv.1.208a27, 209a1, 27, 4.211b6, 212a5, 5.212b24, 213a5, 8.216a31; not infinite, regress, iv.3.-210b27; void, full, iv.6.213b14, 7.214a16, 26, 8.214b32, 215a21, 216a18, 9.216b28, 217a15; greater, vii.8.215b16; time, now, iv.10.218a14, 16, 11.219a9, 220a13, 12.221-a24, 28, b29; vi.1.232a18, 2.232b26, 3.233b33, 234a5, 19, b8, 5.235b33, 37, 236a21, 25, 6.236b23, 237a15; viii.1.251b22, 26, 3.253b26; change, v.1.255a7, b1; vi.-4.234b10, 15, 19, 5.235b7, 20, 21, 6.237a17, 18, 34, b3, 9.240a28, 10.240b23; viii.-1.251b6, 2.252b11, 3.253b27, 7.260b14, 10.267b2; succession, v.3.227a18; continuous, v.3.227a22, 23; vi.1.231a29; vii.2.245a15; viii.6.259b23, 8.262b22, 10.267a21; touch, v.3.227a24; vi.1.231b3; viii.5.256b19; not between, v.3.227a32; rest, stop, pause, endure, v.4.228b4, 6.230a6; vi.1.232a12, 8.238b24, 27, 28, 31; viii.3.253a24,

ANALYTICAL INDEX OF TECHNICAL TERMS   223

b33, 254a20, 6.259b27, 7.261b10, 24, 8.262a18, 66, 17, 26, 263a1, 264a19, 26, 32, 10.267a11; divisible, vi.4.236a21, 6.237a32; solution, vi.9.239b26; alteration, vii.3.246b15, 247a14; viii.7.260a33, b26; magnitude, viii.6.258b25, 26; count, viii.-8.263a31

123a. ἀναγκαῖον, *necessarium, necessitas,* necessary, inevitable: principles, i.6.189a35, 7.191a6; vs. usual, chance, accidental, ii.5.196b20; viii.5.256b10; and nature, ii.8.198b11, 9.200a8, 15; material, ii.9.200a13, 14, 31, b5; ignorance, iii.-1.200b14; movement, iii.2.201b21, 34, 3.202a22; vi.9.239b16, 17; viii.1.250b20, 251-a10, 13, 18, 23, 2.252b31, 253a15, 5.257a33, b34, 7.260a24, 25, 28, b20, 31, 9.265-b9; infinite, finite, iii.4.202b31, 34, 203b29, 5.204a29, 205a32, 8.208a7, 8; place, iv.1.209a30, 2.210a11; void, iv. 7.214b8, 8.214b29; change, vi.6.237a14; viii.1.251-a25, 6.260a16; alteration, vii.3.246a6, 247a6; generation, viii.1.251a17; disappear, viii.3.253b21; mover, moved, viii.5.256b23, 257a14, 258a22, 6.258b13, 10.266a10; be, not be, viii.6.258b18, 7.261a6; pause, viii.8.262a14

123b. ἀναγκάζειν, *cogere,* compel, drive, i.5.288b30, 8.191b30

123c. βία, *vi,* βίαιος, *violentus,* βιαίως, *violenter, vi,* by compulsion, violently: somewhere, iii.5.205b5; not, iv.5.212b31; change, movement, iv.8.215a1-6; v.6.230a-29-32, b4, 24, 25, 231a7; viii.3.253b34, 254a9, viii.4

124. δεῖν, *oportere, opus esse, indigere,* must: proceed, i.1.184a24; viii.5.256b34; principles, beings, 5.188a27, 189a10, 6.189a19, 30, 7.191a4; iii.4.203a30; consider, grasp, i.5.188a31; iii.5.205a7; viii.5.257a28, 6.259b6; nature, i.6.189b20, 9.192a10; viii.6.259a11; change, i.7.190a14; iv.9.216b26; v.2.226b7; viii.1.251b7; being, i.8.-191a31; v.2.226a10, 13; matter, i.9.192a29; explain, ii.2.194b10, 3.195b21, 7.198b4; viii.1.252a22; vapor, ii.8.198b19; becoming, ii.8.199b13, 13, 9.200b1, 2; movement, rest, iii.1.200b13, 2.202a15, 27, 5.205b19; v.4.227b29, 229a6; vi.9.239b12; vii.2.-243b15; viii.1.252a6, 4.254b29, 6.258b10, 259b23, 8.263a6, 9.265a26, 10.267a26, b5, 12; infinite, finite, iii.6.206a13, 30, 207a4, 30; viii.6.259a11; potential, iii.6.206-a18; place, iv.1.208b32, 2.209b34, 210a5, 3.210b10, 4.211a7, 15, 5.212b15, 8.216-b14; void, iv.6.213a19, 31, 7.213b30, 214a5; whole, parts, iv.10.218a7; time, iv.13.222a25; comparable, vii.4.249a4, b5, 15; be capable, viii.1.251a15; be destroyed, viii.1.252a1; alteration, viii.7.260b1; generation, viii.7.261a4, 8.263b31; not be disturbed, viii.7.261b15; δεῖσθαι, *indigere, egere,* need: infinite, iii.7.207-b30; reason, argument, viii.1.252a31, 3.254a31

125. ἐπὶ τὸ πολύ, *frequenter, plerumque,* ἐπὶ πλεῖον, *in plus, latius, ad plus,* usually, for the most part, probable: vs. necessity, chance, ii.5.196b11, 13, 20, 36, 197a4, 19, 35, 7.198b6, 8.198b35, 199b24

126. τύχη, *fortuna,* chance: as factor, ii.4 (denied, ignored, acknowledged, held mysterious), 5 (nonnormal, accidental, indeterminate, unstable), 6 (in relation to automatism, mind, nature, movement), 8.199b23, 25; and ends, ii.8.-198b36, 199a1, b19, 20

126a. τυχόν, *contingens, quodlibet, quodvis,* any chance thing: vs. natural functioning, i.5.188a33, b3, 5, 7; vs. opposite, i.5.188b14; factors, ii.5.197a22; not terminus, ii.8.199b18; not touching, iii.8.208a14; and place, iv.4.211b18, 5.212-b26; change, v.2.225b29; not continuous, v.4.228a23; part, v.4.228b25; not subject, attribute, vii.4.249a2, 3

126b. τυγχάνειν, *sortiri, contingere, attingere,* chance, happen: art, ii.1.192b17; motion, ii.4.196a22; products, ii.8.199b14; direction, iv.1.208b19, 20; contraries, iv.-9.217a25; discuss, iv.10.218a32; place, iv.10.218b12; conditions, vii.3.246b17; one, vii.4.248b19; move, vii.5.250a15, b6; viii.4.254b20, 5.258a8, 9.266a4

126c. ἀποτυγχάνειν, *fallere, aberrare,* fail to achieve, ii.8.199b3

126d. ἀτυχία, δυστυχία, infortunium, infortunium seu calamitas, misfortune, ii.5.197a27, 28, 6.197b8

126e. εὐτυχία, eufortunium, fortunae prosperitas, prosperity, good fortune, ii.5.197a26, 28, 30, 6.197b2, 4, 8, 10

127. αὐτόματον, casus, automatism: and change, ii.4.195b31-36, 196a12, b8, 5.196b30, 197a34, 6.197a36, 27, b13-198a13, 8.198b30; and vortex, ii.4.196a26, 34, b3; and ends, ii.8.198b36, 199a5

127a. μάτην, vanum, frustra, in vain, futile, ii.6.197b22, 26, 27, 30; iii.4.203b5

128. ἁμαρτάνειν, peccare, fall short, fail, ii.8.199b2; iv.6.213a24; viii.1.251a32

128a. ἁμαρτία, ἁμάρτημα, peccatum, mistake, ii.8.199a33, b4

129. κωλύειν, prohibere, vetare, prevent, hinder: nothing to prevent, i.3.187a5 (not an opposite something), 9 (beings many); ii.8.198b17, 23 (necessity); iii.-3.202b8 (unity of actualization), 5.205b10 (naturally in movement); iv.3.210b24 (place) v.3.226b29 (high note), 227a3 (between), 6.230b7 (contrary); vi.10.241-b15 (always movements); vii.2.245a1 (fail to notice), 5.258b, 3 (divisible), 8.264-b17 (circular motion); move without, iv.1.208b11; and actualization, viii.4.255-b4-12, 23, 24, 256a2; movements, viii.8.262a11

129a. ἐμποδίζειν, impedire, prohibere, interfere with, prevent, defeat, resist, ii.8.199a11, b18, b26; iv.7.214b5, 8.215a22, 29, b3, 11; viii.4.255b7, 20, 256a2

130. ἄπειρον, infinitum, immensum, infinite, indefinite; principles, i.2.184b18, 20, 24, 4.187b10, 5.189a12, 16, 20; being, i.2.185a33, b17; viii.3.254b25; and categories, i.2.185a32-b5, 4.187b8, 10, 34; iii.4.203a5, b33, 5.204a10, 17-34, 206a3-8; divisibility, i.2.185b10; iii.1.200b20, 5.200a11, 22-28, 34, 7.207b16, 27; vi.2.233a19, 4.235a37-b5, 5.236a27, 6.237a33, b8, vi.7, 8.239a10, 22; viii.3.253b22, 24, 8.263a28; similar parts, i.4.187a25, 27, 188a2-5; contraries, i.4.187a25; unknowable, i.4.187-b7-9; iii.6.207a26; in kind, i.4.187a9, 10; iii.4.205a22; number of times, ii.1.193-a28 attributes, ii.5.196b28; reasons, ii.5.197a16; and continuity, iii.1.200b17, 19, 4.203a22, 7.207b16; vi.2.233a17-b15, 6.237b15, 8.239a22; viii.8.265a10; being of, iii.4 (as relevant, independent or adjectival, extensive or numerical, explanatory and factual, equivocal), 5 (not actual, whether as a primary being or source or as a simple or composite body in movement or at rest in a place), 6 (potential, a material aspect, not a whole), 7 (ascribed to certain subjects, processes, abstract terms), 8 (nor required as actual by generation, limitation, thought, time, magnitude); iv.1.208a28; subjects of, iii.4.202b31, 32, iii.7; theory of, iii.4.202b35, 203a1, 3, b31, 7.207b28; being, nonbeing of, iii.4.202b35, 36, 203b15, 31, 32, 5.204a23, b3, 6.206a9, 13, 14, b12, 22, 7.207a34, 35, b5, 16-21, 28-34, iii.8; viii.-9.265a18; definition of, iii.4.202b36, 204a2-7, 5.204a14, 6.206a18-b3, 33-207a2, 6, 7; v.233a24; as principle, iii.4.203a3, b4-15, 5.204a21, 3; sensible, body, iii.4.203-a5, 10, b26, 29, 204a2, 5.204a8, b1-206a8, 6.206b25, 7.207b19; universe, worlds, iii.4.203a7, b25, 26, 5.205a1-7, 24, b6, 34, 6.206b24, 7.207b21, 8.208a10; iv.6.213-b24; viii.1.250b18; number, plurality, iii.4.203a10, b24, 32, 5.204a10, 19, b8, 17, 6.206a12, b31, 7.207b1-21; great and small, iii.4.203a15, 6.206b28, 207a30; elements, iii.4.203a16-23, 5.204a15, b13-22; time, iii.4.203b17, 6.206a10, 26, 7.207b15, 22, 24, 8.208a20; iv.10.218a1; vi.2.233a17-b15, vi.7; viii.1.250b25, 252a14, 8.263-a12-14, 19-21, viii.10; mathematical, magnitude, iii.4.203a18, b25, 204a9, 19, 35, 6.206a11, 16, 17, 27, 7.207a34, b, 8.208a22; vi.2.233a17-b15, 7.238a31, b21; generation, destruction, iii.4.203b19, 6.206a23, 30, 8.208a9; viii.3.253b24; limitation, iii.-4.203b21, 8.208a11; thought, iii.4.203b24, 7.207b10, 8.208a17, 21; void, place, iii.-4.203b29, 5.205a7-206a7; iv.4.211b20, 8.215a8, 9; span, iii.4.204a2-6, 5.204a14, b10; vi.2.233a21-b15, vi.7; viii.8.263a5, 6, 9, 15, 19, b4, 9.265a20; addition, subtraction, iii.4.204a7, 6.206a15, b3-33, 207a23, 7.207a33, 35; accidental, iii.5.204a10, 15, 29,

ANALYTICAL INDEX OF TECHNICAL TERMS 225

31, 6.206b22; viii.8.263b8; potential, actual, iii.5.204a21, 25, 206a7, iii.6, 7.207-b11, 12, 18, 29, 8.208a5, viii.8.263a29; change, movement, rest, iii.5.204a7-206a7, 7.207b23, 8.208a20; iv.8.215a21; v.2.225b34; vi.6.237a11, 16, vi.7, 10.241a26, b10, 11, 13, 19; vii.1.242a66, b45, 46, 49, 53, 55-71; viii.2.252b12, 27, 28, 3.253b24, 8.261b27, 265a10, 9.265a17; and explanation, factor, iii.5.205b14, 17, 7.207b34, 35; material aspect, iii.6.206b15, 207a21, 27, 7.20b1, 34-208a4; regress, iii.6.206-b29; iv.1.209a25, 3.210b27; vii.1.242a34; viii.5.256a17, 28, 29; analogy, iii.6.207a2, 18; growth, size, iii.8.208a17; nows, iv.10.218a21; vi.6.237a16; no first in, v.2.226-a4; viii.5.256a18; no ratio, viii.1.252a13; movers, viii.6.258b31, 259a9; and first mover, viii.10.267b19-25

130a. ἀπειρία, *infinitas,* infinity, iii.4.203a12, 7.207b14

131. πέρας, *terminus, finis,* limit: indivisible, i.2.185b18; of bodies, ii.2.193b32; iv.1.209a9; and source, iii.4.203b7; not, iii.4.203b21, 204a6; place, iii.5.206a7; iv.-2.209b2, 5, 9, 10, 4.211b12, 212a6, 13, 20, 26, 30, 5.212b19, 28; end, iii.6.207a15; vi.5.236a13; viii.8.264b27; minimum, iii.7.207b2; and time, now, iv.10.218a24, 11.-220a21-23, i.3.222a12, 18; vi.2.233a14, 3.234a3, 8.239a35; touch, v.3.227a12; and continuity, vi.3.234a9; and movement, change, vi.9.239b23, 10.241a29, 33, b12; viii.2.252b11, 9.265a31, 33

131a. περαίνειν, *inducere, terminari, includere,* limit: number, iii.4.203a12; relative, iii.4.203b21, 22; element, iii.5.204b16; the All, iii.5.205a31, 6.207a16; and touch, iii.8.208a11; time, iv.13.222a26, 14.223b6; vi.2.233b13; vii.1.242b69; line, distance, v.3.226b33; vi.7.238a30

131b. πεπερασμένον, *finitum,* finite, limited, determinate: principles, i.2.184b18, 19, 24, 4.188a17, 6.189a15, 21; being, i.2.185b18; body, i.4.187b25, 26, 33, 34; iv.4.212a30, 5.212b28; vi.7.237b27, 238b4; or infinite, iii.4.202b32, 203-b20, 5.205a21, 22, 31, 6.206a29, 33, b4, 7, 9, 11, 15, 207a24, 7.207b31, 8.208a13; vi.2.233a21-b15, 4.234a37, vi.7.10.241b2; vii.1.242b44, 45, 52, 53, 55-71; viii.6.259-a9-11, viii.10; elements, iii.4.203a18, 5.204b12, 19; universe, iii.5.205a3, 6.207a8-32, 8.208a10; time, iv.8.216a10, 23, 25, 13. 222a29; vi.2.233a21-b15, vi.7; viii.8.-263a14, 19; measure, v.3.226b34; vi.7.238a14; magnitude, vi.2.233b34, 7.238a14, 32; distance, vi.7.237b35, 238a16, 9.239b29; division, vi.7.238a28; movements, mov-ers, viii.5.257a7, 6.259a9-11, viii.10; line, viii.8.261b32, 9.265a20

132. τόπος, *locus,* place: and movement, rest, ii.1.192b14; iii.1.200b21; iv.1-208a32, 2.210a5, 4.211a13, 212a18; v.1.224b5, 12, 225a31, 32, b9, 2.226a32, 35, 36, 4.227b26, 6.229b27, 230a19; vi.7.238b22; vii.1.242a51, b40, 59, 2.243a36-244b2; viii.2.253a15, 3.254a1, viii.4, 5.256b19, 7.260a28, 261a25, 8.262a4, 264b15, 9.265b2, 17, 20, 25, 266a1, 2; change in, iii.1.200b34; iv.2.210a10; v.2.225b19, 3.226b31, 4.227b16, 17; viii.1.251a15, 6.259b19, 7.260b13, 14; and infinite, iii.4.203b29, 5.-205a10, 16, 17, 20, 27, 30-33, 35, b21, 23, 25, 31, 206a1, 2, 3, 5; iv.4.211b21; body in, iii.5.205b31, 206a1; iv.1.4.211a23, 5.212a32; distinctions, iii.5.205b31, 206a2, 3; iv.1.208b12, 4.211a3, 212a29; viii.8.261b35; analysis of, iv.1. (being, status, genus), 2 (not form or material), 3 (not "in" itself or "in" a place), 4 (basic facts, rela-tion to movement, alternative interpretations, definition), 5 (bodies, cosmos; po-tentially, actually; essentially, accidentally; solutions of problems); being of, iv.-1.2.209b16, 3.210b23, 4.210b32, 24b6, 212a5, 20 (defined), 5.213a10; appropriate, resident, iv.1.208b11, 2.210a3, 4.211a5; v.6.230b27; viii.3.253b34, 35, 4.255b14; and void, iv.1.208b25-27, 4.212a13, 6.213b14, 16, 7.213b31, 33, 214a5, 14, 17, 21, 23, 8.214b16-28, 216b9, 10, 14, 9.216b32, 217a4; viii.9.265b26; and explanation, iv.1.-209a19; not of, in place, iv.1.209a25, 21210a6, 9, 3.210b25, 4.211b24; primary, iv.2.209a33, b1, 3.210b24, 4.211a27; v.3.226b22; not form, material, iv.2.3.210b30, 4.211b7, 30, 212a3, 8; and concrete object, iv.2.209b24, 210a7, 4.210a7, 4.210b34,

212a30; "in", iv.3.210b24, 12.221a18, 20, 29; and universe, iv.4.211a24, 29, 5.212-a35, b9, 18; potentially, actually, iv.5.212b3-6; essentially, accidentally, iv.5.212b7-22; solutions of problems concerning, iv.5.212b22-213a11; and time, iv.11.219a15, 12.221a18, 30, 14.223a19; vi.9.240a31; together in, v.3.226b22; contrary, v.3.226b32, 6.230b11; viii.4.255b12, 6.260a9, 8.261b34, 36, 264b14; and continuous, vi.1.231b6; divisible, vi.4.235a18

132a. χώρα, *receptaculum, locus,* location, space: distinct, iv.1.208b7; and beings, bodies, iv.1.208b32, 209a8; and material, iv.2.209b12, 13, 15

132b. χωρίον, *spatium,* place, space: go to, ii.5.197a1; move over, iv.8.216a15

133. κενόν, *inane, vacuum,* empty, void, vacuum: and full (see 133a); "non-being", i.5.188a23; iv.8.215a11; movement in, iii.1.200b21; iv.6.213b4-6, 12-14, 7.214a22-b10; iv.8, 9.216b22-217a20; viii.9.265b24; and place, iv.4.212a14, 6.213-a12-19, 7.213b31, 33, 214a13, 16, 21, 8.214b16-28, 216b8-10, 9.216b31, 217a4, 5; viii.9.265b26; analysis of, iv.6-9; being of, iv.6-9; and natures, iv.6.213b22-27; and explanation, iv.8.214b15, 16, 9.216b35, 217a1, b22; and infinite, iv.8.215a7, 9; and tenuous, iv.9; referent of, iv.9.217b24

133a. πλῆρες, *firmum, solidum, plerum,* full: "being", i.5.188a22; receptacle, iv.6.213a17, 18, 30; and void, iv.6.213b6, 7.214a7, 28, 8.215b20, 26, 216a3, 6, 11, 18

133b. πυκνόν, *densum,* dense, πύκνωσις, πυκνότης, *densitas,* thickening, λεπτόν, μανόν, *subtile, rarum,* rare, tenuous, μάνωσις, μανότης, *raritas,* thinning: denser than fire, rarer than air, i.4.187a14; as contraries, i.4.187a15, 5.188-a22, 189a9, 6.189b9; iv.9.217b10; functioning of, i.6.189a22, 23; and movement, alteration, modification, iv.5.212b3, 7.214a33, 8.215a29-216a11; vii.2.244b7, 3.246-a7; viii.7.260b7-12, 9.265b30; and void, iv.9; fire, iv.9.217a1; and heavy, light, iv.9.217b11-27

134. χρόνος, *tempus,* time: and beginning, i.3.186a14; iv.13.222a12, 33-b7, 14.223b28; and movement, change, iii.1.200b21, 7.207b24; iv.8.215b1-216a11, 10.-218a33-b20, iv.11.12.220b12-222a9, 13.222a29-33, b22-27, 14.222b30-224a2; v.1.224-a35, 2.226b11, 4.227b26, 28, 30, 228a4; vi.2, 4.234b22, 235a10-b5, 5.236a15 (not absolutely first), 17, 30, 36, b4, 5, vi.6, 7, 8.238b27, 239a23, 24, 10.240b22-241a6, 15-23; vii.4, 5; viii.1.251b11-28, 252a31, b5, 3.253b26, 7.260b19, 29, 261b7, 13, 8.-262a3, b21, 264b6, 17, 9.266a6, viii.10; infinite, finite, iii.4.202b31, 203b16, 6.206-a10, 25, b2, 7.207b15, 22, 8.208a20; iv.10.218a1, 25; vi.2.233a21-b15, 10.241b13, 17, 19; vii.1.242b42-53, 69, 70; viii.1.250b25, 252a14, 8.263a12, 13, 19-21, b4, viii.-10; reckoning of, iii.7.207b15; analysis of, iv.10 (difficulties concerning its being and nature), 11 (its dependent, numerical, sequential, diversified, and continuous, character), 12 (its measurement of and by movement and of other things), 13 (the present and related distinctions), 14 (its relativity to processes); alleged parts of, iv.10, 11.220a19, 13.222b8, 13; vi.6 (no partless part), 8, 9.239b8; viii.-3.253b17, 19, 8.263b26-264a6; being of, iv.10.210b30-218a6, 11.218b28, 29, 219a3, 10, 20, b33, 220a1, 12.220a28, 221a5, 10, 20, 25, 13.222b27; nature of, iv.10.218b33, 14.223b21-224a2, 16; and universe, iv.10.218a33-b9, 14.223a17, 18; sense of, iv.11.-218b21-219b2, 13.222b15, 14.223a16-29, b27; and the present, iv.11.218b28, 219-a19, b12, 23, 33, 220a1, 5, 19, 21, 12.220b8, 221a14, iv.13, 14.223a4-13, b1; vi.1.-231b7, 10, 232a19, vi.3, 9.239b8, 31; viii.1.251b19-26, 8.262a30, b2, 263b; contin-uous, iv.11.219a13, 220a5, 26, 12.220b2, 26, 13.222a10, 24, 14.223b1; v.3.226b28, 4.228a30, b2, 3, 8; vi.1.231b8, 10, 19, 232a19, 21, 2.232b24, 26, 233a10, 14; viii.-8.264b7; so much, iv.11.219a14, 13.222a27, 14.223b18; before, after, iv.11.219a14-b2, 9-12, 220a25, 12.220b5-14, 221a14, 14.222b33, 223a4-15, 28; viii.1.251b11, 8.263b9, 9.265a23; and place, iv.11.219a15, 12.221a17, 20, 28, 14.223a19; vi.9.240-a31, 33; defined, iv.11.219b1, 2, 24-26; as number, measure, iv.11.219b1-9, 220-

a3, 15, 25, iv.12, 14.223a16-224a15; viii.1.251b12; same, diverse, iv.12.220b5-14, 13.22a29-b7, 14.223b1-12, 224a2-15; being in, iv.12.220b32-222a9, 13.222b16, 19, 14.222b31, 223a4-15; and coming to a stop, rest, vi.8.238b26-239a10, 18, 26, 31, 34, b3; viii.1.250b29, 252a9, 7.261b24, 8.262b2, 21, 264b4; limit of, vi.8.239a35; viii.3.253b27, 8.262b21; and race-course, vi.9.240a1-17; eternal, viii.1.251b11-28

134a. καιρός, tempus, occasio, occasion, opportunity, i.9.192a36, iv.5.213a5

134b. ἀρχαῖος, antiquus, vetus, old, earlier, ancient: thinkers, i.2.185b26, 6.189b14, 8.190a23; ii.2.194a19; πάλαι, olim, dudum, long ago, former, iv.4.211b32, 33, 13.222b14, 29; v.2.226a23; viii.3.254a16, 8.262b28, 10.267b13; παλαιός, antiquus, ancient, i.6.189b1; ii.4.196a14; μέλλειν, debere velle, futurum esse, be about to, be future: save arguments, i.6.189b1; come into being, i.8.191b23; ii.7.198b7; and time, iv.10.217b34, 218a5, 9, 12.220b7, 222a1, 13.222b2, 8, 9, 14.223a6, 10, 12; vi.3.234a1, 2, 11-18; hope, viii.3.247a13; to be, viii.6.259b24; ἐσόμενος, futurus, future, iv.13.22a11; viii.1.251b21, 6.259b25

134c. ἅμα, simul, together, at the same time: quantitative respect, i.2.185b1; iv.9.216b27; not incompatibles, i.2.185b26, 3.187a4; vi.3.334b4, 6.237a22; viii.5.-257b10, 7.261b5, 9, 15, 26, 8.262a8, b1, 263b11, 25, 29, 264a17, 18, 28, 30, 31, b2, 5, 12, 23, 9.265a18; factor, effect, ii.3.195b18, 21; not potential, actual, iii.-1.201a20; viii.1.251b31; move, be moved, iii.1.201a22, 2.202a7, 9, 3.202a34; v.2.-225b27, 226a14; viii.1.242a58-60, 2.243a, 244b4, 245a9, 5.249b29; viii.8.262b18, 10.266b33, 267a5, 6, 267a11, 18, 26; all things, iii.4.203a27, 30; place, body, iii.4.-203b28, 5.205a34; iv.3.210b18, 4.211b23, 25, 212a29, 30, 5.212a35, 6.213b8, 10, 7.214a29; v.3.226b23; infinite, finite, iii.5.204a33, 205b18, 25; vi.7.238a7; thing, aspects, iv.2.210a7; parts, iv.3.210b3; viii.3.253b21; and time, iv.10.218a, b5, 11.-219a3, 6, 7, b10, 220a1, 12.220b6, 221a5, 13.222b5, 14.223a20, b2, 14.223b1-12; vi.2, 233a10, 6.237a9; viii.1.251b18 (and the heavens), 21; defined, v.3.226b18, 21, 227a23, 33; genus, species, v.4.227b12; subjects, v.4.228a2; movement, v.5.-229a21, 6.230a5, b28; vi.1.231b29, 31, 232a4, 13, 6.236b36, 9.240a9, 13, 14; vii.1.-242b50, 58, 2.243b10; viii.7.261a28, 8.262b11-27, 263a8, 264b10, 10.266b30; contraries, v.6.230b31; touch, vi.1.231a23, 27; now, vi.3.234a4, 14; alteration, vii.4.-249a31; active, passive, viii.4.255b34, 5.257b5; change, viii.5.256b17; movers, viii.-6.258b31

134d. νῦν, nunc, instans, momentum, now, present: reference, ii.5, 9; iii.7; iv.5, 8; v.4, 6; vi.2, 9; viii.1; viii.3, 6-8, 10; in change, iv.4.211b31-36; v.4.228a10; and time, iv.10.218a6-30, iv.11, 12.220b10, 221a14, iv.13, 14.223a4-13, b1; vi.1.231b7, 10, 232a19, vi.3, 5.236a17, 6.237a3-21, 8.239a15, 28, 33-b4, 7, 8, 32, 10.240b33, 241a3, 5; viii.1.251b19-26, 252a15, 8.263b20, 264a13

134e. αἰών, saeculum, aevum, age, viii.1.251a1

135. ἀεί, semper, ever, always: smaller, i.4.187b29; iii.6.206b20; principles, i.6.189a19, b25; subject, i.7.190a15, b3; iii.3.200b33; in becoming, change, i.7.190-b11; iii.4.203a14; iv.1.208b15, 9.216b26, 217a13; vi.6.237b11, 7.238a19, 9.240a29, b1, 5; nature, ii.1.192b34, 8.199b18; prior, ii.3.195b20; not together, ii.3.195b20; explain, ii.3.195b21, 4.196a6, 7.198a35; vs. chance, ii.4.196a21, 5.196b10, 13, 197-a4, 19, 31, 8.198b35, 199b24, 26; factors, ii.6.198a4; convey form, iii.2.202a9; limitation, iii.4.203b20, 21; elements, iii.5.204b14; movement, iii.5; iv.4, 8, 9, 11, 13; vi.7, 9, 10; vii.1, 5; viii.1, 2, 4-9; one after another, iii.6.206a22, 27, 28, 32, 33; take, iii.6.206b10, 17, 207a6, 1, 3, 8; iv.10.2.235a5, 9; viii.6.259a10, 7.260b22, 10.-266b1, 2, 13; greater, iii.7.207b3, 10, 12; place, displacement, iv.2.209b32, 210a8, 4.212a4, 23, 8.216a30, 32, 9.217a18; different, iv.10.218a11; perish, iv.10.218a16; remain, iv.10.218a22; time, present, before, after, iv.11-13; vi.3, 6, 8; viii.1, 8; divisible, iv.12.220a30; vi.1.231b15, 16, 2.232b25, 233a9, 4.235a22, 28, 5.236a33,

228 ARISTOTLE'S PHYSICS

b14; vi.6.237a28, b17, 21, 7.238b7; vii.1.242a47; viii.5.257a34; beings, iv.12.221b3, 4, 222a5-7; accidental, fortuitous, v.1.224b28, 2.225b30; contrary, v.4.229a2; and rest, v.6.230b22, 25, 231a8; vi.9.239b6; viii.6.259a25, 29, 260a12; between, vi.1.-231b9; span, vi.2.233b4; universe, viii.1.250b22; be, occur, viii.1.252a33, 35; movement, rest, viii.3; actualization, viii.4.255b34; and prime mover, viii.6.258b28, 29, 260a15, 10.267b3, 16; stop, viii.8.262a32; traverse, viii.8.263a5; circle, viii.9.265-a34, b6, 7; subtract, viii.10.266a20; power, viii.10.266a26, 267a8

135a. ἀΐδιος, *perpetuus, sempiternus, aeternus,* eternal: element, ii.1.193a26; possibility, being, iii.4.230b30; viii.1.251a17; movement, change, viii.1.251b13, 252-a4, 2.252b9, 253a1, 6.259a6, 7.261a30, 8.263a3, 9.265a25, 266a7, 8, 10.267b24; and principles, viii.1.252b3, 4; mover, viii.6.258b11, 12, 31, 259a7, 8, 13, 14, b33, 260-a15; moved, viii.6.260a1, 7.260b29

136. ἐχόμενον, *consequens, habitum, cohaerens,* next, contiguous, successive, consecutive, dependent: question, i.6.189a11, iv.10.217b29; preference, i.6.189b8; and place, iv.5.212b11; part, iv.9.217a16; vi.5.236b12, 6.237a32, b7; not now, iv.-10.218a18; vi.5.236a17, 25; not in infinite, v.2.226a5; defined, v.3.226b20, 227a6, 10, b1; vi.2.232b24; and movement, change, v.4.227b28, 228a26-b1, 229a4; vi.4.-235a33, 5.235b24; viii.6.258b29, 7.260a22, 24, 26, b21-23, 28, 261a30, 32, b6, 14, 8.263a3, 264b1, 2, 10, 267a7, 15, 24, b15; and indivisible, vi.2.232a24, 233b17, 31; magnitude, distance, vi.2.232a25, 233a11, 13, 6.237a29; time, vi.2.232b24, 26, 233a.0, 13; viii.8.263b30, 264a3; mover, vii.1.242a56, 57; viii.5.257b17, 10.267a9

136a. συνεχές, *continuum,* continuous: unity, i.2.185b9-15; vii.2.244a10; viii.4.255a12, 13, 16, 8.261b28, 264b9, 9.265a28; infinitely divisible, i.2.185b10; iii.-1.200b16-20, 7.207b16, 208a2; vi.1.231a21-b18, 3.234a7, 6.237b11, 15, 8.239a22; viii.5.257b1, 8.263a23-b2; passage, end, ii.2.194a29; movement, change, iii.1.200-b17; iv.11.219a13; v.3.226b20, 4.228a19-b15, 229a1; vi.4.235a6, 5.235b24, 236b17; vii.1.242b60, 62, 2.245a5, 15, 16; viii.2.253a1, 5.257b1, 6.259a5, 16-19, b22, viii.-8.9.265a28, 10.267a13, 21, 24, b16; infinite, iii.4.203a22; and place, iv.4.211a30, 34, b31, 5.212b5, 9.217a3; vi.1.231b4; and void, iv.6.213a34, b2, 7.214a31; and time, iv.10.218a23, 11.219a13, 221a5, 26, 12.220b2, 26, 13.2222a24, 14.223b1; viii.-8.264b6; magnitude, iv.11.219a11, 12; defined, v.3.226b20, 227a10-17, 21, 22, b1; vi.1.231a22, 3.234a8; not, vi.6.237b1; mover, moved, viii.5.258a21, 22, 28; the All, viii.6.259b27; συνεχῶς, *continue, continenter,* continuously: remain, ii.1.193a17; move, change, ii.8.199b16; v.3.226b25, 27; vi.1.232a13, 6.237a12; viii.3.254b3, 6.-259b14, 23, 7.260b20, 8.262a28, b10, 263b6, 264a10, 15, b17, 22, 25, 265a9, 9.-265b8, 10.266b29, 267a20, b10, 17; not increase, decrease, viii.3.253b14; not alteration, viii.3.253b29; going on, viii.6.258b27; συνεχεία, *continuatio,* continuity; and unity, i.3.186a28; of time, iv.13.222a10

136b. ἐφεξῆς, *consequenter, deinceps,* following, successive, neighboring, discrete: consequence, i.8.191a32; events, actions, ii.7.198a35, 8.199a9; terms, iii.1.-200b16; part, iii.1.207a7; elements, iv.5.212b30, 6.213b25; and movement, viii.10.-267a14, b11; now, time, iv.10.218a20; vi.3.234a6; viii.8.264a4; defined, v.3.226b19, 34-227a6, 17-21, 30, b1; vi.1.231a21, 23, b6, 8 ;movement, v.4.228a30; viii.6.259a17, 20, 7.260b21, 22, 261a9, 10, 8.264b7

137. ἀφή, *contactus, tactus,* contact, touch: and continuity, iii.4.203a22; viii.4.255a13; vs. one, iv.5.213a9; body, iv.7.214a7, 8, 8.216b19; unity, v.3.227a17, 20, 26; and alteration, vii.3.247b8

137a. ἅπτεσθαι, *tangere, attingere,* grasp, touch on: persistent nature, i.9.191b35; form, ii.2.194a21; factor, ii.8.198b15; philosophy, iii.4.203a2; body of the All, iii.5.205a21; and limit, iii.8.208a11-14; v.3.227a12, 4.228a25; vi.1.231b2-4, 17, 18; and place, iv.1.211a31, 34, b19, 5.212b6, 19, 31; and motion, iv.8.215a15;

vii.1.242b60, 62, 2.244b1; viii.5.256b19, 258a20, 22, 10.226b30, 227a14; defined, v.-3.226b19, 23, 227a17-33; vi.1.231a21, 22; contiguous, v.3.227a6; and infinite, vi.-2.233a23, 26, 30; ἁπτόν, *tangible, tactile,* tangible, iv.7.214a1, 6, 11, 8.216b20. *See also* 165

137b. προσάπτειν, *adjicere, adjungere,* put in: "is," i.2.185b30

137c. συνάπτειν, *continuare, conjungere, copulare,* connect, be in contact with, link, join, iii.6.207a17; iv.4.212a6, 11.218b25; viii.3.254a16, 8.264b27, 28; σύναψις, *contactus,* mutual contact, v.3.227a15

137d. θίξις, *tactus,* contact, iii.2.202a7,8

138. μέσος, *medius,* intermediate, center, intervening, middle: element, iii.5.205a27; of the universe, iii.5.205b11, 13, 16, 27, 31, 6.207a17; iv.4.212a21, 26, 27, 8.214b15, 215a9; viii.10.267b7; time, iv.11.219a17; viii.8.262b2; and extremes, v.1.224b33, 3.227a10, 5.229b19; of racecourse, vi.9.239b35, 240a6; and mover, moved, vii.2.245b2; vs. continuous, viii.3.253b14; and beginning, end, viii.8.262-a20, 23, 25, b5, 31, 264b32, 9.265a27-b16; μεσότης, *medium, medietas,* mean: present, viii.1.251b20

138a. μεταξύ, *medium, interjectum,* intermediate, means, between: vs. chance, i.5.188b1, 3, 6, 8; and contraries, extremes, i.5.188b23, 24; v.1.224b29, 30, 32, 225b3, 5.229a21, b14-16; vi.1.232a4; element, nature, i.6.189b3, 21; iii.4.203a18; to end, ii.3.194b35; interval, iv.4.211b8, 16, 212a11, 14, 5.212b26; time, iv.10.218-a20, 11.218b26, 28, 219a26; v.4.228b7; vi.3.234a8-10, 6.237a5, 10, 24; viii.1.250b29, 252a9, 7.261b7, 13, 8.263b31; defined, v.3.226b19, 23, 26, 227a10, 30-33; and succession, v.3.227a1; vi.1.231a23, b8, 9, 12; pause, v.4.228b3, 5; distance, vi.6.237-a33; and mover, moved, vii.2; viii.5.257b20; and movement, viii.8.264b31, 265a1

139. μίγμα, *commistio, quod mixtum est,* mixture, μίξις, *commistio, admixtum,* mixture, μειγνύναι, *miscere, admiscere,* mix: one, many, i.4.187a23; everything, i.4.187b1, 188a7; iii.4.203a23; infinite, i.4.187b4; materials, ii.1.192b20; "less," v.4.229a2; ἀμιγής, *immixtus, non mixtus,* unmixed, viii.5.256b25, 27

140. μαθηματικός, *mathematicus,* mathematician, mathematical: and natural scientist, ii.2.193b23, 25, 30; viii.3.253b3; abstractions, ii.2.194a1, 11, 12; and definition, ii.7.198a17; and necessity, ii.9.200a15; and infinite, iii.4.203b18, 25, 5.204-a35, 7.207b28; and place, iv.1.208b23; and time, iv.13.222a15

141. ἀριθμός, *numerous,* number: of principles, i.2,184b20, 7.190b36, 9.192b3; subject, i.7.190a15, b24, 25, 9.192a2; iv.9.217a25; and movement, ii.2.194a4; iv.11.-219b3, 4, 220a2; v.4.227b22, 23, 228a5, 12-15; vii.1.242b65-b41, 4.249b14; viii.2.-252b30; of factors, ii.3.194b17, 195a28, 7.198a14, 15; formal factor, ii.3.194b28, 195a31; and infinite, iii.4.203a7, 13, b24, 5.204a18, 20, b6-10, 17, 6.206a11, b30, 32, 7.207b1-21; viii.8.263a10; and place, iv.2.209b3: and void, iv.6.213b26, 8.215b13; and time, iv.11.219b2, 5, 6, 220a3, 4, 22, 23, 25, iv.12, 14.223a16-b12, 224-a215; vi.9.240a6, 8; viii.1.251b12, 8.263b13; successive, v.3.227a20; and primary being, vii.4.249b23, 24; of parts, vii.5.250a19; vs. definition, viii.8.262a22; 262a22; ἀριθμεῖν, *numerare,* count, iii.5.204b9; iv.11.220a22, 14.223a23-25, b13; viii.8.263a8, 10, 17, 25, 30, b1; ἀριθμητόν, *numerabile,* numerable, iii.5.-204b8, 9; iv.11.219b6, 25, 28, 12.220b19, 14.223a23, 25, 29; ἀριθμούμενον, *quod numeratur,* numbered, iv.11.219b6, 8, 9, 12.220b9; ὧ ἀριθμοῦμεν, *id quo numeramus,* abstract number by which we count, iv.11.219b7-9, 12.220b4, 8

141a. δύο, *duo,* duality, two, being, i.2.185b4; arts, ii.2.194a36; and time, iv.11.219a27, 12.220a29, 32; and point, viii.8.262a21, b6, 25, 263a24, b1, 3; movements, viii.9.265a21; δυάς, *dualitas, binarius,* dyad, i.9.192a10; iv.12.220a27; διχῶς, *dupliciter, duobus modis, bifariam,* twofold: meaning, i.8.191b2; ii.1.193b19, 2.194a12, 35, 3.195b13; iii.1.201a3; iv.3.210a26, 11.219b6; v.2.225b16;

vi.2.233a24, 5.236a7; divided, vi.4.234b21; and movement, viii.1.250b23, 5.257-b17; acted upon, viii.5.256a4; διπλάσιος, *duplex, duplus,* double, twice, ii.-3.195a31; iv.8.215b8, 9; vi.2.233b21, 9.240a1; vii.4.248b14, vii.5; viii.1.251b8, 10.-266b11

142. γεωμέτρης, γεωμετρικός, *geometra, geometricus,* geometer, γεωμετρία, *geometria,* geometry: principles, i.2.185a1, 16, 17; method, ii.2.194a9, 10; teach, viii.5.257a2

142a. μέγεθος, *magnitudo,* magnitude, size, greatness, distance: and being, i.2.185b5, 3.186b12; indivisible, divisible, i.3.187a3; iii.5.204a11, 6.206a11, 27, b1; vi.1.231b19, 21, vi.2, 5.236b10, 6.237a33, b21, 9.239b9, 20, 23; infinite, finite, i.4.187b8; iii.4.202b30, 203b17, 25, 204a1, 5.204a9, 18, 20, b1, 6.206a16, b8, 10, 19, 20, 207a21, iii.7, 8.208a18, 21; vi.7, 8.239a21; viii.10.266a25; possible, i.4.187-b14, 17, 21, 30, 37; no least, i.4.188a12; good, evil, ii.5.197a27; differ in, iii.4.-203b1; place, iv.1.209a16, 18, 2.209b4, 6, 7, 4.211b9, 212a11; and void, iv.8.215-b23, 216a16, 9.217a30, 31, b6, 8; and movement, time, iv.11.219a11, 12, 16, b16, 12.220a29, 32, b25, 27, 28, 221a3; v.2.226a31, 4.228b23, 24; vi.4.234b29, 34, 6.-236b35, 237a2, 29, vi.7, 10.240b10; viii.7.260a27, 261a35, 26, 8.262a5; equal, vi.-9.240a3, 7, 8; and change, vi.9.240a3, 7, 8; and change, vi.10.240b21, 241b1; viii.7.260b15; comparable, vii.4.248b24; path, vii.4.249a19; viii.9.265b4; and movers, viii.6.258b25, viii.10; intermediate, viii.8.264b32

142b. διχοτομία, *decisio, bipartita, divisio, bipartitio, sectio in duas partes,* bisection: argument from, i.3.187a3; vi.9.239b22; and infinite, iii.7.207b11; διχοτομεῖν, *decidere in duo, in duas partes secare,* bisect, vi.9.239b19

143. ἀστρολογία, *astrologia,* astronomy: and natural science, ii.2.193b26, 194a8

143a. οὐρανός, *caelum,* heavens, universe, world: and automatism, ii.4.196a25, 33, b3, 6.198a11; and infinite, iii.4.203a7, b25, 7.207b21; and place, iv.2.209a33, 34, 4.211a14, 24, 25, b29, 212a21, 5.212b7-22, 6.213b23; bulge, iv.9.217a13; and time, generation, iv.10.218b4, 12.221a22, 23, 14.223a18; viii.1.251b18, 19; end movement, viii.6.259b30

144. ἀρμονική, *harmonica, musica,* harmonica, ii.2.194a8

145. ὀπτική, *perspectiva,* optics, ii.2.194a8, 11

146. μετρεῖν, *mensurare, metiri,* measure: part, iv.10.218a6; movement, time, iv.12.220b14-222a9, 14.223b12-224a2; vi.7.238a17; by primary, vii.9.265b10; ἀναμετρεῖν, *metiri,* measure: whole, iv.12.220b3; vi.7.238a22, b11; καταμετρεῖν, *metiri, mensurare,* measure: whole, iv.12.221a2; vi.223.B3, 5, 9, 7.237b28, 238a7, 12, 14, 10.241a13; power, viii.10.266b23

146a. μέτρον, *mensura,* measure: time, iv.12.220b32, 221b7, 8, 23, 25, 14.223b-18, 19, 30; line, v. 3.226b33; rotation, viii.9. 265b8, 10

146b. διάμετρος, *diametrus, dimetiens,* diagonal, diameter, iv.12.221b24, 222a5; viii.8.264b15

146c. σύμμετρον, *commensorabile, habere communem mensuram,* commensurate, ii.7.198a18; iv.12.221b25, 222a6; ἀσύμμετρον *incommensurabile,* incommensurable, iv.12.222a5; συμμετρία, *symmetria,* harmony, vii.3.246b5

147. τὸ μέγα καὶ τὸ μικρόν, *magnum et parvum,* great-and-small: as material, i.4.187a17; and reason, i.5.189a8; as nonbeing, i.9.192a7; as dyad, i.9.192-a10; and infinite, iii.4.203a16, 6.207a30; and place, iv.2.209b35, 210a1, 4.212a7; and void, iv.6.213b10, 11, 9.217a26

147a. μᾶλλον, *magis, potius,* rather, more: being, i.3; element, i.6; and less, i.6.189b10; iv.9.217a5; v.2.226b2, 7, 4.228b18; vii.2.244b5, 4.249b7, 26, 5.250b1; source, i.7; accuracy, i.8; nature, ii.1; automatism, ii.6; factor, ii.9; movement, change, iii.2, 5; iv.8; v.1, 2. 4-6; viii.2, 7-9; void, iii.4; concrete, iii.5; aspect, iii.6 receptacle, iv.4; time, iv.12, 13; viii.1; rest, v.6; colored, vii.4; absurd, viii.1, 5; mover, viii.5, 6; limit, viii.9

147b. μάλιστα, *maxime, potissimum, quam maxime,* chiefly, especially, most: becoming, generation, i.8.191b6; v.1.225a27; differ, ii.6.197b32; vi.10.240b15; factor, ii.6.198a9, 7.198a33; ends, ii.8.199a20, 30; high, ii.9.200a4; infinite, iii.4.203-b16, 22, 204a1; vi.4.235b2; movement, change, iii.5.205b6; iv.1.208a31; vi.5.235-b30; viii.2.252b8, 17, 253a7, 4.254b24, 33, 7.261a23, 8.262a12; "in", iv.3.210b13; place, iv.4.211a14, 212a23; resist, iv.8.215a30; time, iv.10.218b9, 11.220a16, 13.-222b24, 14.223b19; alteration, vii.3.245b6, 247b2; viii.8.265a5; natural, excellent, vii.3.246a14, 15

148. ὑπερέχειν, *excedere, superare, excellere, exsuperare,* preponderate, exceed, ὑπεροχή, *excellentia, exsuperantia, excedentia,* excess: and deficiency, i.4.187a16, 6.189b10, 12; iii.1.200b29, 8.208a15, 16; in mixture, i.4.187b3; and void, iv.8.215-a28, b12-19, 31, 216a13; time, iv.12.221b30; and quality, vii.4.249b26; vs. equal, viii.7.261b20

148a. λείπειν, *relinquere, restare,* remain, fall short: movement, iii.2.201b35; v.2.226a24; vi.2.232b18; viii.4.254b25; in power, iii.5.204b15; infinite, potential, iii.6.206a18; material, iv.2.209b10; contrariety, v.5.229a27 stage, vi.10.240b28; ἀπολείπειν, *deficere, restare, relinquere,* leave behind, fail to reach, fall short: place, iv.4.211a2; vi.2.232a30; movement, change, vi.4.235a4, 5.235b8-22; διαλείπειν, *deficere, intermittere, interrumpere,* cease, be discontinuous, leave, iv.9.217a5, 11.219b8; v.3.226b28, 29, 4.227b31, 228a11, b4, 8; viii.6.258b10, 8.264-b18; ἐλλείπειν, *deficere, deminuere,* be less than, vi.2.233b3; viii.10.266-b3; ἔλλειψις, *defectus, defectio,* deficiency: and excess (see 58f. 148); ἐπιλείπειν, *deficere,* fail, iii.6.206b3, 8.208a8; παραλείπειν, *relinquere, prae-termittere,* pass by: chance, ii.4.196a20; ὑπολείπειν, *deficere,* cease, stop, run out, not, iii.4.203b19, 24; iv.13.222a29, b6; vi.5.236b14

149. κόσμος, *mundus, cosmos,* world: shape of, ii.2.193b30; and automatism, ii.4.196a25; and infinite, iii.4.203b26, 6.206b23; viii.1.250b18, 20; and void, iv.8.-216b18; and movement, viii.2.252b26, 27

149a. διακοσμεῖν, *componere, disponere,* reduce: viii.9.265b32

149b. κοσμοποιία, *mundi creatio, cosmopoeia,* cosmogony, ii.4.196a22; κοσμοποεῖν, *mundum facera,* construct a cosmology, viii.1.250b16

149c. κατασκευάζειν, *parare, constituere, probare, astruere, efficere,* constitute, establish: nature of beings, i.6.189a26; void, iv.8.216a22; generation, destruction, viii.9.265b31

149d. κατακλυσμός, *diluvium, inundatio,* cataclysm, iv.13.222a23, 26

150. πᾶν, *omne, universum, quodlibet,* all, any, universe: indiscriminately, i.1.184b13; science common to, i.2.185a2; and movement, change, rest, i.2.185-a13, 3.186a18, 4.187a32, b2, 8.191b12; ii.3.194b21; iii.4.203a29, 5.205a6; iv.5. 212-a34, 6.213b13; vii.4.248a10; viii.1.252a21, 2.252b26, viii.3, 6.259a24, 260a12, 8.265-a8; and alleged unitary being, i.2.185a20-186a3, 3.187a1, 7, 5.188a20, 6.189b2; and starting-point, i.3.186a13; together, i.4.187a30, b1, 22, 31; iii.4.203a25, 26, 32; viii.1.250b25; and principles, factors, i.5.188a28, 6.189a16; ii.6.198a13; be not, i.8.191b26; ii.7.198b3; ending, ii.2.194a32; and order, ii.4.196a28; viii.1.252a12; convertible by limitation, ii.6.197a37; v.3.227a19, 25; movers, iii.201a24, 25, 2.-202a3; viii.6.258b30-32; infinite, finite, iii.4.202b32, 5.205a2, 21, 23, 6.207a18-20, 8.208a11; vi.7.237b26; element, iii.5.204b31; and part, iii.5.205a11, b22, 6.206b19, 207a12; place, iii.5.205a33; iv.1.208b30, 33, 4.211a25, 5.212b14-18; beyond, iv.6.-213b1; void, iv.8.214b19; and time, iv.10.218b7, 12.221a28, 32, 14.223a17; viii.8.-

263b18-264b6; one, vii.4.248b25; vs. some, viii.1.252b1, viii.3; and first principle, viii.6.259b27; *et passim;* πανταχῇ, πάντη, *undique,* in every direction, in every part, iii.5.204b20, 21; iv.7.214b8, 8.215a23, 24; viii.5.257b2; πάντως, *penitus, omnino,* comprehensively, in a general way, under any and all conditions: explain, ii.7.198b5; void, iv.9.217b22; motion, vii.1.242b64; act, viii.1.251b2; παντελῶς, *penitus, omnino,* (not) at all, altogether, completely, in general: and time, viii.7.261b22; principles, i.7.190b36; differ, i.9.192a8; immovable, ii.7.- 198b2; πανταχοῦ, *ubique,* everywhere: mass, iii.4.203b28; and infinite, iii.5.- 205a18, 19; and time, iv.10.218b13, 12.220b5, 14.223b11, 12; πάμπαν, *penitus, totum,* altogether: identical, iii.6.207a13; rest, viii.2.252b23; *et passim*

V. HUMAN NATURE AND SOUL

151. ἄνθρωπος, *homo,* man: and being, primary being, i.2.185a7, 24; iii.6.- 206a30; as subject, i.2.185b29; v.4.228a1; vi.5.236b4, 5; viii.8.262a3; animal, i.3.- 186a21; ii.3.195b1; iv.3.210a18; definition of, i.3.186b14-35; ii.2.194a6; natural, ii.- 1.193b6, 8-12; generates, ii.2.194b13, 7.198a27; as species, ii.3.195a36, b23; iii.2.- 201a11, 12; vii.4.249b20; and chance, ii.5.197a10; and necessity, ii.9.200b3; and infinite, iii.6.206a26, 30, b2, 7.207b6, 7; whole, iii.6.207a10; iv.3.210b2, 4; and the void, iv.6.213a23, 27; and time, iv.12.220b12; and time, iv.12.220b12; and change, movement, v.1.225a23, 24, 2.225b17, 22, 25, 4.227b25; viii.5.256a7, 13, 23; and love, strife, viii.1.252a29

151a. ἀνθρώπινος, *humanus,* human: intelligence, ii.4.196b6; affairs, iv.14.- 223b25; ζῷον, *animal, animans,* animal: man *(see* 151); being, i.3.186b16; ii.- 1.192b9; parts of, i-4-187b13-21; ii.4.196a24; viii.4.254b23; generation of, i.7.190- b4, 8.191b19-25; ii.4.196a29, 35, 8.199b8, 13; vii.4.249b21; automatism, ii.6.197b14; behavior, ii.8.199a20; and movement, viii.2.252b22, 25, 253a12, 13, 18, 20; viii.4, 6.259b3, 8, 10, 7.261a16, 9.265b34;

152. ζωή, *vita,* life, viii.1.250b14; ζωτικόν, *vitale,* animal, viii.4.255a6; ζῆν, *vivere,* live, viii.7.261a16

153. ἀθάνατος, *immortalis,* deathless, undying, iii.4.203b12; viii.1.250b13, 6.259b25

154. ψυχή, *anima,* soul, living being, self: and primary being, i.2.185a24; and knowledge, iv.3.210b1; and place, iv.5.212b12; and time, iv.11.218b31, 219a6, 27, 14.223a17-29; and change, alteration, v.2.226a12; vii.3.246a10, 247a1, b17, 248a8; and movement, viii.9.265b32

154a. ἔμψυχος, *animatus,* animate, vii.2.244b9, 13, 14, 245a1; vii.2.252b17, 253a9, 4.255a7, 18, 6.259b2, 9.265b34

154b. ἄψυχος, *inanimatus,* inanimate, ii.6.197b7, 14; vii.2.244b8, 13, 14, 245a10; viii.2.252b14, 21, 4.255a17

155. ὁρμή, *impetus,* tendency, ii.1.192b18; ὁρμᾶν, *movere, discedere, procedere,* start, viii.8.262a17, b14

156. ὀρέγεσθαι, *desiderare, appetere,* have an impulse, i.9.192a18, 19; ὄρεξις, *appetitus, appetitio,* desire, viii.2.253a17

157. ἔθων, *solere,* be accustomed to, iv.13.222b25

157a. εἰωθός, *consuetum, vulgo haberi solere,* accustomed, iv.14.223b24, vi.2.233a13

157b. ἠθικός, *moralis,* moral, vii.3.247a7

158. τιμᾶν, *honorare,* value, ii.6.197b11

159. φιλότης, *amicitia,* love, friendship, viii.1.252a26

159a. φιλία, *concordia, amicitia,* love, friendship, νεῖκος, *discordia, contentio, lis, dissidium,* strife: contraries, i.5.188b34, 6.189a24; factors, ii.4.196a18, 8.198-b16; iii.4.203b13; and one, many, viii.1.2.50b28; alternate, viii.1.252a29; and movement, viii.9.265b21

160. διώκειν, *persequi,* follow, pursue, ii.5.197a17; vi.9.239b16, 25

161. φεύγειν, *fugere,* avoid, flee, ii.5.197a17; vi.9.239b17; viii.1.252a29

162. ἡδύς, *dulcis, jucundus,* pleasant, v.6.230b8; ἡδονή, *voluptas,* pleasure, vii.3.247a8, 14, 16; ἥδεσθαι, *laetari,* experience pleasure, vii.3.247a12

162a. λυπηρός, *tristis, molestus,* painful, v.6.230b8; λύπη, *tristitia, dolor,* pain, vii.3.247a8, 16

162b. ταραχή, *turbatio, perturbatio, tumultus,* disturbance, excitement, vii.3.-247b18, 248a1; ταράττεσθαι, *turbari, conturbari,* be disturbed, viii.7.261b16

162c. χαίρειν, *gaudere, valere,* bid farewell, ii.8.198b15

162d. εὐδαιμονία, *felicitas, beatitudo,* happiness, ii.6.197b4, 5

163. χρῆσθαι, *uti,* use, take recourse to, work with, apply: in our interest, ii.2.194a34; art, ii.2.194b2, 3; chance, ii.4.196a20; infinite, iii.4.203b17, 6.206b30, 7.207b25, 30, 208a3; as receptacle, iv.4.212a17; point, iv.11.220a12, 18; viii.8.262-b25, 263a24; intermediate, v.1.224b31, 5.229b16; viii.8.262b5; demonstration, vi.-2.233a7; knowledge, vii.3.247b16; viii.1.251a33; movement, viii.3.253b2

163a. χρῆσις, *usus,* use, vii.3.247b7, 248a5

163b. χρήσιμος, *utilis, conducere,* useful, important, significant, ii.8.198b26; viii.7.261b10, 15

163c. χρῆμα, *res,* thing, iii.4.203a25

163d. προσχρῆσθαι, *congruere, adhiben, indigere, adjungere,* take recourse to: in definition, i.2.185b2; iii.1.200b19

164. κρίνειν, *judicare, dijudicare, discernere,* discriminate: what we ascertain directly, indirectly, ii.1.193a5; more, less, iv.11.219b4; by time, iv.14.223b27; rest, vi.8.239a16; by sense, viii.3.248a1; same, different, vii.4.249a27; fail to, viii.3.254a32; κρίσις, *discretio, judicari,* discrimination, iv.8.216b19

164a. ἀποκρίνειν, *segregare, secernere, disgregare, separare,* separate, i.4.-187b29; ἀποκρίνεσθαι, *congregari, secerni,* be drawn, ii.4.196a21; ἀποκρινόμενον, *dscretum, separatum,* independent, iv.8.216a24, 9.217b20

164b. ἐκκρίνειν, *segregare, secernere,* separate: contraries, i.4.187a20; out of mixtures, i.4.187a23, b23, 24, 29, 31, 188a1; emit, vii.2.243b14

164c. σύγκρισις, *congregatio, confusio, concretio, aggregatio,* combination: and separation, i.4.187a31; vii.2.243b8, 11; viii.9.265b19, 31; συγκρίνεσθαι, *congregari, conerescere,* combine: love, viii.9.265b22; atoms, viii.9.265b29

164d. διάκρισις, *segregatio, secretio, disgregatio,* separation: and combination (see 146c); διακρίνειν, *segregare, secernere, discerne, distinguere, disgregare, disjicere,* separate, distinguish: not attributes, i.4.188a5, 7, 10; and order, ii.4.196-a27; and mind, viii.1.250b26, 9.265b22; strife, viii.1.252a27, 9.265b21; distribute, viii.6.259b13; and movement, viii.7.260b11-13, 9.265b29 (atoms)

164e. ἀκρίβεια, *certitudo,* accuracy: potential, actual, i.8.191b29; first philosophy, i.9.192a35

165. αἰσθάνεσθαι, *sentire,* sense: change, time, iv.11.218b32, 219a4, 30 (See also 137a)

165a. αἴσθησις, *sensus,* sense perception: acquaintance, i.1.184a25; and reasoning, i.5.188b32, 189a5, 6, 9; viii.3.253a33, 8.262a18; not evident to. iv.8.216-a29; viii.3.253b11, 254a26; and time, iv.11.219a25; and alteration, vii.2.244b10-245a2, 3.247a10, 19

165b. αἰσθητικός, *sensitivus, sensibilis,* sensitive, perceptive, vii.2.244b10, 3.247a6, 17, 248a8; viii.2.253a19

165c. αἰσθητόν, *sensible, sensile,* sensible: differences, i.6.189b7; vii.2.244b5; infinite, iii.4.203a6, 10, 204a2, 5.204a8, b2, 7, 32, 205a8, 9, 6.206b25, 207a29, 7.-207b19, 208a2, 8.208a9; and place, iii.5.205a10, 26, b31; iv.1.208b28, 209a17; bulk, iv.9.217a8; and the void, iv.6.213a29, 7.214a7, 8; and alteration, vii.2.245a3, 3.245b3, 5, 247a7, 10, 14, 248a7; and movement, viii.8.265a4

165d. ἀναίσθητος, *insensibilis, sensu carens,* imperceptible, insensitive, i.4.-187b1; iv.13.222b15; vii.2.245a10

165e. ἀναισθησία, *insensibilitas, sensu non percipere,* insensibility, iv.11.218b26

166. βλέπειν, *respicere,* look vii.4.249b15; ἀνάβλεψις *respectio, aspectus,* look, vii.3.247b8; ἀποβλέπειν, *aspicere, respicere,* look, ii.2.194a19; ἐπιβλέπειν, *intendere, animum advertere, considerare, respicere,* re-examine, i.7.190a14; viii.6.259a21; ἰδεῖν, *videre, cognoscere, scire, perspicere,* see, ii.5.197a17, 8.199b27; iii.2.202a2; iv.2.209b22, 10.218a11; vi.10.240b15; viii.1.251a6; observe, i.4.187b2, 6.189a29, 8.191b33; ii.4.196b2, 5.196b10; iii.1.201b4, 4.203a24, 204a4, 5.204a13, 16, 17, 6.206b5; iv.3.210b8, 8.215a25, 216a13; v.6.230a28; vii.1.242b61; viii.2.252b12, 21, 253a11, ὁρᾶν, *videre, cernere, conspicere,* see, 3.254a6, 35, 4.255-a17, 5.256b21, 6.259a1, 5; παρορᾶν, *despicere, neglectere,* overlook, i.3.186a32; συνορᾶν, *videre,* command a view of both, i.3.186a32; ὄψις, *visus, aspectus,* sight, iv.4.211b2; vii.2.245a7

166a. ἀκρόασις, *auditus, auscultatio,* treatise (title); ἀκοή, *auditus,* hearing, vii.2.245a7

166b. ὄσφρησις, *adoratus,* smelling, vii.2.245a8

166c. γεῦσις, *gustus,* taste, vii.2.245a9; χυμός, *sapor,* flavor, vii.2.245a9

167. ἀνάμνησις, *recordatio,* remembering, v.2.225b32; μνεία, *memoria, mentio,* attention, ii.4.196a17; viii.9.265b18; μεμνῆσθαι, *memoria, recordari, reminiscere,* remember, vii.3.247a9, 12

167a. ἐλπίζειν, *spes, sperare,* hope, vii.3.247a11; ἐλπίς, *spes,* hope, vii.3.-247a11

168. ἐμπειρία, *experientia,* experience, vii.3.247b21

168a. ἀπειρία, *infirmitas, imperitia,* inexperience, i.8.191a26

VI. MIND

169. νοῦς, *intellectus, mens,* mind, rational being: Anaxagores on, i.4.188a9; iii.4.203a31; viii.1.250b26, 5.256b25, 9.265b22; factor, ii.4.196a18, 30, 6.198a6, 10, 12, 8.198b16; iii.4.203b13; animals, ii.8.199a22; and time, iv.14.223a26

169a. νοεῖν, *intelligere, cognoscere, concipere,* think, apprehend, conceive: fail to, i.1.193a9; mind, iii.4.203a32; number, iii.7.207b10; infinite, iii.8.208a16, 19; time, iv.11.219a27; viii.1.251b20; color, v.224b18, 20; point, viii.8.262b7; κατανοεῖν, *intelligere,* keep in mind: place, iv.4.211a12

169b. νοητός, *intelligibilis,* intelligible: and infinite, iii.5.204b1, 6, 6.207a29, 30; elements, iv.1.209a18

169c. νόησις, *intellectus, cogitatio,* conception, thought: abtract, ii.2.193a34; infinite, iii.4.203b24, 8.208a14, 16, 20; and time, iv.13.222a16

169d. νοητικός, *intellectivus,* supposed, mental: infinite, iii.8.208a22; state, vii.3.247b1

169e. ὁμονοητικῶς ἔχειν, *concorditer intelligi, in eadem sententia convenire,* agree, viii.1.251b14

170. διάνοια, *intellectus, animus, mens, intelligentia, cogitatio,* mind, intelligence, design, thought, design, sense: serious, i.9.192a16; obscure to, ii.4.196b6; to some end, i.5.196b22, 197a2, 7, 8; good, evil, ii.5.197a29; factors, ii.6.198a4; and change, time, iv.11.218b22; and movement, viii.2.253a17, 3.253a34

171. τέχνη, *ars,* art: and nature, ii.1.192b18, 193a16, 31, 32, 34, 35, 2.194a21, 8.199a12-20, b27, 30, 9.200b1; and use, ii.2.194a33, b7, 8.199a17, b1; technician, ii.3.195a31; and animals, ii.8.199a21; mistake in, ii.8.199a33

172. φρονεῖν, *prudens, esse,* be intelligent, vii.3.247b11; φρόνιμος, *prudens,* intelligent, vii.3.247b18

173. φαντάζεσθαι, *imaginari, videre,* seem, i.9.192a15

173a. φαντασία, *phantasia, speciem prae se ferre,* experience, iv.4.211b34; imagination, viii.3.254a28, 29

173b. φαίνεσθαι, *videri, apparere,* be apparent, appear, be observed, seem: one, i.3.186a4, 9-192a1; different, i.4.187b2; ii.1.192b12; subject matters, ii.2.193-b28; good, ii.3.195a26; chance, ii.4.196a7; ends, ii.8.199a24; movement, change, iii.2.201b35; iv.14.222b33; v.4.228a9; viii.1.251a31, 10.267a14, 19; elements, iii.5.-204b35; possible, iii.6.206a12; infinite, iii.6.206b6, 7.208a2; place, iv.1.208a33, 4.212-a10, 13; contrast, iv.6.213b16; void, iv.7.214a6; present, iv.10.218a9, 11.218b31; absurd, viii.1.251a21; time, viii.1.251b15; love, strife, viii.1.252a30; being, viii.3.254-a26; generation, viii.7.261a13

173c. παρεμφαίνεσθαι, *apparere, speciem referre, videri,* appear with, be observed with, iv.4.212a8, 14. 224a1

174. ὑπόληψις, *existimatio,* apprehension, v.4.227b14

175. δόξα, *opinio, sententia,* opinion: of natural philosophers, i.4.187a28, 35, 8.191a34; concerning principles, i.6.189b12; of the void, iv.6.213a21; and movement, viii.3.253b1, 254a27, 29, 6.259b3

175a. δόγμα, *dogma,* teaching, iv.2.209b15

175b. δοκεῖν, *videri,* appear, be held: nature of a thing, i.4.187b6; ii.1.193a9; different, i.5.188b37; vii.4.249a25; viii.3.254a28; prior, i.6.189a32; element, i.6.189-b3, 17; contraries, i.6.189b27; v.5.229b2; becoming, i.8.191b17; matter, ii.2.194a19; chance, ii.4.196b5, 5.197a10, 11, 6.197b4, 8.199a1; slight miss, ii.5.197a30; end, ii.8.198b28, 199a3; movement, iii.1.200b16, 201a25, 2.201b25, 28, 31; v.4.228a15, b16, 5.229a10, 23, 6.229b23, 230b25, 26, 30; vii.2.244a11; viii.2.252b7, 253a7, 11, 3.254a12, 30, 6.259b6, 8.264a9; philosophers, iii.4.203a1; infinite, iii.4.203b11, 24, 26, 5.204b5, 6.206b28, 7.207a34, 8.208a5; place, iv.1.208b5, 29, 2.209b2, 6, 18, 28, 32, 4.210b33, 211a9, 28, b11, 17, 30, 212a7, 12, 22, 28; void, iv.6.213a16, b5, 19, 7.213b31, 214a21, 8.214b16, 215a23, 216b18; dense, rare, iv.9.217b17; viii.7.-260b10; time, iv.10.218a3, 8, b6, 9, 11.218b23, 29, 219a6-8, 14, 30, 32, 13.222b4, 14.223a17, b21, 29; viii.7.261b23; mover, vii.1.242b54; alteration, vii.3.246a4; science, viii.1.252a30; generation, viii.7.261a3

175c. ἐοικέναι, *videri, visum esse, assimilari,* seem: infinite, i.4.187a26; principles, i.6.189b11; chance, ii.4.196a10; nature, ii.8.199b31; "all things together," iii.4.203a25; movement, v.4.228b18; viii.1.252a5, 7, 4.254b30

175d. οἴεσθαι, *arbitraria, putare, opinari, sentire,* claim, think, conceive, expect: comprehension, i.184a12; ii.3.194b18; infinite, finite, i.3.186a11, 4.187a27; iii.6.207a15; derivation, i.6.189a16; becoming, i.8.191b11; explanation, ii.3.194b34; to meet, ii.4.196a4; chance, ii.4.196a10, 17, 19; ends, ii.8.199b26; necessity, ii.9.-199b35; place, iv.4.211a14, 8.214b20; being, iv.6.213a29, 7.213b32, 214a1; void, iv.7.214a24, 216a23, 9.216b22; time, iv.11.218b29; vi.9.240a1; and alteration, vii.-3.247b8; movement, viii.9.265b28

175e. ὑποπτεύειν, *concipere, suspicari,* suspect, iv.10.217b33

175f. νομίζειν, *putare, censere, existimare, arbitrari, opinari,* hold, accept: becoming, i.4.187a36; contraries, i.6.189a35; necessity, ii.9.200a2; place, iv.1.208-b32; movement, rest, viii.1.252a6; as principle, viii.1.252a32; finite, viii.6.259a9

175g. νόμος, *lex, forma,* convention, ii.1.193a15

176. λογισμός, *ratiocinatio,* practical reasoning, ii.9.200a23, 24; λογιστικόν, *ratiocinativum, facutas rationandi,* training, iv.3.210a30

177. βούλεσθαι, *velle, placere,* want, mean, wish, tend, insist: impossible separation, i.4.188a9; completion, ii.2.194a32; to meet, ii.4.196a4, 5.197a17; infinite, iii.7.207b31; place, iv.4.212a18; void, iv.6.213a23, 28, 7.214a20, 9.217b22; be, not be, viii.6.258b17; to show, viii.6.259a28

177a. βουλεύειν, *deliberare, consultare,* decide, advise, deliberate, ii.3.194b30, 195a22, 8.199a21, b27, 28

177b. ἐθέλειν, *velle, debere,* tend, ii.3.195a25

178. προαίρεσις, *propositum, praeelectio,* intention, choice, plan, ii.5.196b18, 19, 197a6, 7, 6.197b8, 22; προαιρεῖν, *proponere, praeeligere,* choose, ii.5.177a2, 3, 6.197b21

179. ἐπιστήμη, *scientia,* science: order of procedure in, i.1. (analysis of situations into their elements, of the generic into the specific); of nature, i.1.184a15; iii.4.202b30, 36; universal, i.2.185a2; vii.3.247b20; matter, form, ii.2.194a22; and soul, iv.3.210b1; and movement, iv.4.211a22; v.1.224b13, 2.225b33, 5.229b5; viii.-3.253b1, 4.255b2, 5.257a14; species, genus, v.4.227b13, 14; get, use, vii.3.247b10, 16; of contraries, viii.1.251a30, 33

179a. ἐπίστασθαι, *scire, scientia, (refertur ad demonstrationem),* getting scientific knowledge, reflect: conditions of, i.1.184a10; iv.3.210a30; vii.3.247b11; of universal, vii.3.247b6, 20; and movement, viii.1.251a22, 4.255a33, b3

179b. ἐπιστητόν, *scibile, sub scientiam cadere,* knowable: being, i.6.189a13; persistent nature, i.7.191a8

179c. τὸ ἐπιστῆμον, *sciens,* knowing, vii.3.247b2, 4, 5, 15, 18, 4.255b23

180. σοφός, *sapiens,* sage, wise, ii.4.196a9; iv.13.222b17

181. γιγνώσκειν, *cognoscere,* comprehend, i.1.184a12

181a. γνωρίζειν, *cognoscere, cognitio, notitia, nosse, considerare,* master, know, find out: facts, principles, i.1.184a12, 13; matter, form, ii.2.194a18, 26, b1, 6; iv.-2.209b21; place, infinite, iv.1.208a28, 2.209b18; time, movement, iv.11.219a22, b17, 24; number, iv.12.220b21

181b. γνώριμος, *notus, cognotus,* familiar, conclusive, intelligible, determinate, known, familiar: in scientific procedure, i.1.184a16-18, 21, 22, 25; particular, universal, i.5.188b32, 189a4, 6; directly, indirectly, ii.1.193a6; movement, iii.3.202b26; iv.14.223b20; present, iv.11.219b29

181c. γνωριστικός, *cognoscitivus,* knowing: art, ii.2.194b4

182. εἰδέναι, *intelligere, cognoscere, scire, noscere, (refertur ad definitiones),* getting knowledge, mean: conditions of, i.1.184a10; of composites, i.4.187b10-13; astronomy, ii.2.193b27; matter, form, ii.2.194a22, b10; and explanation, ii.3.194b17, 18, 22, 7.198a22, b4; chance, ii.5.196b17, 34

182a. οὐκ εἰδότως, *non scite,* inadvertently, i.4.188a5

183. πιστεύειν, *credere, scire,* rely on, confidently conclude, be sure: thinking, iii.8.208a15; movement, rest, viii.3.254a3, 32, 6.259a20, b20, 8.264a7

183a. πίστις, *fides, probatio,* conviction, certitude: infinite, iii.4.203b15; void, iv.6.213a15; change, v.1.224b30; move, rest, viii.3.254a35; endure, viii.7.261b25; pause, viii.8.262a18

183b. ἀπιστία, *incredulitas, refutatio,* disbelief: void, iv.6.213a15

184. μανθάνειν, *addiscere, intelligere, discere,* understand, learn: being itself, i.3.187a8; be taught, iii.3.202a33, b3, 5, 11, 17, iv.12.221b1; v.2.226a22, 5.229b4; vii.3.247b19; viii.4.255a33-b3, 5.257a13, b5

184a. μάθημα, μάθησις, *doctrina, disciplina, disputatio, doctrinatio,* discipline, learning, definition in, ii.2.194a8; and movement, iii.1.201a18, 3.202a32, b2, 16, 20; v.2.226a15, 4.227b13, viii.3.253b3

184b. μαθητικός, *disciplinativus, discere posse,* being taught, viii.5.257a21

184c. ἀμαθέστατος, *indisciplinabilis, insipientissimus,* most stupid, iv.13.222b18

185. διδάσκειν, *docere,* teach, iii.3.202b3, 4, 10, 17; viii.3.202b3, 4, 10, 17; viii.5.257a12, 9, 10, 13, b5

185a. διδασκαλικός, *docens, quod docendi vim habet,* teacher, iii.3.202b7

185b. δίδαξις, *dactio,* teaching, iii.3.202a32, b2, 7, 16, 19

186. θέα, *speculatio, inspectio,* attention, iv.2.209b20

187. θεωρεῖν, *speculari, contemplari, perspicere, dispicere, considerare,* view, see, develop a theory, show, consider: concerning "becoming," i.7.189b32; principles, i.7.191a22; natural science, ii.2.193b23, 194a14; infinite, iii.4.202b35, 5.204-b10; place, iv.1.208a33, 4.211a6; void, iv.6.213a12; alteration, vii.3.245b5; move, rest, viii.3.254b4; actually, viii.4.255b2, 4, 22

187a. θεωρία, *consideratio, inspectio, speclatio,* theory: special, common, iii.-1.200b24; of the infinite, iii.4.203a1, b3, 31, 7.207b28; of generation, destruction, viii.1.250b17; of nature, viii.1.251a6

188. πράττειν, *agere, facere,* actio, act, conduct: by design, ii.5.196b22; affairs, ii.6.197b6, 12, 13; to some end, ii.8.199a9-11, b21; and time, iv.13.222b23; and pleasure, pain, vii.3.247a9

188a. πρακτός, *practicus,* practical, ii.6.197b3

188b. πρᾶξις, *actus, actio,* praxis, conduct of life, overt action, ii.6.197b2, 9.200a23, 24; vii.3.247a10

188c. πρᾶγμα, *res,* thing, state of affairs, event, path, actuality: starting-point of, i.3.186a14; nature of, i.4.187b7; and movement, 6.260a7; iii.1.200b33; v.3.226-b26, 30, 4.227b28; viii.1.251a11, 5.256b5; and infinite, iii.8.208a15; viii.8.263a17; and place, iv.1.209a22, 2.209b23, 30, 31, 210a7, 9, 4.211a1, 212a1, 4, 29; and form, iv.4.211b13; and material, iv.7.214a15; in time, iv.12.221a16, 22, 14.223b31; viii.8.263b11, 15, 16; and number, iv.12.221b15; human, iv.14.223b25; and love, strife, viii.1.252a9; and generation, viii.7.261a4

188d. εὐπραξία, *eupraxia, recta actio,* doing well, ii.6.197b5

189. ποιητής, *poeta,* poet, ii.2.194a30

189a. ποίησις, *factio, effectio,* doing, iii.3.202a23, 26, 28, 32, b3, 20

189b. ποιητικός, *factivus, effectrix, efficiens, activus, effectivus,* constructive, active, capable of acting: art, ii.2.194b2, 4; relative, iii.1.200b30, 201a23, 3.202-a23, b26; viii.4.255b34; things in contact, iv.5.212b32; motion, iv.9.217b25. *See also* 34

190. ἀγνοεῖν, *ignorare,* be unaware, be unclear, ἄγνοια, *ignorantia,* ignorance, perplexity: becoming, i.8.191b11, 34; ii.4.196a13; movement, nature, iii.1.200b14; v.2.225b33; vs. knowledge, viii.4.255b5

190a. ἄγνωστος, *ignotus,* unknowable: infinite, i.4.187b7-9; iii.6.207a25, 31

191. θαυμάζειν, *mirari, admirari,* find mysterious, strange: coming from "what is-not," i.8.191b16; automatism, ii.4.196a28; θαυμαστόν, *mirabile, admirabile,* strange, astonishing, ii.4.196a11; iv.1.208b34; viii.3.253b32

192. μυθολογεῖν, *fabulari,* tell legends, iv.11.218b24

193. φιλοσοφία, *philosophia,* philosophy: and scientific problems, i.2.185a20, 8.191a24; first, i.9.192a36; ii.2.194b14; work on, ii.2.194a36; iii.4.203a2

194. ζητεῖν, *quaerere, inquirere,* search, undertake, inquire, look for: beings, principles, i.2.184b22-25, 8.191a24; viii.1.252b1; the impossible, i.4.188a9; explanation, ii.3.194b23, 195b22; and animals, ii.8.199a21; infinite, iii.5.204a14; place, iv.-1.209a4, 24, 2.210a4, 4.211a12; void, iv.7.214a15; time, iv.11.219a2; movement, viii.2.253a4, 3.253a33, 254a31, 4.255b13

194a. ζήτησις, *quaestio*, inquiry: infinite, iii.5.204a34

195. σκέπτεσθαι, *intendere, considerare*, examine, consider, look to: "all things one," i.2.185b7; contraries, i.5.188a31; movement, iii.1.200b23; vi.10.241b14; viii.-1.251a5, 5.257a31, 7.260a21; infinite, iii.4.204a1; time, iv.10.218b10; path, vii.4.-249a21; generation, destruction, vii.4.249b20

195a. ἐπισκέπτεσθαι, *considerare*, consider, examine, reflect further: generation, ii.1.193b21; factors, ii.3.194b16; chance, automatism, ii.4.195b36; motion, viii.-7.260b15

195b. σκέψις, *respectus, consideratio, intentio*, inquiry: having philosophic import, i.2.185a20; place, iv.4.211a7; void, iv.6.213a16; movement, v.4.228a20; viii.3.253a22

195c. σκοπεῖν, *intendere, considerare*, consider, investigate: being, i.2.184b6; nature, i.2.184b6; ii.2.194a13; unity, i.2.185a5; mathematician, ii.2.193b25, 194a10; events, ii.7.198a34; movement, iii.2.201b19; viii.2.252b8, 3.254a30, 5.256b3, 257a28, 6.258b16; infinite, iii.4.203b16, 204b4, 5.205a7; place, iv.2.209b5, 18, 3.210-b8, 4.211b30; void, iv.8.216a26

195d. ἐπισκοπεῖν, *intendere, considerare, dispicere*, consider, investigate, reflect: elements, i.6.189b17; becoming, i.7.190b3; infinite, iii.5.204a2; void, iv.8.214b30; time, iv.10.218b9, 14.223a16; motion, viii.8.264a8

196. πειρᾶν, *tentare, conari*, try: to determine principles, i.1.184a15; ii.3.194b23; to prove, ii.1.193a3; iv.6.213a22; to define, ii.2.194a2; iii.1.200b15; to inquire, iv.-4.211a7

197. πραγματεύεσθαι, *negociari, tractare*, occupy oneself with, explore: mathematics, ii.2.193b31; nature, ii.4.202b35

197a. πραγματεία, *negotium, tractatio*, inquiry, science, enterprise: aim at knowledge, ii.3.194b18; and movement, ii.7.198a30; viii.3.253a32

198. μέθοδος, *scientia, methodus, ars*, scientific exploration, search, analysis: systematic, i.1.184a11; iii.1.200b13, 5.204b3; principle, viii.1.251a7; and movement, viii.3.253b7, 7.261a30

199. μαρτύριον, *testimonium*, evidence, μαρτυρεῖν, *testari*, attest, iv.6.213b-21; viii.9.265b17

200. ἀπορία, *defectus, dubitatio, opinio*, difficulty, problem, difficulty: not. i.2.185a11, 3.186a9; physical, i.2.185a18; unity of part and whole, i.2.185b11; number of elements, i.6.189b29; of the ancients, i.8.191a24, b30; necessity, end, ii.8.198b16; movement, rest, iii.3.202a21; v.4.227b14, 228a19, 6.230b21, 28; viii.3.-253a23, 4.255a1, 5.258a27, 8.262b9, 10.266b28, 267b9; infinite, iii.4.203b23, 31; place, iv.1.208a33, 209a2, 24, 5.212b23; void, iv.7.214b6; ἀπόρημα, *oppositum, quod dubitatur*, perplexing question, iv.4.211a10; πρόβλημα, *problema*, problem, iv.6.212b3

200a. ἀπορεῖν, *deficere, ambigere, dubitare, opponere, indubitationem revocare*, διαπορεῖν, *opponere, dubitare*, become perplexed, ask, debate, raise problems: one, many, i.2.186a1; contraries, i.6.189a22, 28; natural science, ii.2.194a15; chance, automatism, ii.4.195b36, 196a8, 5.197a22; necessity, end, ii.8.198b33; animal behavior, ii.8.199a22; movement, change, rest, iii.3.202a13; v.1.224b13, 4.227-b14, 6.230a12; 18, 231a5; vii.4.248a10; viii.2.253a6, 8, 3.253a31, 4.254b33, 6.260-a12, 10.266b27, 267a17; place, iv.1.209a30, 2.210a12, 3.210a25, b23, 31, 4.211-a8; void, iv.7.214a9, 9.217a11; time, iv.10.217b30, 218a30, 14.223a22, 29

200b. δυσχερής, *inconveniens, absurdus,* troublesome, v.1.225a30

200c. δυσκολία, *difficultas, negotium,* perplexity, difficulty: place, iv.4.211a10; movement, vi.9.239b11

200d. χαλεπός, *difficilis, gravis,* difficult: not, i.2.185a12, 3.186a5, 10; iii.6.-206a17; viii.2.242b7, 3.253b13; movement, iii.2.201b33, 202a2; place, iv.2.209b18, 22, 3.210b23, 4.212a8

200e. ῥάδιος, *facilis,* easy: not, ii.2.201b17 (movement); iv.2.209b20 (matter, form), 10.218a10 (present); void, iv.7.214b11

201. προηπορημένον, *praedubitatum, dubitatum,* problem, προηυπο ρημένον, *exquisitum, expositum,* solution: place, iv.1.208a35

202. ὅτι, *quod, quia, quoniam,* that, because, i.1.184a14, 3.186b14, 187a10, 4.188a11, 12, 5.188a26, 6.189a12-14, 7.191a20, 9.192b2; ii.3.195b23, 6.198a10, 8.198-b17; iii.5.204b30, 32, 6.206b28, 7.207b23; v.1.225a27; *et passim*

203. διότι, *quia, ob hoc quod, ideo quia, cur, quoniam,* why, because, i.4.187b1; ii.7.198b8, 8.198b10; iii.7.207b26; iv.4.211a24, b35, 14.223b30; *et passim;* διό, *unde, idcirco, ideoque,* therefore, i.1.184a23, 2.185b27, 4.187b1, 6.189b5, 7.190-a29; ii.1.192b26, 193b9, 2.193b33, 4.196a16, 5.197a6, 27; iii.1.200b18, 6.207a25; iv.3.210b3, 4.211b33, 7.214a21, 9.216b35, 12.221a28; viii.5.256b24; *et passim;*

203a. διά τί, *quare, (causa) cur, (sit) propter quid, quam ob causam,* why: change, i.3.186a16, 18; v4.228a10; and knowledge, ii.3.194b19 ; final factor, ii.3.194b33, 9.200a10; chance, ii.4.196a8; fourfold, ii.7.198a15, 16, 19, 23, 31, b5; and movement, iii.5.205b8; v.6.230a19; viii.2.254a6, 3.253a23, 4.255b14; place, iv.2.209b34, 8.216b10; stop, iv.8.215a19; time, iv.14.223a17; incomparable, vii.-4.298b7, 20

204. ἀφιέναι, *abire, dimittere, omittere,* dismiss, v.1.224b27

205. λανθάνειν, *latere, non animadvertere, ignorare,* overlook, do covertly, be inattentive to, take no notice of: pattern, i.5.188b11; advocates of "ideas," ii.2.-193b35; movement, iii.1.200b13; iv.7.214a18, 11.218b22, 28; vii.2.244b15, 245a1; viii.3.235b11, 32; differences, change, vii.4.249a22

205a. ἐπιλανθάνειν, *oblivisci, oblivionem inducere,* cast into oblivion, forget, iv.12.221a32, 13.222b18

205b. λήθη, *oblivio,* forgetting, v.2.225b32

206. λύειν, *solvere, amovere,* resolve, refute, ransom, free argument, i.2.185a8, b14, 3.186a5; vi.9.239b11; viii.2.252b7; problem, i.7.190b33, 8.190a23, b30, 34; iv.3.210b23, 4.211a8, 5.212b23, 7.214b7, 10; viii.8.263a12, 10.267a17; from hindrance, viii.4.256a2; prisoner, ii.8.199b20

206a. ἀπολύειν, *resolvere, sejungere,* separate: categories, i.2.185a28

206b. διαλύειν, *dissolvere, resolere,* refute, analyze, resolve: in geometry, i.2.185a17; composite being, i.7.190b22; iii.5.204b33

206c. λύσις, *solutio,* solution: of argument, i.3.186a23; vi.9.239b26; of problems, viii.3.253a31, 8.262b28, 263a15, 22

207. τάξις, *ordo, ordinatio,* arrangement, order: as a genus of contraries, i.5.188a24; pattern, i.5.188b15, 20; form, i.7.190b28; of universe, ii.4.196a28; and nature, viii.1.252a12, 13, 16, 21

207a. ἐπιτάττειν, *instituere, imperare,* specify: form, ii.2.194b6

207b. ἀταξία, inordinatio, the unordered: an "opposite," i.7.190b15

207c. ἄτακτον, inordinatum, ordine vacare, unordered, viii.1.252a11

VII. THE DIVINE

208. ἥρως, *heros,* hero, iv.11.218b24

240 INDEX OF NAMES

209. δαιμόνιος, felix, numen excellens, mysterious, ii.4.296b7
210. θεῖον, divinum, numen, divine, i.9.192a17; ii.4.196a33, b6; iii.4.203b13
211. ὁ Ἑρμῆς, Mercurius, Hermes, i.7.190b7; Ζεύς, Juppiter, Zeus, ii.8.-
198b18; θεός, deus, a god, viii.8.262a3

INDEX OF NAMES

Anaxagoras, principles, contraries, (i.2.184b22); viii.1.252a10; infinite parts, i.-
4.189a17; iii.4.203a20, 23, 5.205b1-24; (iv.3.210a25); mind, i.4.188a9; (ii.4.196-
a18, 30, 8.198b16); iii.4.203a31; viii.5.256b24, 9.265b22; chance, necessity, (ii.4.-
196b5, 9.199b35); no void, iv.6.213a24; generation, destruction, (vii.2.243b10)

Anaximander, one, contraries, i.4.187a21; infinite, iii.4.203b14, (5.204b23-29;
viii.1.250b18); earth, (iv.8.214b31)

Anaximenes, air, (i.2.184b17, 6.189b6; ii.1.193a21; iii.4.203a18, 5.204a31, 205-
a27); infinite, (viii.1.250b18); dense, rare, (viii.9.265b30)

Antiphon, squaring the circle, i.2.185a17; nature, material, (viii.9.265b30)

Archytas, eternal, (iii.4.203b28-30)

Aristotle, De caelo, (iv.8.214b24); Metaphysics, i.(2.186a2, 3.187a1, 5.189a1),
8.191b29, 9.192a36; ii.2.(194a19), b14, (ii.3, 5.196b21-25, 197a5-14, 25-27, 6.198-
a7-13; iii.1-5, 7; v.1-3; vi.1.232a22); On Generation and Corruption, (i.3.187a1;
ii.1.193b21); iv.5.213a5, (7.214b3); On Philosophy, ii.2.194a36; Posterior Analytics,
(i.1.184a25; ii.1.193a3)

Coriscus, iv.11.219b21; v.4.227b32, 33

Democritus, principles, infinite, i.2.184b21; iii.4.203a20, 33; (viii.1.250b18); being, nonbeing, i.(3.187a1-3), 5.188a23; plenum, vacuum, i.5.188a22; (iii.1.200-
b21, 4.203b26); iv.6.213a34; (viii.9.265b23); differences, i.2.184b21, 5.188a24; iii.-
4.203b1; form, ii.2.194a20; chance, necessity, (ii.4.196a1, 24, 9.199b35); animal
intelligence, (ii.8.199a22); generation, destruction, (vii.2.243b10); viii.1.251b16;
explanation, viii.1.252a34; microcosm, (viii.2.252b26)

Diogenes of Apollonia, air, (i.2.184b17, 6.189b6; ii.1.193a21; iii.4.203a18, 5.204-
a31); infinite, (viii.1.250b18)

Eleatics, contrasted with "natural philosophers," i.2.184b17, 3.186a20, 4.187a12;
see also Melissus, Parmenides, Zeno of Elea

Empedocles, four elements, i.(2.184b19), 4.187a22, 25, 26, 188a18, 6.189a15;
(ii.1.193a22; iii.4.203a18); love, strife, (1.5.188b34, 6.189a24; ii.4.196a18, 8.198-
b16); viii.1.252a7, (10.265b19); form, ii.2.194a20; chance, ii.4.196a20, 24, 8.198-
b32, (199b5); generation, destruction, (vii.2.243b10); order, viii.1.252a20

Heraclitus, argument, i.2.185a7; good, bad, i.2.185b20; fire, (i.6.189b3; ii.1.193-
a21, 4.196a18); iii.5.205a3

Hippasus, (i.6.189b3; ii.1.193a21, 4.196a18)

Hippo, water, (i.2.184b17; ii.1.193a21)

Hippocrates, four elements, (i.2.184b19)

Hippocrates of Chios, squaring the circle, (i.2.185a16); division, (iii.6.206b12)

Homer, iv.12.221b32

Leucippus, plenum, vacuum, iv.6.213a34, (b4-22; viii.9.265b23); infinite, (viii.-
1.250b18)

Lycophron, "is" (copula), i.2.185b28

Melissus, principle, one, immovable, i.2.184b16, 3.186a10-22; (viii.3.254a25);
arguments, i.2.185a9, 10, 3.186a6, 8, 11; being, infinite, i.2.185a32, b17; iii.6.207-
a15-17; (viii.3.254a25); no void, iv.6.213b12, 7.214a27; see also Eleatics

Metrodorus of Chios, void, (iv.6.213b1)

Parmenides, principle, one, immovable, i.2.184b16; arguments, i.2.185a9, 3.-186a7, 22, 32, 187a1; being, limited, i.2.185b18; iii.6.207a15-17; meaning of "being', i.3.186a22-187a11; contraries, i.5.188a20, (b33); becoming, i.9.192a1; *see also* Eleatics

Paron, time, iv.13.222b18

Plato, nonbeing, (i.3.187a2); great and small, i.4.187a17, (6.189b15, 9.192a7, 11); iii.4.203a16, 6.206b27-33; iv.2.209b35, 210a1; formal principle, (i.9.192a34); ransomed, (ii.8.199b20); movement, (iii.2.201b20; viii.10.266b27-267a20); infinite, iii.4.203a4-16, (5.204a8), 6.206b27-33; "ideas,' iii.4.203a8, 10; contrasted with "natural philosophers," iii.4.203a16; place, iv.2.209b11-16, 33-210a2; *Timaeus*, iv.2.209-b12, 210a2; (viii.1.251b17, 10.267a16); "unwritten teachings," iv.2.209b14; "participation," iv.2.209b35, 210a1; matter, iv.2.210a1; *Parmenides*, (iv.3.210a25); earth, (iv.8.214b31); time, (iv.10.218b33); viii.1.251b17; *Phaedrus*, self-mover, (vii.1.-241b39; viii.9.265b32); bodily excellences, (vii.3.246b4); *see also* Platonists

Platonists, nonbeing, (i.9); "ideas," (ii.2.193b35-194a12, 3.194b26); indivisible lines, (iii.6.206a17); points, units, (v.3.227a27); *see also* Plato, Pythagoreans

Polyclitus, ii.3

Protarchus, altar stones, ii.6.197b10

Pythagoreans, contraries, (i.5.188b34; iii.2.201b25); infinite, iii.4.203a4-16, (5.-204a8, 31, 34); contrasted with "natural philosophers," iii.4.203a16; number-system, (iii.6.206b32; viii.4.249b24); void, iv.6.213b(1), 22-27; time, (iv.10.218b1); points, units, (v.3.227a27); *see also* Platonists

Socrates, v.4.228a3

Thales, water, (i.2.184b19, 6.189b3; ii.1.193a21; iii.4.203a18, 5.205a27)

Xenophanes, contraries,)i.5.188b33); infinite, (viii.1.250b18)

Xuthus, universe, iv.9.216b26

Zeno of Elea, bisection, (i.3.187a3); place, iv.1.209a23, 3.210b22; infinite, vi.-2.233a21; viii.8.263a5; paradoxes concerning motion, vi.9; noise, vii.5.250a20; *see also* Eleatics

RICHARD HOPE,

born in Pueblo, Colorado, in 1895, was educated at
the University of Southern California (M.A., B.D.,
1923, 1926) and at Columbia University (Ph.D.,
1930). After teaching at Concordia Collegiate Institute and Columbia, he joined the faculty of the
University of Pittsburgh in 1930 and at the time of
his death was Chairman of the Department of
Philosophy. His writings include *The Book of
Diogenes Laertius* (1930), *A Guide to Readings in
Philosophy* (1939), *How Man Thinks* (1949), and a
much-praised translation of Aristotle's *Metaphysics*.
Professor Hope's work on the translation of the
Physics was completed shortly before his untimely
death in 1955.